Volume 12
Applied Physics
and Engineering
An International Series
Thyristor Physics

Thyristor Physics

Adolph Blicher

Springer–Verlag
New York   Heidelberg   Berlin
1976

To Ann and Peter

Adolph Blicher
Formerly with RCA Corp.
Advanced Devices and Applications Dept.
Solid State Technology Center
Somerville, New Jersey 08876

Library of Congress Cataloging in Publication Data

Blicher, Adolph.
  Thyristor physics.

  (Applied physics and engineering; v. 12)
  Bibliography
  Includes index.
  1. Thyristors. I. Title.
TK7871.99.T5B58   621.3815'28   75-46523

© 1976 by Springer-Verlag New York Inc.
Softcover reprint of the hardcover 1st edition 1976

ISBN-13: 978-1-4612-9879-3   e-ISBN-13: 978-1-4612-9877-9
DOI: 10.1007/978-1-4612-9877-9

Preface

In this volume I attempt to present concisely the physical principles underlying the operation and performance characteristics of the class of semiconductor p-n-p-n switches known as thyristors. The semiconductor controlled rectifier (SCR), the triode AC switch (Triac) the gate turn-off switch (GTO), and the reverse conducting thyristor (RCT) are some of the most important devices belonging to this device family.

This book is aimed both at semiconductor-device physicists, designers, and students and at those electronic circuit designers who wish to apply thyristors creatively without the limitation of considering them as "black boxes," described only by insufficiently understood electrical ratings.

The book endeavors to present an up-to-date account of the progress made in understanding the operation, potentialities, and limitations of thyristors as switching circuit elements. It assumes some basic knowledge of transistor physics and stresses the phenomenological aspects of thyristor theory with the use of mathematics not going beyond calculus and differential equations. The first two chapters discuss basic thyristor operation theory. The subsequent chapters are devoted to the study of the static and dynamic properties of the SCR, the RCT, the GTO, and the triac; they include discussions of forward voltage drops, maximum voltage-blocking capabilities, turn-on and turn-off transients, current and voltage rise rates, and desirable and undesirable triggering effects.

A chapter on silicon surface phenomena has been incorporated to facilitate understanding of the existent approaches to surface-stable high-voltage planar and mesa devices, as reviewed in two chapters. The final part of the book is devoted to basic thyristor circuits.

It will become apparent that some thyristor properties have not been fully analyzed or even fully understood at this time. In some other cases, the available analysis may not lead to a closed-form solution. One should not be particularly surprised at this, realizing that a p-n-p-n switch consists of three p-n junctions and that even a single junction is still not fully understood, especially at high current densities or high reverse voltages. Very important refinements of the first-order theory of forward- or reverse-biased junctions are constantly being made. For example, there is on-going discussion as to injection efficiency at high current levels and/or high doping concentrations. Even the junction depletion approximation appears to be inadequate for the field characterization of the reverse-biased junctions with one side only very lightly doped.

Despite these difficulties, a solid foundation for the understanding and prediction of the thyristor behavior has been established. This book's purpose is to make this information accessible to all those interested.

In its major part, this work was supported by the RCA Laboratories, Princeton, N.J. and RCA Solid State Division, Somerville, N.J. to provide necessary background material for the training of RCA personnel engaged in semiconductor-devices studies and applications.

I wish to express here my thanks for the support and encouragement I have received from Mr. G. B. Herzog and Dr. W. M. Webster of RCA Laboratories and Messrs. J. Assour, D. Baugher, D. Burke, J. W. Gaylord, B. A. Jacoby, and C. R. Turner of the RCA Solid State Division. I wish also to express my appreciation for the many very useful discussions I had while writing this book to Mr. J. M. Neilson as well as to Messrs. H. Becke, T. C. McNulty, J. Olmstead, P. Smith, and J. E. Wojslawowicz of the RCA Solid State Division.

My work has been greatly facilitated by the expert typing of Mrs. B. J. Peterson, the thorough literature searches of Ms. E. Jankovics, and the prompt graphic services of Mr. D. Russell, Engineering Publications, RCA, Somerville, N.J.

For permission to use numerous illustrations special thanks are due to the Bell System Technical Journal; to the Electrochemical Society, New York City, N.Y.; to the Electronic Industries Association; to the Institute of Electrical and Electronic Engineers, New York City, N.Y.; to Pergamon Press, G.B.; to RCA Engineering Publications, Somerville, N.J.; and to Taylor and Francis, Ltd., G.B.

List of symbols

M_n	Electron avalanche multiplication factor, 1	P_{20}	Acceptor concentration in the p_2-base, 7
MT_1	Triac terminal No. 1	Q	Quantity of heat, 16
MT_2	Triac terminal No. 2	Q_M	Total charge on the metal electrode, 13
M_p	Hole avalanche multiplication factor, 1	Q_s	Total charge in the semiconductor, 13
m	Constant, 3	Q_{SS}	Fixed surface charge, 13
m	Mass, 16	q	Electronic charge $=$
N	Impurity concentration, 7		1.6×10^{-19} Coulomb
N_A	Acceptor concentration, 13		
N_B	$C_B =$ Impurity concentration, 15	q_B	Total excess base charge per unit area, 5
N_D	Donor concentration, 14	q_{B_1}	Total excess charge in the n_1-base
N_{SS}	Number of fixed surface states, 14	q_{B_2}	Total excess charge in the p_2-base, 5
N_{10}	Donor concentration in the n-base, 7	q_{CS}	Charge injected into the collector body
N_{20}	Donor concentration in the n+-emitter, 7	q_E	Total excess emitter charge per unit area, 5
n	Excess, injected minority-carrier density, 4	q_N	Total excess base charge per unit area in a
n(x)	Injected carried density at x, 9		transistor (active mode), 5
n_e	Injected electron density at the emitter—base junction, 9	q_{S_1}	Excess n-base charge due to collector injection, 5
n_i	Intrinsic carrier concentration, 13	q_{S_2}	Excess p-base charge due to collector injection, 5
n_0	Equilibrium electron concentration, 13	R	Contact resistance per unit area, 7
P	Power dissipation, 16	R_C	Ohmic contact resistance, 7
P_j	Power dissipated in a p-n junction, 16	R_T	Total thyristor ohmic resistance per unit area, 7
P_n	Power dissipated by electrons, 16	T_C	Instantaneous device temperature during cooling cycle, 17
P_p	Power dissipated by holes, 16	T_H	Instantaneous device temperature during heating cycle, 17
P_R	Instantaneous reverse power, 16	T_j	Junction temperature, 10
p_0	Equilibrium concentration of holes in bulk silicon	T_1	$t_{c1} = p_1-n_2-p_2$ transistor minority-carrier base transit time, 8
P_{00}	Acceptor concentration in the p+ region, 7		
P_{10}	Acceptor concentration in the p_1-emitter, 7	T_{1I}	$p_2-n_1-p_1$ inverse transistor

	minority-carrier base transit time, 8	V_h	Anode voltage at the holding point, 1
T_2	$t_{c2} = n_2 - p_2 - n_1$ transistor minority-carrier base transit time, 8	V_j	Total SCR junction voltage drop, 7
T_{2I}	$n_1 - p_2 - n_2$ inverse transistor minority-carrier base transit time, 8	V_r	Reverse anode potential for turn off, 8
t	time, 17	V_s	Anode voltage at the switching point, 1
t	p–base width, 6	V_T	SCR total voltage drop in the on state, 7
t_d	delay time, 5	V_{WT}	Voltage drop in the SCR effective base, 7
t_{f_1}	SCR first fall time, 8		
t_{f_2}	SCR second fall time, 8	$V_{(n_1)}$	Voltage drop in the n_1-base, 7
t_p	Minority carrier transit time in the GTO p-base, 9	$V_{(n_2)}$	Voltage drop in the n+-emitter, 7
t_q	Circuit commutated turn-off time, 1	$V_{(p_1)}$	Voltage drop in the p_1-emitter, 7
t_r	Rise time, 5	$V_{(p_0)}$	Voltage drop in the p+ emitter region, 7
t_{rr}	Junction recovery time		
t_s	Spreading time, 5	$V_{(p_2)}$	Voltage drop in the p_2-base, 7
t_s	Focusing time (storage) of a GTO, 9	V_1	Potential of junction J_1, 1
t_{s1}	SCR first storage time, 8	V_2	Potential of junction J_2, 1
t_{s2}	SCR second storage time, 8	V_3	Potential of junction J_3, 1
		v	charge propagation velocity (GTO), 9
t_{gt}	Delay plus rise time, 5		
U	Dissipated energy, 10	v_s	Plasma propagation velocity, 5
V	Potential		
V_A	Applied potential	W	Metallurgical base width, 1
V_B	Junction avalanche breakdown voltage, 3	W_0	Electric base width (W less depletion layer width), 1
V_{BO}	Thyristor turn-on (breakover) voltage, 3		
$V_{(BR)B}$	Thyristor maximum reverse-blocking capability, 3	W_p	GTO p-base width, 9
		x_0	Silicon dioxide thickness, 13
V_{CB}	Collector–base voltage, 2	Z	Ratio of the p_2-base sheet conductance to the n+-emitter conductance, 7
$V_{CEO(sus)}$	Transistor open base breakdown voltage, 3		
V_{DX}	Off-state voltage in a specified circuit, 8	*Greek*	
V_F	Forward voltage drop, 16	α	Common-base DC current gain, 1
V_{FB}	Flat band potential, 13		
V_{gt}	Gate triggering current, 17	α_E	Effective common-base DC current gain, 2

X

α_e	Effective common-base small-signal current gain, 2	γ_{2I}	$n_1-p_2-n_2$ transistor emitter efficiency, 8
α_N	n-p-n transistor common-base DC current, 1	ϵ_0	Permittivity of free space $= 8.854 \times 10^{-12}$ F/m
α_n	n-p-n transistor common-base small-signal current gain at some frequency f, 2	ϵ_1	Dielectric constant of silicon, 14
α_P	p-n-p transistor common-base DC-current gain, 1	ϵ_2	Dielectric constant of a dielectric coating, 14
α_{PI}	$p_2-n_1-p_1$ transistor DC common base current gain, 16	η	Intrinsic stand-off ratio of a unijunction transistor, 17
α_P	p-n-p transistor common-base small-signal current gain at some frequency f, 2	Θ_C	Conduction angle, 17
		Θ_F	Firing angle, 17
		Θ	ωt, 17
α_1	p-n-p transistor common-base small-signal low-frequency current gain, 1	θ	Thermal resistance, 16
		λ	Space constant, 14
		μ	Ambipolar mobility, 4
α_2	n-p-n transistor common-base small-signal low-frequency current gain, 1	μ_{cc}	Carrier-carrier scattered mobility, 4
		μ_n	Electron mobility, 4
		μ_p	Hole mobility, 4
α_{11}	$p_2-n_1-p_1$ inverse transistor common DC current gain, 8	μ_0	Low current density mobility, 4
		ρ	Resistivity
α_{2I}	$n_1-p_2-n_2$ inverse transistor DC gain, 8	ρ	Electric charge density, 14
β_{off}	GTO turn-off gain, 9	ρ_S	Sheet resistance, 12
β_T	Transistor base transport factor, 2	σ	Average electrical conductivity in the p-base, 6
β_1	$p_1-n_1-p_2$ transistor transport factor, 8	σ	Surface charge, 14
β_{1I}	$n_2-p_n-n_1$ transistor transport factor, 8	τ	Lifetime, 5
		τ_0	Effective lifetime in the depletion region, 2
β_{2I}	$n_1-p_2-n_2$ transistor transport factor, 8	τ_C	Minority carrier transit time, 5
γ	Emitter efficiency, 2	τ_{C_1}	Hole transit time in the SCR n-base, 5
γ_1	$p_1-n_1-p_2$ transistor emitter efficiency, 8	τ_{C_2}	Electron transit time in the SCR p-base, 5
γ_{1I}	$p_2-n_1-p_1$ transistor emitter efficiency, 8	τ_E	Minority carrier lifetime in the emitter, 5
γ_2	$n_2-p_2-n_1$ transistor emitter efficiency, 8	τ_{eff}	Effective lifetime of minority carriers, 4

xi

Contents

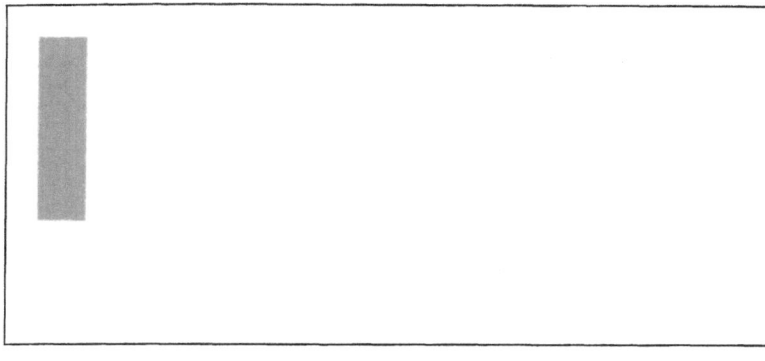

Device basics

Summary

Thyristors are three-terminal solid-state switches whose bistable state depends on a regenerative feedback mechanism occuring in a p-n-p-n structure.

The silicon-controlled rectifier (SCR) is a device belonging to the class of thyristors with the ability of unidirectional conduction. After defining the switching and holding points of an SCR, the gate turn-on conditions are examined, and it is shown that the $V\text{–}I$ coordinates of the switching point are dependent on small-signal current gains of the p-n-p and n-p-n SCR sections. The turn-on occurs when the sum of small-signal alphas is equal to unity.

In the case of a shorted emitter, the turn-on condition is simplified to a product of small-signal current gain and avalanche multiplication factor $M\alpha_1 = 1$.

At zero potential applied to the center junction the current flowing through the device is I_1, and at this current level the sum of DC alphas α_N and α_P is equal to unity.

The holding current I_h is greater than I_1 and, therefore, the $V\text{–}I$ coordinates of the holding point are dependent on the sum of DC alphas.

1.1
Introduction

A thyristor is any semiconductor switch whose bistable action depends on p-n-p-n structure regeneration feedback. The SCR (silicon-controlled rectifier) is one of the thyristor devices and has three terminals—an anode, a cathode, and a control electrode (gate) (see Figure 1.1). It is known also as a reverse-blocking triode thyristor. A thyristor that blocks the forward potential but conducts in the reverse direction is known as the reverse-conducting thyristor (RCT). A p-n-p-n diode with anode and cathode, but without

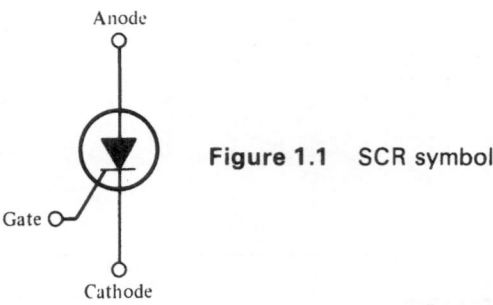

Anode

Gate

Cathode

Figure 1.1 SCR symbol.

any control electrode is known as Shockley diode. It is a reverse-blocking diode thyristor. The triode AC switch (triac) is a thyristor that can conduct current in both directions.

A thyristor that can be turned off by the action of the gate only is known as a GTO. The light-activated silicon-controlled rectifier is known as LASCR. There are several other devices (not all of them discussed in this book) that belong to the thyristor family such as the silicon unilateral switch (SUS), the complementary SCR (CSCR), the programmable unijunction transistor (PUT), and the silicon-controlled switch (SCS).

Thyristor electrical characteristic terminology differs considerably from the one used for transistors. In the following we list some of the most important SCR and Triac electrical definitions:

On-state. The condition of the thyristor corresponding to the low-resistance, low-voltage portion of the device anode–cathode (*principal*) voltage–current characteristic.

Off-state. The condition of the thyristor corresponding to the high-resistance, low-current portion of the anode–cathode $V–I$ characteristic.

Reverse-blocking state. The condition of a reverse-blocking thyristor corresponding to the portion of the anode-to-cathode $V–I$ characteristic for reverse currents of lower magnitude than the reverse-breakdown circuit.

Principal voltage. The voltage between the main terminals.

Forward voltage. A positive anode-to-cathode voltage.

Off-state voltage. The principal voltage when the thyristor is in the off-state.

Critical rate of rise of the off-state voltage (dV/dt). The minimum value of the rate of rise of principal voltage which may cause switching from the off-state to the on-state.

Reapplied rate of rise of voltage (reapplied dV/dt). In a reverse-blocking thyristor, the rate of rise of forward voltage following turn-off (testing condition).

Breakover voltage. The principal voltage of the breakover point.

On-state voltage. The principal voltage when the thyristor is the on-state.

Principal current. The current through the collector junction.

On-state current. The principal current when the thyristor is in the on-state.

Forward current. For a reverse-blocking thyristor, the principal current for a positive anode-to-cathode voltage.

Surge on-state current. An on-state current of short time duration.

Critical rate of rise of on-state current. The maximum value of the rate of rise of on-state current (di/dt) which a thyristor can withstand without deleterious effect.

Off-state current. The principal current when the thyristor is in the off-state.

Holding current. The minimum principal current required to maintain the thyristor in the on-state.

Latching current. The minimum principal current required to maintain the thyristor in the on-state immediately after switching from the off-state to the on-state has occurred and the triggering signal has been removed.

Reverse voltage. A negative anode-to-cathode voltage.

Reverse current. The current for negative anode-to-cathode voltage for a reverse-blocking thyristor.

Circuit-commutated turn-off time (t_q). The time interval between the instant when the principal current has decreased to zero after external switching of the principal voltage circuit, and the instant when the thyristor is capable of supporting a specified principal voltage without turning on.

Critical rate of rise of commutation voltage of a bidirectional thyristor. The minimum value of the rate of rise of principal voltage which will cause switching from the off-state to the on-state immediately following the on-state current conduction in the opposite quadrant.

EIA-NEMA[1] 1972 *Recommended Standard for Thyristors* contains a complete list of thyristor definitions.

1.2
SCR current–voltage characteristic

An SCR is a four-layer device that can be considered as an n-p-n transistor with an added p-layer (Figure 1.2).

The n emitter of such a structure is the device cathode, the added p layer is the hole emitter and device anode. The p-base contact of

---

[1] Electronic Industries Association, 2001 Eye Street, N.W. Washington, D.C. 20006.

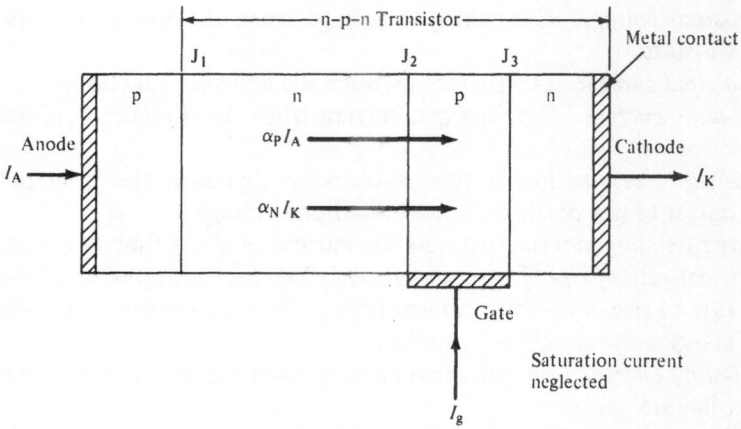

Figure 1.2 Current flow in an SCR.

the n-p-n transistor portion is the control electrode—the gate. The p-n-p-n structure has three p-n junctions, J_1, J_2, and J_3, and can be also visualized as a combination of an n-p-n and p-n-p transistor sharing a common collector, junction J_2 (Figure 1.3). Each transistor section is supplied with a base driving current by the collector of the other transistor. The gate is usually connected to the p-base of the n-p-n transistor. When a positive potential is applied to the anode, a small saturation current starts flowing since the collecting junction J_2 is reverse-biased and blocks the flow of any substantial current despite the fact that the two other junctions are forward-biased. The device is in the forward-blocking state. If the polarity of the applied potential is reversed, junctions J_1 and J_3 become reverse-biased and J_2 is forward-biased. J_1 and J_3 will now block any substantial current flow. The forward- and reverse-blocking capability will exist as long as the applied potentials are below the breakdown potentials of J_2 or the sum of breakdown potentials of J_1 and J_3 junctions, respectively, and no forward bias is applied to the gate.

We consider now the situation when the device is at first in the forward-blocking condition so that J_2 junction is reverse biased. If sufficiently high positive (forward) bias is applied to the gate, the n-type cathode starts emitting. The n-p-n transistor conducts and provides the base current of the p-n-p transistor. Because of the existence of the regenerative feedback when the total loop gain exceeds unity, both transistors are driven after a short delay into saturation. This means that all three junctions become forward-biased, its impedance reaches its lowest value, and the device is turned on.

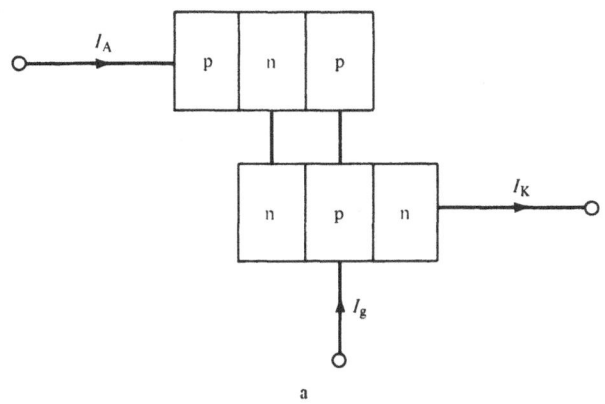

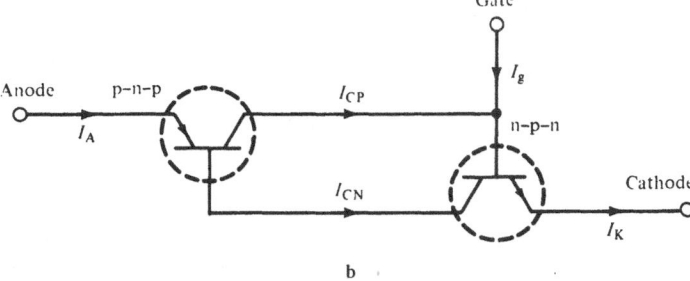

Figure 1.3 Two-transistor equivalent circuit.

The switching (breakover) point is a sensitive function of the gate-current magnitude, so that a family of current–voltage characteristics may be plotted with the gate current I_g as a parameter (Figure 1.4).

1.2.1
Switching and holding points

The region in the first quadrant of the current–voltage characteristic of Figure 1.5 is the forward region. The switching point 1 is defined by high-voltage, low-current coordinates V_s and I_s and a slope [1.1]

$$\frac{dV}{dI}\bigg|_{I_g = \text{constant}} = 0$$

The other point of interest is the holding point 2 characterized by low-voltage, high-current V_h, I_h coordinates, and a slope [1.1]

$$\frac{dV}{dI}\bigg|_{I_g = \text{constant}} = 0$$

5

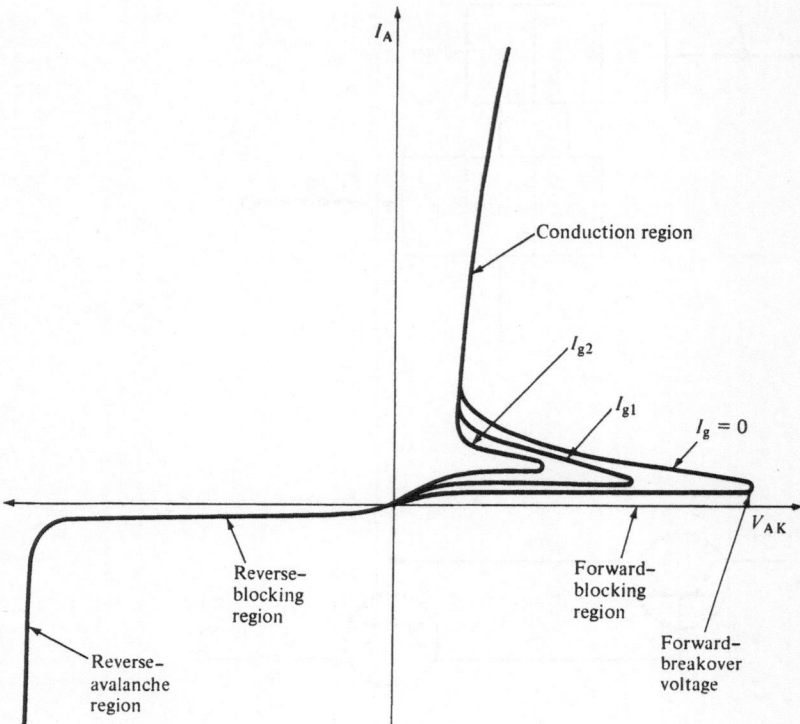

Figure 1.4 Family of SCR V–I characteristics.

Figure 1.5 Current–voltage characteristics of a p-n-p-n device.

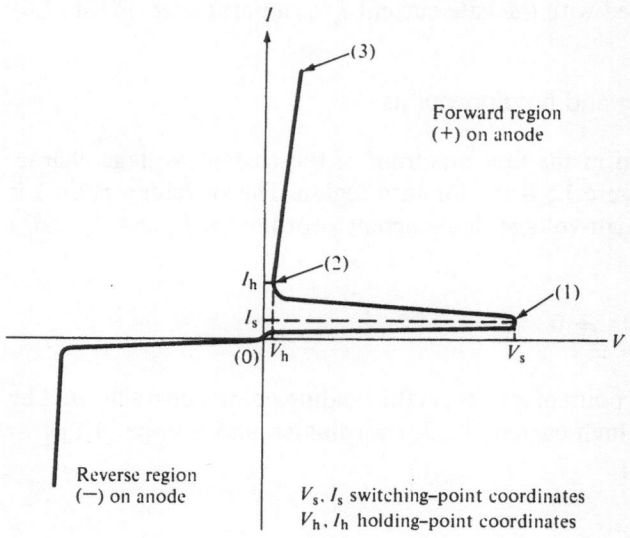

The current gain in a device is usually not entirely uniform over the whole conducting area of the device. Also, the gate current is applied usually in a form of a short-duration pulse. Because of these two prevailing conditions, the minimum latching current required for a positive turn-on will be higher than the holding current.

1.3
Basic SCR construction features

Figure 1.6a represents a top view of the device with a centrally located gate. This arrangement has many advantages which will become obvious from the understanding of the device physics.

Figure 1.6b represents device cross section. This particular device has the cathode metallization extended beyond the cathode itself,

Figure 1.6 SCR basic construction.

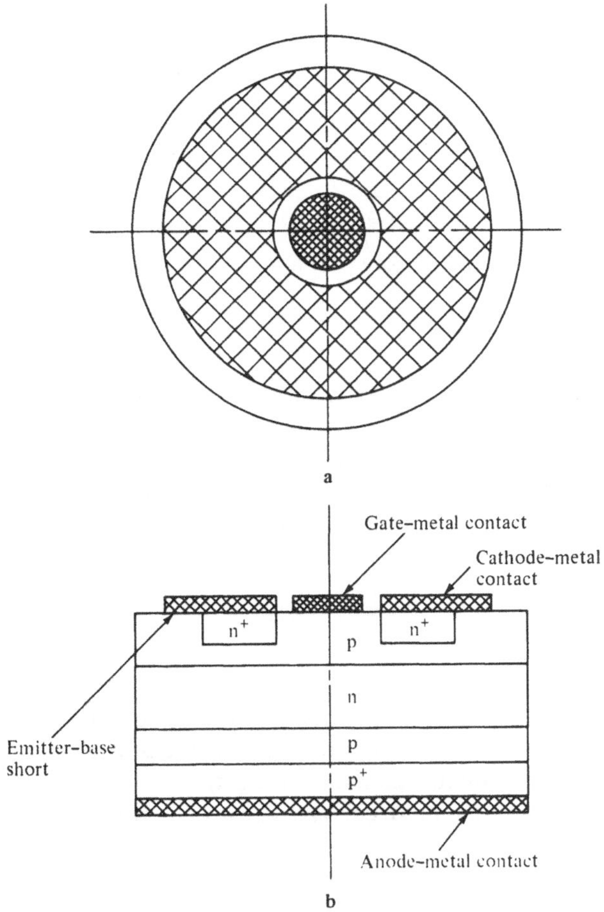

a

Gate–metal contact

Cathode–metal contact

n$^+$ p   n$^+$

n

Emitter–base short

p

p$^+$

Anode–metal contact

b

7

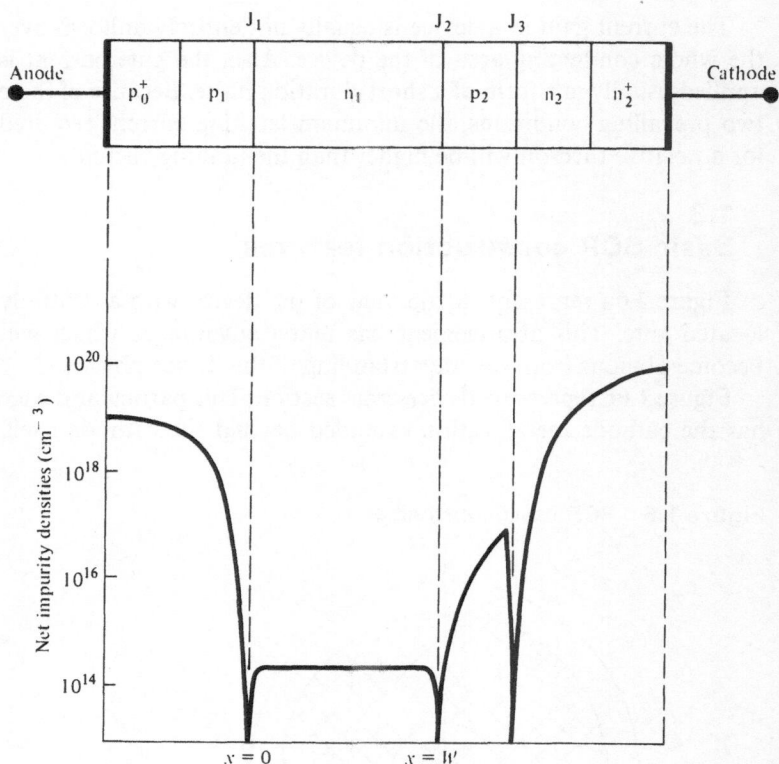

Figure 1.7 Typical impurity density profile of a p-n-p-n device.

so that the entire cathode perimeter is shorted to the p base. This "shorted" emitter improves device performance at elevated temperatures and at fast-varying anode–cathode potentials.

A typical impurity doping profile for a diffused SCR is given in Figure 1.7. The n-base region is usually uniformly and lightly doped for high-voltage devices.

1.4
Gate triggering

There are several mechanisms by which an SCR can be triggered. In this section we limit our discussion to triggering the device by the application of forward potential to the device gate.

The DC alpha (common base current gain) of a transistor is defined by the equation

$$I_C = \alpha I_E + I_{CBO} \tag{1.1}$$

where I_C and I_E are transistor collector and emitter currents, respectively, and I_{CBO} is the collector-saturation current.

Small-signal alpha α_1 is derived by differentiating Equation (1.1) in respect to I_E and assuming constant collector voltage:

$$\alpha_1 = \alpha + I_E \frac{d\alpha}{dI_E} \tag{1.2}$$

Small-signal alpha is the sum of the DC alpha and the product of the emitter current and the DC alpha slope.

At very low current levels the slope is positive so that the small-signal alpha is always larger than the DC alpha until a sufficiently high current level is reached at which the DC current gain reaches its peak. Above this current level the small-signal alpha is smaller than the DC alpha.

If we neglect the saturation current, which is acceptable at room temperature, we can express the anode current as the sum of two currents reaching the collector J_2 (Figure 1.8):

$$I_A = \alpha_P I_A + \alpha_N I_K \tag{1.3}$$

with

$$I_K = I_A + I_g \tag{1.4}$$

I_A, I_K, and I_g are the anode, the cathode, and the gate currents, respectively. α_N is the n-p-n transistor current gain while α_P is the p-n-p transistor gain.

From Equations (1.3) and (1.4)

$$I_A = \frac{\alpha_N I_g}{1 - (\alpha_P + \alpha_N)} \tag{1.5}$$

The breakover point is defined by $dV/dI = 0$. However, the condition

$$\left. \frac{dI_A}{dI_g} \right|_{V = \text{constant}} \to \infty \tag{1.6}$$

can be used instead [1.2].

From (1.2) and (1.5), we can derive (see Appendix)

$$\frac{dI_A}{dI_g} = \frac{\alpha_2}{1 - (\alpha_1 + \alpha_2)} \tag{1.7}$$

where α_1, α_2 are small-signal alphas of the p-n-p and n-p-n transistors, respectively.

When

$$\alpha_1 + \alpha_2 = 1 \tag{1.8}$$

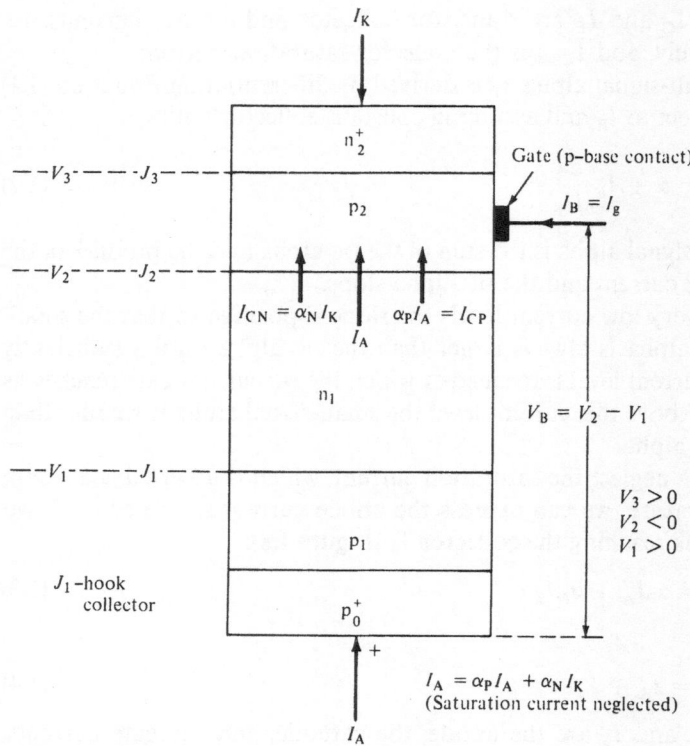

Figure 1.8 SCR schematic structure at low current levels, in the off condition.

then $dI/dI_g \to \infty$ and the device is turned on. Thus the condition for turning the device on requires that the sum of small-signal alphas be at least equal to unity.

Current gains are current and voltage dependent. At voltages sufficiently close to the avalanche-breakdown potential of the collecting junction J_2, carrier-multiplication effects should be taken into account. If M_p and M_n are avalanche-multiplication factors for holes and electrons, respectively, then the requirement (1.8) should be replaced by

$$M_p \alpha_1 + M_n \alpha_2 = 1 \tag{1.9}$$

M_p and M_n are functions of the potential across J_2 junction. The approximate empirical expression for the multiplication factor as a function of the applied voltage V is

$$M = \frac{1}{1 - (V/V_B)^m} \tag{1.10}$$

where V_B is the junction breakdown potential and m is a constant which has different values for holes and electrons but for simplification may be assumed the same for both and equal or greater than four.

1.5
Holding current

The holding-point coordinates (Figure 1.5) are given by I_h and V_h. The holding point is defined as the low-voltage, high-current point at which the slope $dV/dI = 0$ at $I_g = $ constant.

An accurate analysis of the $V–I$ relationship at the holding point is not available due to the complexity of the problem. There is another point on the $V–I$ characteristic, however, that can be determined more easily [1.3]. When an SCR is triggered from the off condition into the on condition, the potential V_2 across the center junction J_2 goes from the reverse to forward bias and at some current level becomes equal to zero. This current I_1 is called the turn-off current. For currents larger than I_1 the center junction is forward-biased. At this newly defined point, there is no avalanche multiplication and there is no saturation current flowing since the potential is assumed to be zero.

Since $I_A = I_1$, we obtain from (1.5)

$$\alpha_N + \alpha_P = 1 - \frac{\alpha_N I_g}{I_1} \tag{1.11a}$$

therefore

$$\alpha_N + \alpha_P < 1 \tag{1.11b}$$

At the turn-off point the sum of DC alphas is less than unity according to expression (1.11). Often there is no gate current flowing when the device is turned off and $I_g = 0$. Equation (1.11) is reduced then to

$$\alpha_N + \alpha_P = 1 \tag{1.12}$$

The holding point occurs for positive potentials for current levels above I_1. It follows then from (1.11) or (1.12) that the holding current will depend on DC alphas.

It is very difficult to find a general relationship between the holding and switching currents. In an idealized, special case of a symmetrical structure [1.3], it was found that

$$I_h \simeq 2.5 \, I_1 \tag{1.13}$$

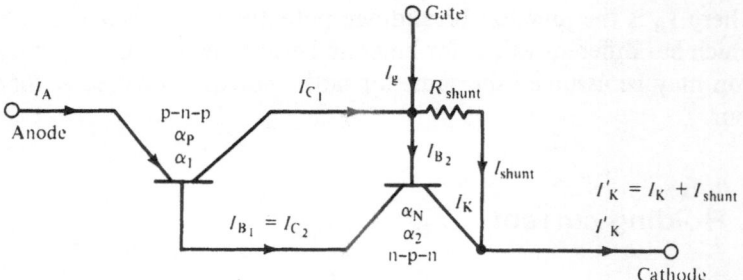

Figure 1.9 Two-transistor analog of a p-n-p-n switch with shunted emitter.

1.6
Triggering a shorted emitter SCR

A shorted emitter is widely used in the design of power thyristors in order to achieve a significant improvement in the high-temperature blocking capability and the device ability to withstand rapidly rising potentials without the undesired switching effect due to the capacitive, displacement current. Figure 1.9 represents a two-transistor analog of a p-n-p-n switch with a shorted (shunted) emitter. In practice, the shorting is accomplished by extending the cathode metallization over the cathode edges[2] onto the p^+-base region as shown in Figure 1.6b. This way the total or part of the emitter periphery is shorted to the p base.

We assume that the device is in the forward-blocking condition, i.e., the anode is positively biased and so are the junctions J_1 and J_3, while the center junction J_2 is reverse biased. No external gate potential is applied. At room temperature and in the absence of sudden potential change, only a small saturation current will be collected by J_2. The hole-current component of this saturation current will flow laterally in the p base toward the cathode-shorting contact. As long as this current is small, there will be only a very small lateral potential present in the base if the p-base resistivity is not very high and lateral dimensions not very long. At high temperatures or when a sudden potential variation occurs, an increased hole-current flow will create sufficiently high forward bias for the n emitter, so that appreciable injection will take place and the device will turn on. This will happen, however, at much higher temperatures or displacement currents than in the absence of the shorted emitter. In the latter case, the junction J_2 collected holes will forward-bias

---

[2] Another technique used for the same purpose consists of a number of shorted dots distributed over the major part of the cathode area.

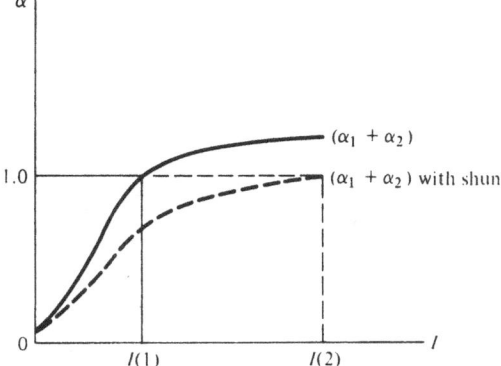

Figure 1.10 Sum of alphas versus current with and without shunt.

the n emitter much sooner (Figure 1.10) and the condition $\alpha_1 + \alpha_2$ = 1 will be satisfied much sooner, i.e., at lower current than with the short (shunt) present.

For the shorted-emitter case, we can define a new DC alpha dependent on the shunting current I_{shunt} magnitude [1.4]:

$$\alpha_{shunt} = \alpha_N\left(\frac{I_K}{I_K + I_{shunt}}\right) \tag{1.14}$$

Since $I_{shunt} > 0$, α_{shunt} is smaller than α_N. The same situation will exist for small-signal "alpha shunt." Before the turn on the ratio of the shunting current I_{shunt} to the cathode current I_K is always much larger than unity, so that α_{2shunt} of the n-p-n transistor will be very small and usually much below the value of α_1 of the p-n-p section of the SCR, and it can be neglected in the turn-on condition (1.9), which, assuming[3] $M_n = M_p = M$ becomes

$$M\alpha_1 = 1 \tag{1.15}$$

APPENDIX

Differentiating (1.3) we get

$$\frac{dI_A}{dI_g} = \alpha_P\frac{dI_A}{dI_g} + I_A\frac{d\alpha_P}{dI_g} + \alpha_N\frac{dI_K}{dI_g} + I_K\frac{d\alpha_N}{dI_g} \tag{1.A}$$

From relationship (1.2) for small-signal alphas we obtain

$$\frac{d\alpha_P}{dI_A} = \frac{\alpha_1}{I_A} - \frac{\alpha_P}{I_A} \tag{1.B}$$

---

[3] For silicon M_n and M_p have in reality different values; therefore, Equation (1.15) is only a reasonable approximation.

and

$$\frac{d\alpha_N}{dI_K} = \frac{\alpha_2}{I_K} - \frac{\alpha_N}{I_K} \tag{1.C}$$

Noting that

$$\frac{d\alpha_P}{dI_g} = \frac{d\alpha_P}{dI_A}\frac{dI_A}{dI_g} = \frac{\alpha_1}{I_A}\frac{dI_A}{dI_g} - \frac{\alpha_P}{I_A}\frac{dI_A}{dI_g} \tag{1.D}$$

$$\frac{d\alpha_N}{dI_g} = \frac{d\alpha_N}{dI_K}\frac{dI_K}{dI_g} = \frac{\alpha_2}{I_K}\frac{dI_K}{dI_g} - \frac{\alpha_N}{I_K}\frac{dI_K}{dI_g} \tag{1.E}$$

substituting Equations (1.D), (1.E), and (1.4) into (1.A) we obtain

$$\frac{dI_A}{dI_g} = \frac{\alpha_2}{1 - \alpha_1 - \alpha_2} \tag{1.F}$$

References

1.1 W. Fulop. Three terminal measurements of current amplification factors of controlled rectifiers. *IEEE Trans. Electron Devices, 10*: 120–135, 1963.

1.2 F. E. Gentry. Turn on criteria for pnpn devices. *IEEE Trans. Electron Devices, ED-11*: 74, 1964.

1.3 J. F. Gibbons. A critique of the theory of pnpn devices. *IEEE Trans. Electron Devices, ED-11*: 406–413, 1969; and Graphical analysis of the *I–V* characteristics of generalized pnpn devices. *Proc. IEEE, 55* (8): 1366–1374, 1967.

1.4 R. W. Aldrich and N. Holonyak, Jr. Two-terminal asymmetrical and symmetrical silicon negative resistance switches. *J. Appl. Phys., 30*: 1819–1824, 1959.

1.5 F. E. Gentry, F. W. Gutzwiller, N. Holonyak, Jr., and E. E. Von Zastrow. *Semiconductor Controlled Rectifiers.* Englewood Cliffs, N.J.: Prentice-Hall, 1964.

Also of interest

W. Fulop, W. Fong Yan, M. R. Joadat-Ghasabi, and C. A. Hogarth. Three-terminal current gain measurements of high-power thyristors. *Int. J. Electron., 33* (6): 601–609, 1972.

2

Current gain

Summary

The functional dependence of current gain on current before the turn on of the SCR is reached is determined from the theoretical considerations under somewhat idealized conditions.

The transport factors in both device bases are functions of the electric fields and diffusion. The internal field in the n base depends on the magnitude of the current flowing; consequently, the transport factor is strongly current dependent. The field in the p base has approximately a constant value, so the transport factor is independent of the current level within certain limits.

Because of the existence of the recombination-generation currents at the n emitter (cathode) and p emitter (anode), the emitter efficiencies at low-current levels, occurring before the turn-on, are strongly current-dependent and increase with increasing current density.

If the SCR is considered before the turn-on as an n-p-n transistor with a so-called hook collector, then it is possible to determine experimentally the values of current gains of both the n-p-n and the p-n-p transistors using only the three available SCR terminals.

2.1
Variation of current gain with current

The turn on and holding points can be determined from the knowledge of the small-signal and DC current gain values, respectively. The functional relationship between the alphas and current, before the turn on is reached, can be determined from the theoretical considerations under some idealized conditions [2.1]. It is assumed that the SCR is in the off condition (Figure 1.7) and its anode is positively biased in respect to the cathode, so that junctions J_1 and J_3 are forward biased, but J_2 is reverse biased and blocks the current flow.

J_1 and J_2 are emitter junctions of the p-n-p and n-p-n transistors, respectively, and J_2 is the collector junction for both transistors. The collected current density for either transistor is

$$J_C = \alpha J_E + J_{CBO} \tag{2.1}$$

where α is the DC current gain:

$$\alpha = \beta_T \gamma \tag{2.2}$$

β_T is the transport factor, γ is the emitter efficiency, and J_{CBO} is the collector saturation current density.

For the p-n-p transistor

$$\gamma = \frac{J_p}{J_p + J_n + J_{rg}} \tag{2.3}$$

where J_p and J_n are injected hole and electron current densities flowing into the base and into the emitter regions, respectively, and J_{rg} is the recombination current density of the space-charge region of the emitter junction.

$$J_{rg} \propto \frac{d}{\tau_o} \exp\left(\frac{qV}{2kT}\right) \tag{2.4}$$

In this expression d is the depletion layer width (a function of the applied potential), τ_o is the effective lifetime in the depletion region, and V is the applied emitter-base voltage. For the case when the recombination centers are lying very close to the energy gap center, τ_o is about equal to the geometric mean of two terms. One is the limiting value of lifetime in the highly doped n-type silicon and the other one is the lifetime in the highly doped p-type silicon.

Gold and copper impurities are among the recombination centers lying very close to the energy gap center.

More accurate computation of τ_o may be made using appropriate Sah–Noyce–Shockley expressions [2.2].

Figure 1.6 shows the approximate doping profiles in various device regions. It is usually the case that the doping of the p base is higher than the doping level of the n base. Also, the p-emitter doping is larger than that of the n base, and the n-emitter concentration higher than that of the p base. Since in the off condition the current flow is very low, we can neglect any conductivity modulation effects.

There is a current density J_A flowing in the n base consisting of holes and electrons. If the resistivity ρ of the n base is uniform, then the electric field will be

$$E = \rho J_A \tag{2.5}$$

The distribution of holes in the n base is determined using the one-dimensional equation for the steady-state hole-current flow in the n base and two boundary conditions (Figure 1.7). At $x = 0$ the injected current density is proportional to $\exp(qV/kT)$. At $x = W_0$, i.e., at the collector edge, the excess charge is equal to zero (W_0 is the electric base width).

Knowing the distribution of holes in the n base, it is possible to compute the hole-current densities at the collector and at the emitter. The ratio of these two currents is by definition the transport factor β_T [2.1]:

$$\beta_T = \frac{B \exp(A W_0)}{A \sin h(B W_0) + B \cos h(B W_0)} \tag{2.6}$$

where

$$A = \frac{1}{L_c} \tag{2.7}$$

$$B = \left(\frac{1}{L_c^2} + \frac{1}{L_p^2}\right)^{1/2} \tag{2.8}$$

L_p is the diffusion length and L_c is a characteristic length defined by

$$L_c \equiv \frac{2kT}{qE} \tag{2.9}$$

Since $E = \rho J_A$, it is evident that the transport factor β_T is a function of current density J_A.

When $E \to 0$, $L_c \to \infty$ and (2.6) is reduced to the familiar expression for the transport factor in the absence of an electric field:

$$\beta_T = \frac{1}{\cos h\left(\dfrac{W_0}{L_p}\right)} \tag{2.10}$$

The concept of the characteristic length L_c may be further clarified by the realization that the current is carried both by drift (field) and diffusion. These two processes occur simultaneously and, from the equivalent circuit point of view in parallel. When the field is large, the quantity L_c becomes smaller than the diffusion length L_p. In this case $B \simeq 1/L_c$ from (2.8), and the conduction takes place by drift. The concept of the critical length L_c is thus helpful in determining whether the conduction is mostly governed by drift or by diffusion.

Due to the assumption that the p-emitter concentration is much higher than that of the n base, the electron component J_n in the expression for the emitter efficiency may be neglected, and (2.3) becomes

$$\gamma = \frac{J_p}{J_p + J_{rg}} \tag{2.11}$$

17

This expression shows that the space-charge recombination current is responsible for the lowering of the emitter-injection efficiency at the very low current levels. The emitter efficiency improves, however, when the current density increases. It is so because the injected current density J_p is proportional to $\exp(qV/kT)$ while the recombination-generation current density is proportional to $\exp(qV/2kT)$. The recombination-generation current increases at a slower rate than the injected current density. In addition, τ_o increases with the injected current density.

Figures 2.1, 2.2, and 2.3 represent the transport factor β_T, the emitter efficiency γ, and the alphas of a typical p-n-p transistor in function of current density [2.1]. The current gain of the n-p-n section of the SCR can be determined basically from similar considerations. There are, however, some differences in the physical parameters. The n-p-n transistor has usually a much narrower p-base width than the n base of the p-n-p transistor.

The built-in field created in the p base by diffusion may be assumed to be constant (this is exact only for an exponentially decaying impurity concentration). Due to the field constancy, the values of A and B in Equations (2.6), (2.7) and (2.8) become independent of current so that the n-p-n transistor current gain is a function of the n-emitter efficiency only.

Figure 2.1　Transport factor β_T as a function of current and base width for a typical transistor. (After Yang and Voulgaris [2.1].)

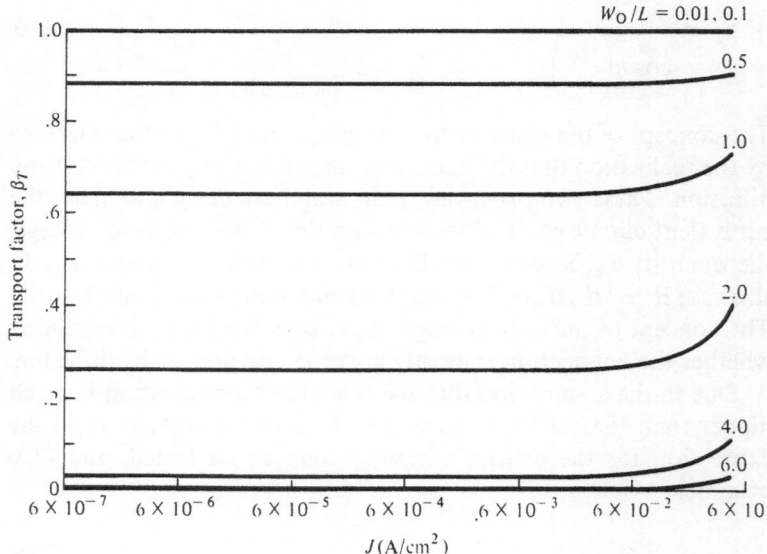

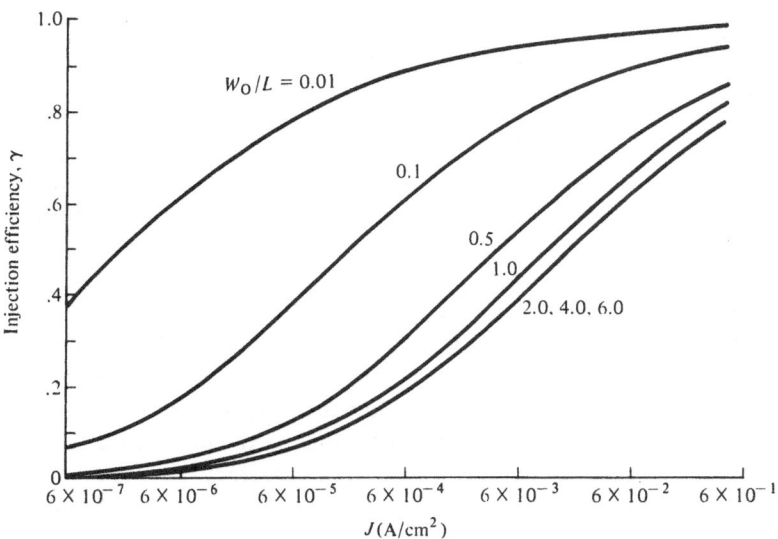

Figure 2.2 Variation of injection efficiency, γ, with current and base width for a typical transistor. (After Yang and Voulgaris [2.1].)

Figure 2.3 Small-signal alpha and DC alpha as functions of current and base width for a typical transistor. (After Yang and Voulgaris [2.1].)

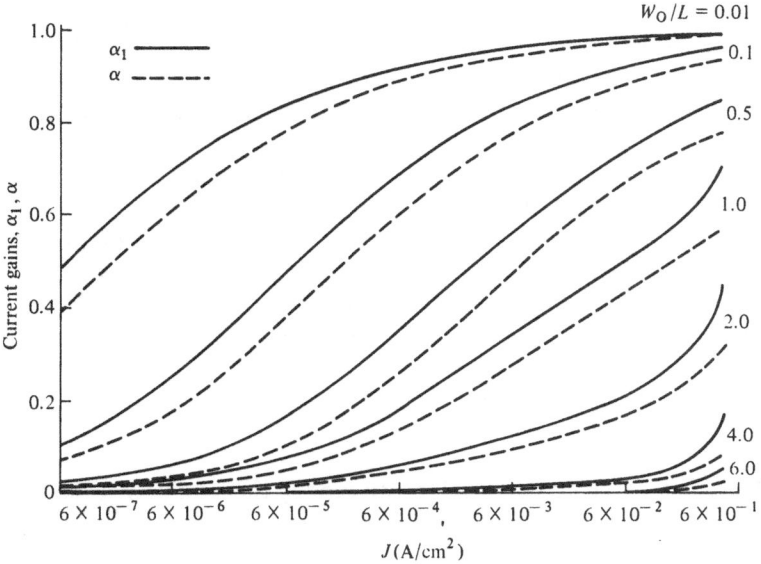

The expressions for β_T and γ used for the p-n-p transistor can be used for the n-p-n transistor by introducing appropriate values for electron (rather than for hole) lifetimes, mobilities, etc.

The diagram of Figure 2.3 shows that at the very low current densities the transport factor of the p-n-p section is very small but starts increasing rapidly, when appreciable high current densities are reached. The small-signal and DC current gains are functions of W_O/L ratios for both types of transistors. For narrow base widths, the transport factor may become essentially independent of current density and dependent only on transport by diffusion. For very wide bases, β_T depends primarily on the electric field and consequently on the current density. Also for the very wide bases the emitter efficiency becomes almost independent of the base width.

An actual device differs from an idealized one in that the true current density is not known with great accuracy. An actual device is never uniform, and the emission from the emitter junctions may occur at some small region rather than be uniformly distributed over the entire emitter area. Also, when the current gain is very low the base current is comparatively large, so that there is a transverse voltage drop in a base, which may cause edge injection. For these reasons, all analytical expressions are true for rather narrow emitters, a few thousandths of an inch wide, so that the injection nonuniformities are less probable.

2.2
Current gain measurement

Generally, only three terminals are available in an SCR—the cathode, anode, and p-gate contacts. There is no contact available for the access to the n base. In view of this, only an indirect method of alpha measurement can be utilized.

When an SCR is in the forward-blocking condition, it may be considered as an n-p-n transistor with an added p layer. The additional n-p junction J_1 can emit holes when forward-biased. The additional structure is the so-called hook collector.

In a transistor with a hook collector, the collector current in the common emitter operation is, neglecting the saturation current,

$$I_C = \frac{\alpha_N I_B}{1 - (\alpha_N + \alpha_P)} \tag{2.12}$$

The effect of the hook collector is to modify the DC alpha α_N, so that the effective alpha becomes [1.1 and 1.6]

$$\alpha_E = m_{PO}\alpha_N \tag{2.13}$$

with

$$m_{PO} = \frac{1}{1 - \alpha_p} \tag{2.14}$$

Similar expressions are also valid for small-signal alphas

$$\alpha_e = \alpha_n m_p \tag{2.15}$$

with

$$m_p = \frac{1}{1 - \alpha_p} \tag{2.16}$$

α_n and α_p are small-signal current gains of the n-p-n and p-n-p transistors at some measurement frequency f.

From the transistor theory, the frequency variation of small-signal alphas is such that

$$\alpha_n = \frac{\alpha_2}{1 + jf/f_n} \tag{2.17}$$

$$\alpha_p = \frac{\alpha_1}{1 + jf/f_p} \tag{2.18}$$

α_1 and α_2 are small-signal alphas at very low frequencies and f_p and f_n are corresponding alpha cut-off frequencies.

From Equations (2.15)–(2.18)

$$\frac{\alpha_e}{\alpha_{eo}} = \frac{\alpha_e}{m_{po}\alpha_2} \tag{2.19}$$

where

$$\alpha_{eo} = m_{po}\alpha_2; \qquad m_{po} = \frac{1}{1 - \alpha_1} \tag{2.20}$$

Figure 2.4 is a theoretical plot of the normalized effective alpha α_e/α_{eo}, in decibels, versus the log of the normalized frequency f/f_p. The study of the theoretical plot shows several regions of interest. We limit our discussion here to the plateau region c-d only. In this region α_e is frequency-independent.

The plateau region occurs when the ratio of the cut-off frequencies f_p/f_n is much smaller than unity, i.e., when there is an appreciable difference between the two characteristic frequencies. This region is determined by the condition $f/f_p \gg 1$ on one side and by $f/f_n \ll 1$ on the other. In the plateau region the expression (2.19) simplifies [1.1] to

$$\frac{\alpha_e}{\alpha_{eo}} = \frac{1}{m_{po}} = 1 - \alpha_1 \tag{2.21}$$

21

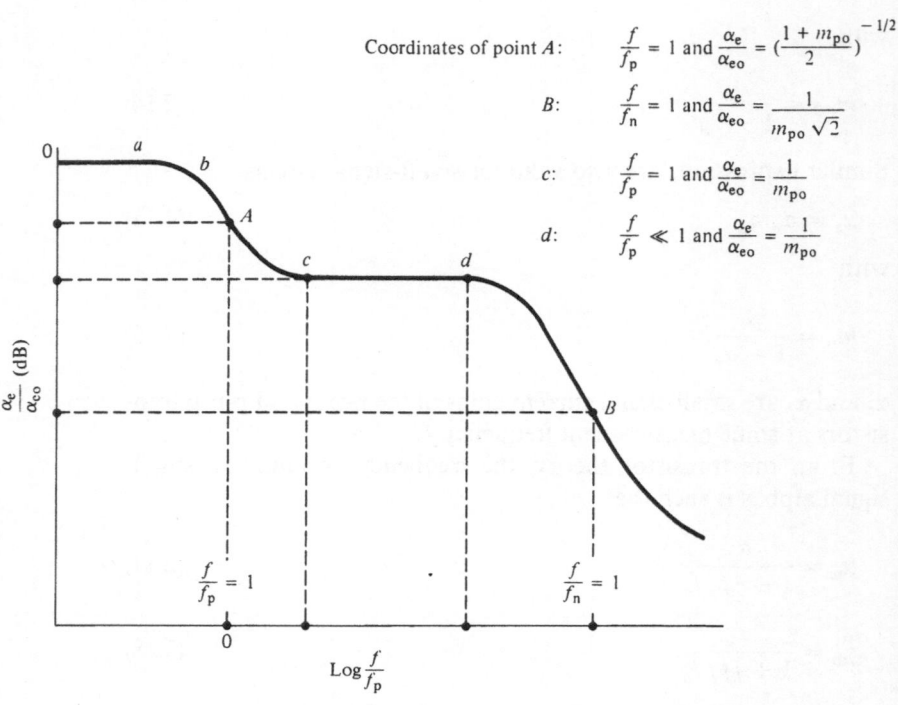

Coordinates of point A: $\quad \frac{f}{f_p} = 1$ and $\frac{\alpha_e}{\alpha_{eo}} = \left(\frac{1+m_{po}}{2}\right)^{-1/2}$

B: $\quad \frac{f}{f_n} = 1$ and $\frac{\alpha_e}{\alpha_{eo}} = \frac{1}{m_{po}\sqrt{2}}$

c: $\quad \frac{f}{f_p} = 1$ and $\frac{\alpha_e}{\alpha_{eo}} = \frac{1}{m_{po}}$

d: $\quad \frac{f}{f_p} \ll 1$ and $\frac{\alpha_e}{\alpha_{eo}} = \frac{1}{m_{po}}$

Figure 2.4 Current gain versus frequency. (After Fulop [1.1].)

Figure 2.5 Device No. 1: measured α_e variation with respect to frequency. $V_{CB} = +10V$, parametric in I_A. (After Fulop [1.1].)

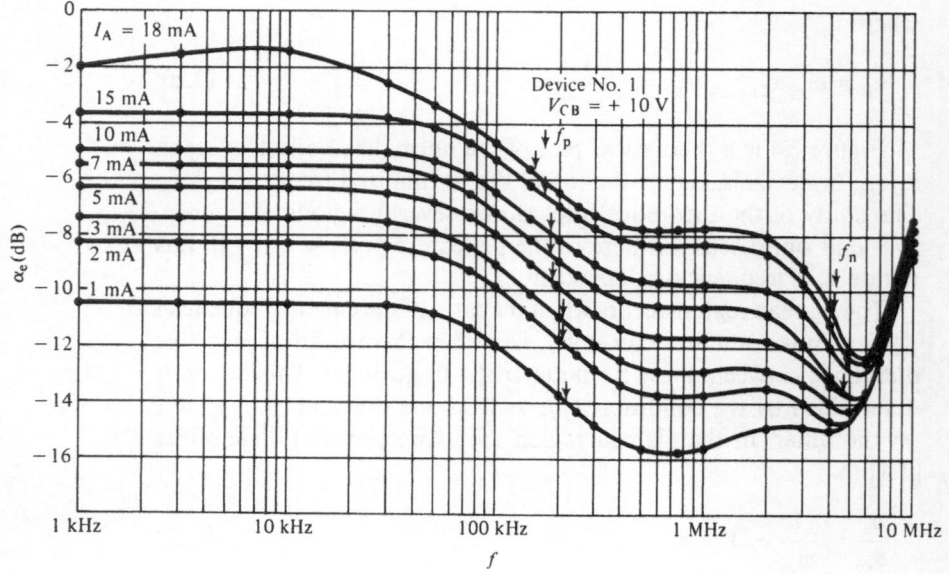

22

Referring now to the experimental plot of Figure 2.5 with anode current as a parameter [1.1], we determine the α_e value within the plateau region at the desired anode current I_A. Next, we find from the same curve the value of low frequency α_e, i.e., α_{eo}, by extrapolating back to $f = 1 \, kHz$, and from Equation (2.21) we find α_1 and m_{po}. α_2 can be found from the expression (2.20).

Remembering the relation (1.2) and knowing the small-signal alphas and their current dependence, it is possible to find the DC alphas by numerical integration.

References

2.1 E. S. Yang and N. E. Voulgaris. On the variation of small signal alphas of a pnpn device with current. In *Solid State Electronics*, Vol. 10. Pergamon Press, Oxford, England: 1967, pp. 641–648.

2.2 C. T. Sah, R. N. Noyce, and W. Shockley. Carrier generation and recombination in p-n junctions and p-n junctions characteristics. *Proc. IRE, 45*: 1228, 1957.

Thyristor maximum voltage-blocking capability

Summary

The maximum forward- and reverse-voltage–blocking capability of power thyristors are dependent on the small-signal current gains of the p-n-p and n-p-n transistor sections, on the junction type, and on silicon resistivity.

For small values of current gains, the maximum forward- and reverse-blocking potentials are equal. In the shorted cathode emitter case, both the maximum forward breakover point and maximum reverse-blocking voltage are determined by the condition, $\alpha_1 = 1/M$, which states that the breakdown occurs when the small-signal common base current gain of the idealized symmetrical p-n-p thyristor section is equal to the reciprocal of the avalanche-multiplication factor M. This relationship permits the approximate computation of the maximum blocking capability since the multiplication factor and current gain are known voltage functions.

Improved breakover and/or other characteristics can be obtained in the reverse conducting thyristor (RCT) in which both the cathode and anode emitters have shorted regions.

3.1
Introduction

The evaluation of the maximum reverse- and forward-blocking capability of a thyristor will be carried out for a somewhat idealized model. The very exact solutions are useful only in those cases when surface breakdown effects are negligibly small, as in a device with perfectly beveled mesas or provided with field plates. Only then the bulk breakdown is the limiting factor and not the usually imperfect surface.

The simplification consists of the assumption that both p regions of the thyristor are diffused in exactly the same way and have the same thicknesses. This approximation is good enough for many devices [3.1]. As a result, the p-n-p section of a thyristor will be considered to be entirely symmetrical, i.e., the p_1-n_1-p_2 and p_2-n_1-p_1 transistors are considered to be exactly the same. Figure 3.1 shows schematically the thyristor structure and an approximate impurity distribution in the various thyristor regions.

When the device is in the reverse-blocking state, the J_1 and J_3 junctions are reverse biased while the central junction J_2 is forward biased. The applied potential is supported, however, almost entirely by the junction J_1 since the breakdown of J_3 is usually very low, on the order of 10 V.

For the reverse-blocking case, we have to consider then the p_2-n_1-p_1 transistor. With p_2 acting as an emitter and p_1 as a collector, the depletion layer (Figure 3.1a) extends almost entirely into the n_1 base and somewhat into the p_1 collector.

Figure 3.1 Depletion regions (cross-hatched areas) for the reverse- and forward-blocking voltages.

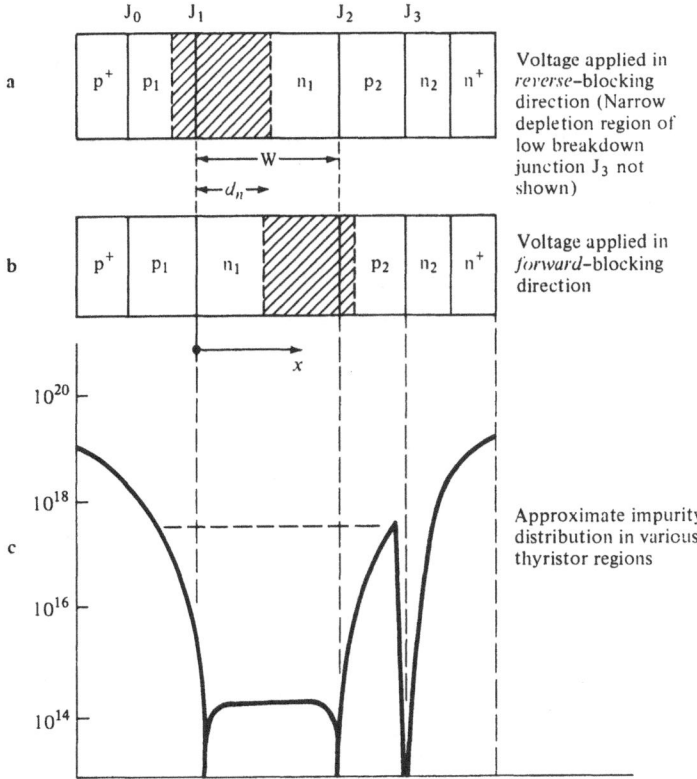

In the forward-blocking condition the roles of the p regions are reversed; the p_1 region becomes the emitter and the p_2 region the collector. The depletion region is shifted now to the junction J_2, as per Figure 3.1b, and extends mainly into the n_1 region, which is usually very lightly doped with a concentration of about 10^{14} atoms/cm^3.

Because of the assumed symmetry, the p_1-n_1-p_2 and p_2-n_1-p_1 transistors have exactly the same DC and small-signal current gains.

The very narrow depletion regions of the forward-biased junctions or low-breakdown junctions are not shown in Figure 3.1.

3.2
Maximum forward-blocking capability V_{BO}

Since we are interested here in the maximum blocking capability, the discussion is limited to the special case when there is no potential applied to the thyristor gate and the gate current is zero. Under these conditions, the device behaves essentially as a Shockley p-n-p-n diode. For any forward potential applied to the gate, the breakover point will be lower of course than the maximum blocking capability.

3.2.1
Nonshorted emitter case

For a device with a nonshorted emitter, the condition for breakover is (see Chapter 1)

$$M_p \alpha_1 + M_n \alpha_2 = 1 \tag{3.1}$$

where M_p and M_n are avalanche multiplication factors for holes and electrons, respectively, and α_1 and α_2 are small-signal current gains of the p-n-p and the n-p-n thyristor sections, respectively.

The multiplication factors M_p and M_n are related to the applied voltage V and the breakdown voltage V_B of the particular type of junction considered. The empirical relationship for the multiplication factor is

$$M = \frac{1}{1 - \left(\dfrac{V}{V_B}\right)^m} \tag{3.2}$$

The constant m has different values for p and n materials. It is quite common to consider for simplification just one value of m. This value varies in the literature between 4 and 8. For more accurate computations of M, see [3.2].

26

An approximate expression for the step junction avalanche break-down in silicon [3.3] is

$$V_B \simeq \left(\frac{10^{16}}{C_B}\right)^{3/4} \times 60 \tag{3.3}$$

where C_B is the background doping per cm^3. For the linearly graded junctions with an impurity gradient a per cm^4

$$V_B \simeq \left(\frac{3 \times 10^{20}}{a}\right)^{2/5} \times 60 \tag{3.4}$$

Both formulae are valid for avalanche breakdown of plane junctions at room temperature only. At higher temperatures the breakdown voltage increases. Figure 3.2 is a plot of step junction breakdown voltage versus doping density of the lightly doped side. The same chart provides the information about the depletion width [3.4].

Figure 3.3 [3.5] represents breakdown voltages of plane-diffused junctions with different depths and for a surface concentration of 10^{19} cm^{-3}. Figures 3.4 and 15.2 give similar information for planar (cylindrical) junctions.

In the case of the nonshorted emitter, the condition for the breakover in the forward-blocking direction is

$$M_p \alpha_1 + M_n \alpha_2 = 1 \tag{3.5}$$

For simplification, we assume $M_p \simeq M_n = M$ and the expression (3.5) becomes

$$M(\alpha_1 + \alpha_2) = 1 \tag{3.5a}$$

From (3.2) and (3.5a) the breakover voltage is

$$V_{BO} = V_B(1 - \alpha_1 - \alpha_2)^{1/m} \tag{3.6}$$

3.2.2
Shorted-emitter case

The shorted emitter is presently in general use in most thyristor types with the exception of the sensitive gate devices.

When the n emitter of the thyristor is shorted, the small-signal current gain α_2 of the n-p-n transistor remains very small until a sufficiently large current density is reached; only then it starts increasing very rapidly. Below this current level, the p-n-p section current gain α_1 will be usually much larger than the n-p-n section current gain α_2 due to the emitter-shorting effect. If we neglect α_2, in the limiting case, the expression (3.5a) is reduced to a simple

27

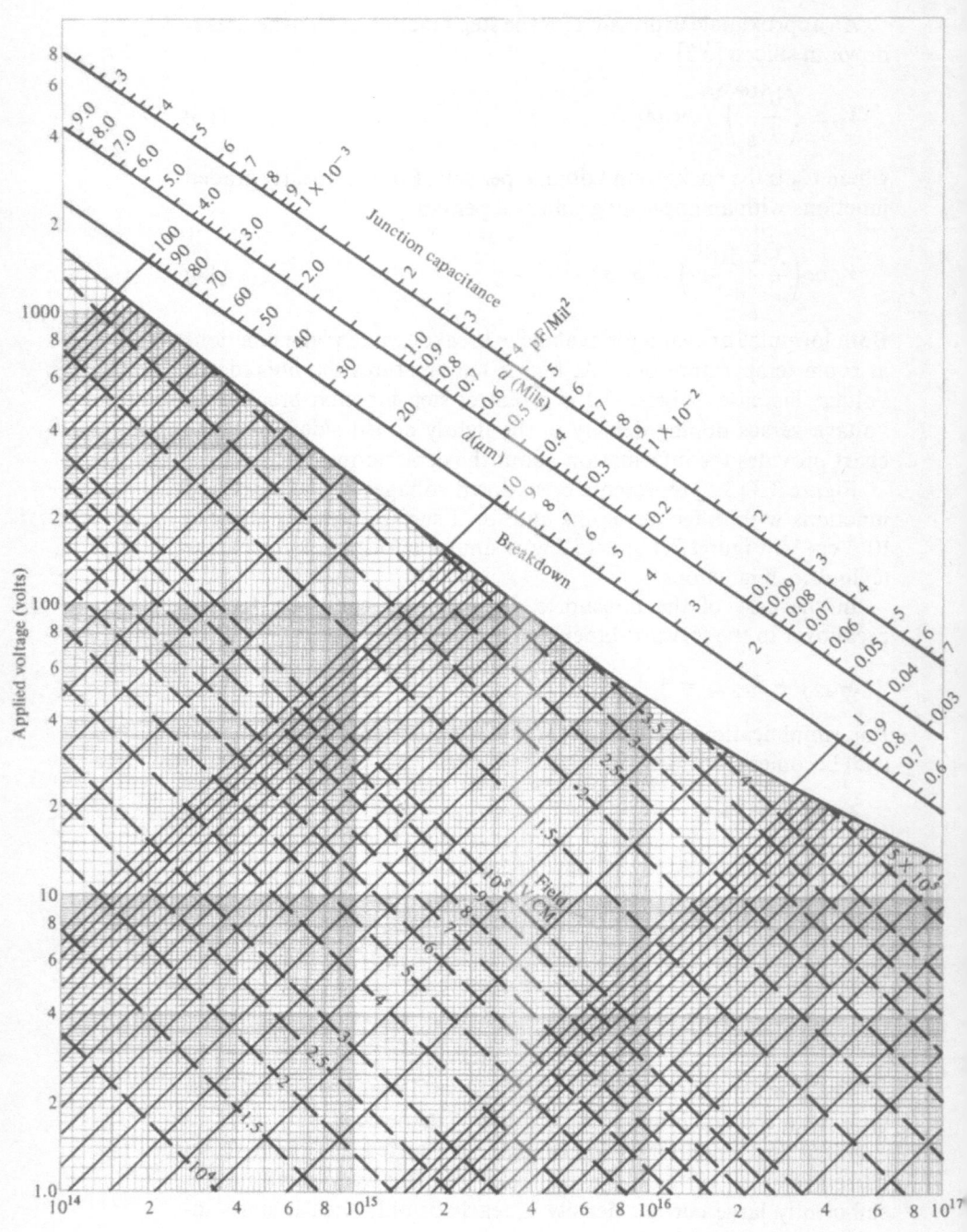

Figure 3.2 Abrupt junction breakdown in silicon (see Appendix 1). (After Olmstead [3.4].)

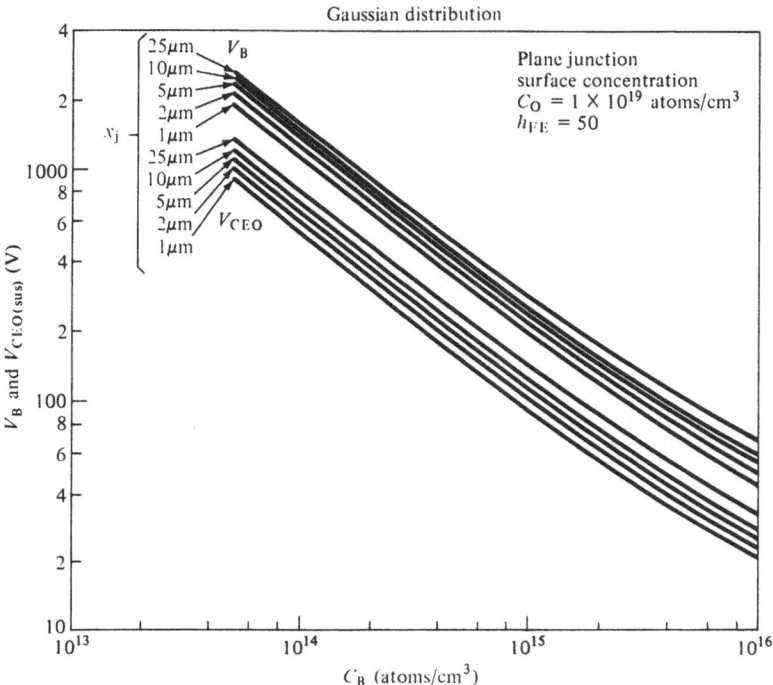

Figure 3.3 Dependence of breakdown voltage V_B, sustaining voltage $V_{CEO(sus)}$ on background doping C_B and junction depth x_J. (After Kannam [3.5].)

Figure 3.4 Dependence of breakdown voltage V_B on background doping C_B. (After Kannam [3.5].)

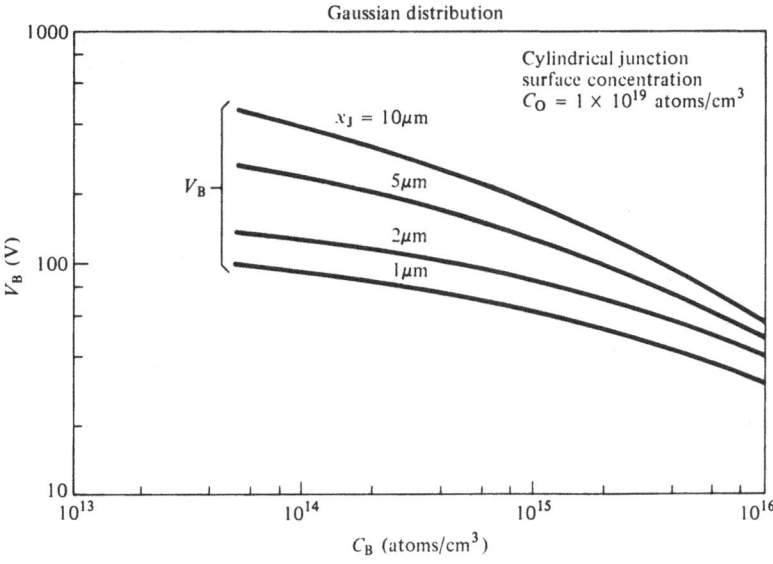

29

condition (valid for moderate temperatures)

$$M\alpha_1 = 1 \qquad (3.5b)$$

and the breakover voltage becomes

$$V_{BO} = V_B(1 - \alpha_1)^{1/m} \qquad (3.7)$$

3.3
Maximum reverse-blocking capability

When the anode potential of the thyristor is negative, the depletion region is located at the junction J_1 (Figure 3.1). The device behaves like a p-n-p transistor with the small-signal current gain α_1, the same as considered for the forward-blocking voltage because of the assumed symmetry.

The p-n-p transistor is an open-base transistor so that the collector breakdown is characterized by the sustaining voltage $V_{CEO} = V_{(BR)R}$. The potential applied between the emitter and the collector of the p-n-p transistor is V_{CEO} and Equation (3.2) becomes

$$M = \frac{1}{1 - \left(\dfrac{V_{CEO}}{V_B}\right)^m} \qquad (3.8)$$

When the collector avalanche multiplication effects cannot be neglected, the collector current in a transistor is

$$I_C = M(\alpha_1 I_E + I_{CBO}) \qquad (3.9)$$

where I_E is the emitter current and I_{CBO} the saturation current. When the base is open, the base current $I_B = 0$, $I_E = I_{CEO}$ (open-base collector current) and relation (3.9) yields

$$I_{CEO} = \frac{M I_{CBO}}{1 - M\alpha_1} \qquad (3.10)$$

When $M\alpha_1 = 1$, $I_{CEO} \rightarrow \infty$ and breakdown occurs.

The condition

$$M\alpha_1 = 1 \qquad (3.11)$$

for the maximum reverse breakdown $V_{(BR)R}$ is then the same as the condition obtained for the maximum forward-blocking voltage. From (3.7) and (3.11), we derive the expression for the breakdown voltage

$$V_{CEO(sus)} = V_{(BR)R} = V_B(1 - \alpha_1)^{1/m} \qquad (3.12)$$

At small current densities α_1 is a functiion of several variables such as current, voltage, and temperature. For simplification we are assuming here that the temperature is constant and $T = 300°K$.

In principle it is possible to determine α_1 knowing the current density and the applied potential. The value of the potential determines the effective electrical base width W_0, which is equal to the geometrical base with W less the depletion layer width d_n in the n base region (Figure 3.1a):

$$W_0 = W - d_n \tag{3.13}$$

For very crude simplified computations, we may assume that we have reached that current level at which the emitter efficiency γ approaches unity and the transport factor β_T is almost independent of current. The smaller the ratio W_0/L_p of the electrical base width W_0 to the diffusion length L_p of holes in the n-base region, the more accurate this assumption is.

Under these conditions, the expression for small-signal current gain simplifies to the well known relationship

$$\alpha_1 = \frac{1}{\cosh(W_0/L_p)} \tag{3.14}$$

The depletion region width for an abrupt junction can be obtained from the expression

$$d_n = \left(\frac{2K\varepsilon_0 V}{qN_D}\right)^{1/2} \tag{3.15}$$

where q is the electronic charge (1.6×10^{-19} C); N_D is the doping density of the n_1-base region; K is the dielectric constant (12, for silicon); V is the applied potential; and ε_0 is the free space permittivity ($\cong 8.9 \times 10^{-14}$ F/cm).

Expression (3.15) is valid only for one-sided step junction, i.e., for the case when N_D is much smaller than the doping density of the p^+ region. The value of d_n can be obtained directly from Figure 3.2, which gives the relationship between carrier concentration, applied voltage, and the depletion-layer thickness. For $V = 500$ V, $N_D = 2 \times 10^{14}$, $d_n \simeq 55$ µm.

For graded junctions (gaussian or erfc) Lawrence and Warner [3.6], charts should be used. Two of these charts for gaussian distribution are reproduced in Figures 3.5 and 3.6. In most practical cases, the step-junction approximation will be quite satisfactory.

For a thyristor with n-base width $W = 100$ µm, n-base concentration $= 2 \times 10^{14}$ cm^{-3}, $d_n \simeq 50$ µm, $W_0 = 60$ µm. The hole lifetime in the n base is assumed to be 2 µsec. If $D_p = 13$ cm^2/sec, then $L_p = 50$ µm, $\alpha_1 = 0.55$, and $V_{(BR)R} \simeq 0.90 \, V_B$, with $m = 4$.

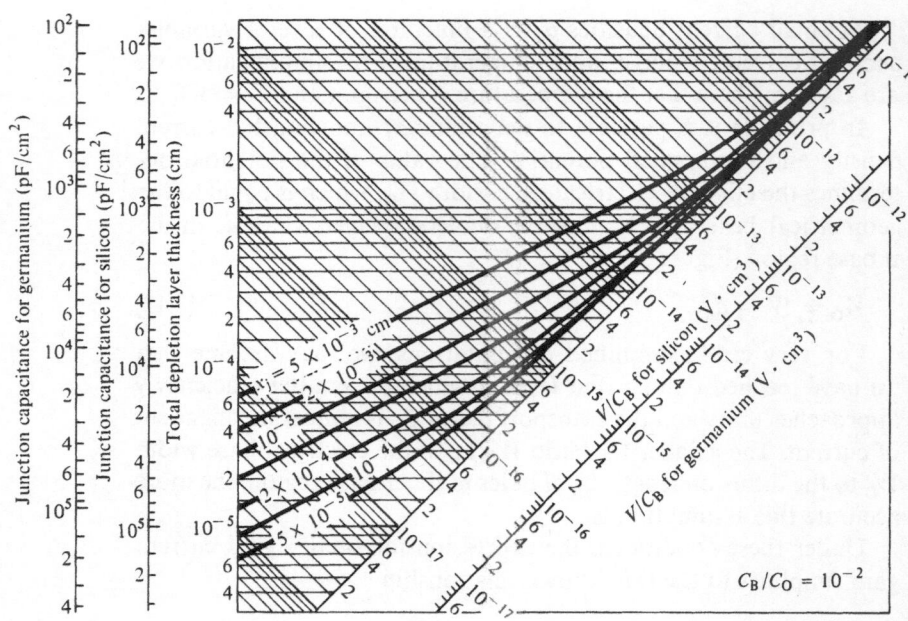

Figure 3.5   Chart for use in range 3×10^{-3} to 3×10^{-2}, gaussian distribution. (After Lawrence and Werner [3.5].)

3.4
Punch-through condition

It was assumed so far that the depletion region width was always smaller than the n-base width, and no punch-through occurred. When this condition is not fulfilled, then $W = d_n$ and $W_O = 0$. For one-sided step junction, the punch-through voltage can be determined from Figure 3.2, making $d_n = W$. The graph of Figure 3.2 is plotted on the basis of Equation (3.15) which can be rewritten for the punch-through potential as

$$V_{PT} = \frac{qW^2 N_D}{2K\varepsilon_0} \tag{3.16}$$

For the n_1-base width, $W = 110\ \mu m$, $N_D = 2 \times 10^{14}\ cm^{-3}$, the punch-through potential $V_{PT} \simeq 1800$ V. For a base width of 50 μm the punch-through would occur at only 500 V and the thyristor breakdown or breakover will occur much sooner than could be expected from the n-type silicon doped to 2×10^{14}. A step junction, with this doping density on the lightly doped side, would break down at 1000 V.

3.5
Temperature dependence

The avalanche breakdown voltage of a p-n junction increases slightly [3.7] with increasing temperature, as shown in Figure 3.7. The effect is more pronounced for higher resistivities.

For an impurity concentration of 10^{14} atoms/cm^3, the change may be by about 20 percent for a temperature variation between 300°K and 400°K. Accordingly, the reverse-blocking potential will increase with the increasing temperature. In the reverse direction, alphas remain small even at moderately elevated temperatures, so that the temperature effect on breakdown prevails. The same is true for the forward-blocking potential, but only for a very small temperature increase over the room temperature. Further increase will result in a lower breakover since current gains α_1 and α_2 increase with increasing current, and at some elevated temperature their sum will become close enough to unity to cause the breakover to happen. This assumes that the breakover had not happened prior to that due to the increase in the saturation current.

Figure 3.6 Chart for use in range 3×10^{-3} to 3×10^{-2}, gaussian distribution (see Appendix 2). (After Lawrence and Werner [3.5].)

Reprinted with permission from *The Bell System Technical Journal,* © 1960 by The American Telephone and Telegraph Company.

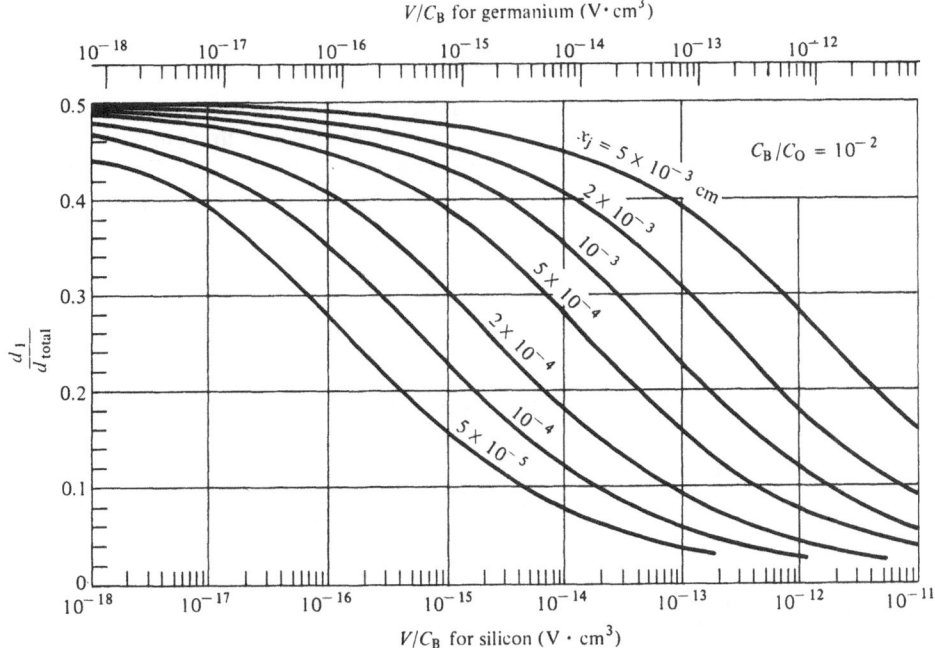

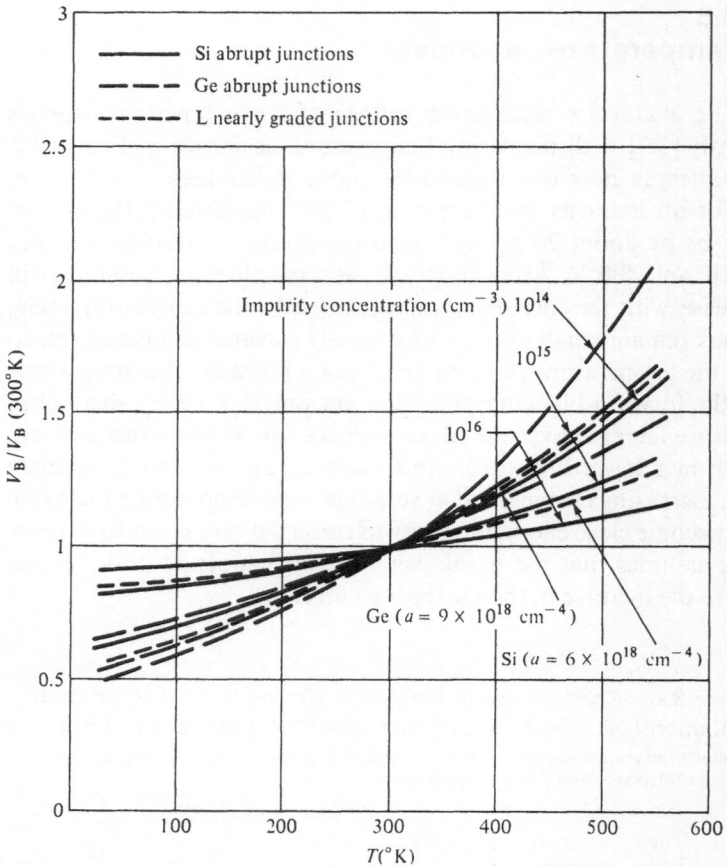

Figure 3.7 Normalized avalanche-breakdown voltage versus lattice temperature. The breakdown voltage increases with increasing temperature, i.e., positive temperature coefficient for breakdown due to avalanche multiplication. (After Crowell and Sze [3.7].)

The current crossing the J_2 junction when the device is in the reverse-blocking mode is

$$I_A = \alpha_P I_A + \alpha_N I_K \qquad (3.17)$$

if the saturation current and avalanche multiplication effects are neglected. I_A and I_K are anode and cathode currents, respectively, in the absence of gate current.

If these effects are not negligible then

$$I_A = M(\alpha_P I_A + \alpha_N I_K + I_{SC}) \qquad (3.18)$$

where I_{SC} is the J_2 junction space-charge–generated current. For the shorted emitter case, we assume $\alpha_2 = 0$ and

$$I_A = M(\alpha_P I_A + I_{SC}) \tag{3.19}$$

From (3.2) and (3.19)

$$\frac{V}{V_B} = \left[1 - \left(\alpha_P + \frac{I_{SC}}{I_A} \right) \right]^{1/m} \tag{3.20}$$

Expression (3.20) shows the high sensitivity of the maximum forward-blocking voltage to the temperature-dependent I_{SC}.

3.6
Surface breakdown

If special care is not taken of the thyristor surfaces, then the high-voltage capability will be limited in all probability by the surface breakdown, which is lower than the Si bulk breakdown; this may lead to the device's destruction and inoperativeness. For this reason, and also to avoid life problems caused by surface ion migration, it is necessary to take appropriate steps to minimize the field at the device surface.

One of the most common methods of surface field minimization consists of junction shaping or beveling introduced by Davies and Gentry [3.7] and modified recently by Cornu [3.9].

3.7
Reverse-conducting thyristor (RCT)

In most current inverter systems, i.e., in circuits for changing a DC into an AC current, an antiparallel rectifier is connected across the thyristor structure and, therefore, the thyristor reverse-blocking potential is not utilized [3.10–3.12]. As a result, the device designer has more latitude in the optimization of such thyristor parameters as the forward-blocking voltage at room and high temperatures, the forward voltage drop at full conduction, and the device speed.

3.7.1
RCT structure

A cross-sectional view of an RCT structure is illustrated by Figure 3.8 [3.10]. The RCT differs from the ordinary SCR in that it has shorting regions not only in the cathode but also in the anode emitter. The antiparallel rectifier is incorporated in the device and consists effectively of the anode metallization, the n$^+$-n-p structure, and the cathode metallization. The metallizations also serve to short selected areas of the cathode and anode emitters. The principal purpose of the additional anode shorts is to prevent the p-n-p transistor from amplifying the thermally generated leakage current of the forward-voltage-blocking junction J_2. The high conductivity

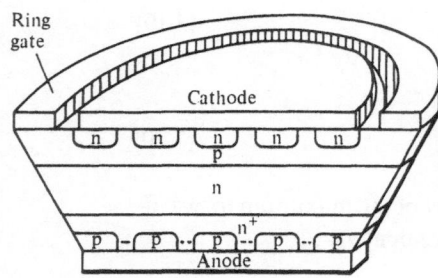

Figure 3.8 Cross section of the RCT structure. (After Kokosa and Tuft [3.10].)

n^{++} layer in the structure serves to foster good ohmic contact to the anode metallization. A high-resistivity n region would form Schottky barriers with anode metallization, making the anode shorts less effective.

Between the n-base region and the p anode there is a highly doped n^+ region which stops the spreading into the p emitter of the forward-blocking junction space-charge region and prevents a punch-through condition.

3.7.2
Maximum forward-blocking voltage of an RCT

In the reverse-conducting thyristors both the anode and the cathode emitters contain shorted regions. *At low current levels* and moderate junction temperatures below 100°C the current gains α_1 and α_2 are small enough to be neglected in expression (3.6) which simplifies, therefore, to

$$V_{BO} \simeq V_B \qquad (3.21)$$

The breakover voltage is, therefore, approximately equal to the breakdown voltage of the central junction J_2. Furthermore, since the device includes an n^+ region for punch-through prevention, the n-base width can be made just large enough for the requisite avalanche breakdown and the forward-blocking voltage. Since in the moderate- and high-voltage devices the n-base concentration is very low (on the order of 10^{13} to 10^{14} cm^{-3}), the device may be designed from the breakdown point of view as a p-i-n diode, which has the advantage of about a 40-percent smaller depletion region than a regular abrupt p-n junction diode for the same avalanche breakdown voltage. In a p-i-n diode there is zero net impurity density in the intrinsic i region so that the electric field has a constant value determined by the ratio of the applied potential and the i-region width. As a result, the avalanche-breakdown voltage is almost proportional to the width of the intrinsic region thickness [3.13]. The proportionality constant is about 20 V/μm.

An RCT can be designed, therefore, with considerably smaller n-base width than a standard SCR and a smaller forward-voltage drop in the conducting state (see Chapter 7).

The full advantage of the p-i-n breakdown improvement can be utilized only when there is no surface breakdown occurring prior to the bulk breakdown. Kokosa and Taft [3.10] used a positively beveled mesa junction (see Chapter 13) in order to minimize the surface fields and to fully use the bulk silicon capability. Their structure was able to forward block a potential of 1500 V at a 150°C junction temperature and exhibited a voltage drop of 3.0 V at a forward current of 1500 A at 90°C.

APPENDIX 1

To obtain the breakdown voltage for a one-sided (p^+-n or n^+-p) junction, start with the concentration of the lightly doped side as an abscissa and follow the ordinate until the *intersection* with the curve labeled "breakdown" and read the breakdown voltage on the ordinate (volts) axis. For example, for a concentration of 1.9×10^{14} atom/cm^3 the breakdown reading is 1000 V.

The depletion layer width on the lightly doped side may be determined by extending the continuous 45-degree inclined line beyond the point of *intersection* until the depletion width line (d microns or d mils) is reached. In our example $x = 80\,\mu m = 3.3$ mils. By further extension of the 45 degrees line, one may obtain junction capacitance per unit area. In our example, junction capacitance is 8.2×10^{-3} pF.

The lines of constant peak electric fields have been plotted by the interrupted 45-degree inclined lines sloping from the left to the right, with the coordinates being the doping density and the applied voltage.

APPENDIX 2

In Figures 3.5 and 3.6 C_B denotes the background concentration; C_O the diffused impurity surface concentration; V, the applied voltage; x_j, the junction depth; $d_{total} = d_1 + d_2$, the total depletion-region thickness; and d_1, that part of the depletion layer width that extends into the more heavily doped junction region.

Figure 3.5 shows the total depletion layer thickness versus the ratio of voltage to background impurity concentration for gaussian distribution and C_B/C_O ratio between 3×10^{-3} and 3×10^{-2}.

As an example, for $C_B/C_O = 3 \times 10^{-3}$ with $C_B = 10^{14}$, when the applied voltage is 100 V, then $V/C_B = 10^{-12}$. If $x_j = 2 \times 10^{-3}$ cm,

37

one obtains from Figure 3.5 a total depletion-layer thickness $d_{total} = 4 \times 10^{-3}$ cm. From Figure 3.6 one obtains $d_1/d_{total} = 0.18$; therefore, $d_1 = 0.72 \times 10^{-3}$ cm and $d_2 = 3.28 \times 10^{-3}$ cm.

References

3.1   A. Herlet. The maximum blocking capability of silicon thyristors. *Solid State Electronics*, *8*: 655–671, 1968.

3.2   J. L. Moll, J. L. Su, and A. C. M. Wang. Multiplication in collector junctions of silicon n-p-n and p-n-p transistors. *IEEE Trans. Electron Devices*, *ED-17* (5): 420–423, 1970.

3.3   S. M. Sze and G. Gibbons. Avalanche breakdown voltages of abrupt and linearly graded p-n junctions in Ge, Si, GaAs, GaP. *Appl. Phys. Lett.*, *8*: 111, 1966.

3.4   J. Olmstead. RCA, Solid State Technology Center, Somerville, N.J., private communication, 1970.

3.5   P. Kannam. RCA, Solid State Technology Center, Somerville, N.J., private communication, 1973.

3.6   H. Lawrence and R. M. Werner, Jr. Diffused junction depletion layer calculations. Monograph 3517, Bell Telephone System Technical Publications, and *BSTJ*, *39*: 389–404, 1960.

3.7   C. R. Crowell and S. M. Sze. Temperature dependence of avalanche multiplication on semiconductors. *Appl Phys. Lett.*, *9*: 242–244, 1966.

3.8   R. Davies and F. Gentry. Control of electric fields at the surface of p-n junctions. *IEEE Trans. Electron Devices*, *ED-11*: 313–323, 1964.

3.9   Jozef Cornu. Field distribution near the surface of beveled p-n junctions in high-voltage devices. *Trans. Electron Devices*, *ED-20* (7): 347–352, 1973.

3.10 R. A. Kokosa and B. R. Tuft. A high voltage, high temperature reverse conducting thyristor. *IEEE Trans. Electron Devices*, *ED-17* (9): 667–672, 1970.

3.11 L. S. Greenberg, and E. F. McKeon. ITR—A new reverse conducting thyristor for horizontal deflection. 22nd Electronic Components Conference, Washington, D.C., May 1972.

3.12 T. Matsuzawa and Y. Usunaga. Some electrical characteristics of a reverse conducting thyristor. *IEEE Trans. Electron Devices*, *ED*-17 (9): 816, 1970.

3.13 J. L. Moll. *Physics of Semiconductors*. New York: McGraw-Hill, 1964.

Some high-injection-level effects

Summary

At high injection levels occurring in power thyristors, the injected carrier concentration may exceed several times the impurity concentration for the lightly doped regions. As a result, semiconductor and device properties are significantly modified as compared to those at moderate or low injection levels.

Due to the presence of a large and equal number of mobile carriers, it is convenient to use the concept of an ambipolar mobility and diffusivity.

At high-carrier-injection levels the mobilities and diffusivities are functions of current density.

Lifetime may be very much affected due to the Auger recombination mechanism, if the injected carrier density is on the order of 10^{18} carriers cm^3

The two other effects of importance at high injection levels are the gain fall-off with current and the presence of a non-negligible voltage drop across p$^+$p or n$^+$n junctions.

4.1
Introduction

Power thyristors operate under steady-state conditions at current densities on the order of several hundred amperes per cm^2. Some of the ratings for device characterization may require current densities on the order of several thousands of amperes per cm^2. Also, during transient conditions current densities may exceed values of 10^4 A/cm^2. At these high current densities and, consequently, at large carrier-injection levels, several silicon crystal properties are considerably modified as compared to the properties at low injection levels. As a result, semiconductor device properties are also changed and are not adequately described by small-signal theories.

A high injection level is defined as one at which the injected minority carrier density approaches or exceeds the existing equilibrium value of majority carrier concentration. At room temperature, all impurities are assumed to be ionized so that the carrier concentration equals the doping impurity concentration. Those regions of the p-n-p-n structure which are not very highly doped are likely to be the most affected by the injection.

4.2
Ambipolar mobility and diffusivity

In the study of semiconductor devices at high injection levels, it is no more possible to consider the hole and electron transport separately since the holes move in a cloud of electrons, and vice versa.

A convenient way to handle this situation [4.1] is to combine the continuity equation for holes with that for electrons and to define two new parameters, the ambipolar mobility, μ, and ambipolar diffusion constant, D, which are algebraic functions of hole and electron mobilities or diffusivities. The resulting single equation for both types of carriers allows concentrating our attention on the excess minority density, while the presence of majority-carrier concentration is taken care of automatically by the newly defined ambipolar parameters. The expressions for the ambipolar mobility and diffusivity are

$$\mu = \mu_n \mu_p (p - n)/(n\mu_n + p\mu_p) \tag{4.1}$$

$$D = D_n D_p (p + n)/(nD_n + pD_p) \tag{4.2}$$

p and n are total hole and electron concentrations, respectively. μ and D are not connected by the Einstein relationship as are the normal mobilities and diffusion constants.

4.3
Mobility and diffusivity versus current density

At low current levels the carrier mobility is limited mainly by two mechanisms: crystal-lattice and doping-impurity scattering. The probability of impurity scattering depends on the total concentration of ionized impurities present in the silicon crystal. Lattice scattering is a result of thermal vibration of the atoms at the lattice sites.

Due to the presence of a large number of injected carriers, a third mechanism, the carrier–carrier scattering becomes significant. If μ_0 is the low-level carrier mobility and μ_{cc} is the carrier–carrier scat-

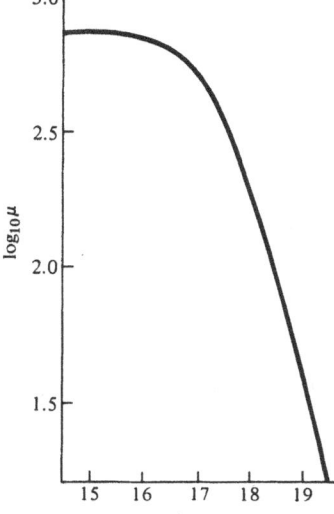

Figure 4.1 Ambipolar mobility μ against carrier density n in silicon at 300°K and $/N_B/\,10^{15}$ cm^{-3}. (After M. Lietz [4.2].)

tered mobility, then the resulting high-injection level mobility μ is obtained from

$$\frac{1}{\mu} = \frac{1}{\mu_0} + \frac{1}{\mu_{cc}} \qquad (4.3)$$

For injected carrier densities smaller than 10^{19} carriers cm^3, the high injection-level ambipolar mobility can be approximated [4.2] by

$$\mu = \frac{A}{n + B} \qquad (4.4)$$

Figure 4.2 Effect of carrier–carrier scattering on carrier diffusion coefficients. (After Howard and Johnson [4.4].)

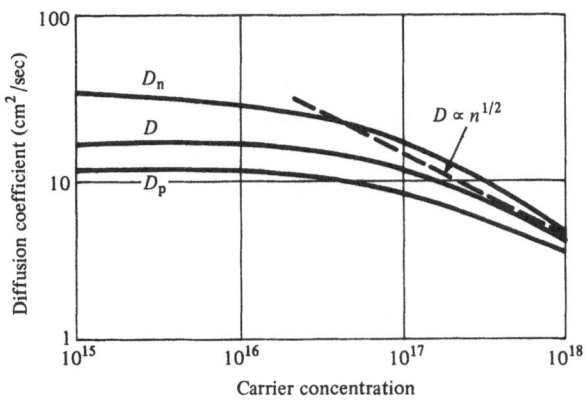

41

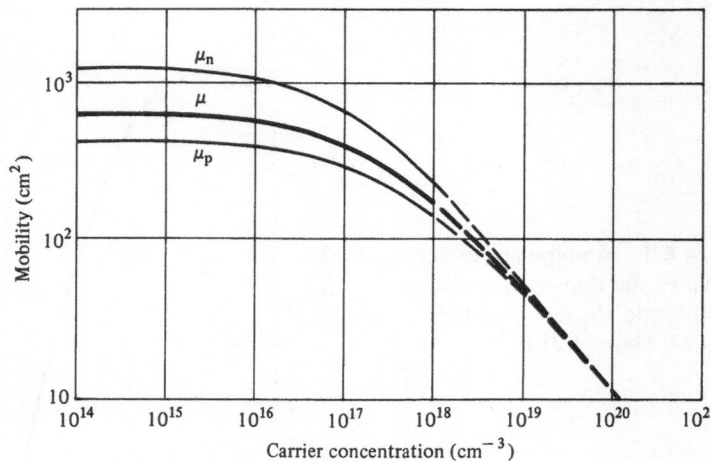

Figure 4.3 Electron, hole, and ambipolar mobilities versus carrier concentration. (After Burtscher *et al.* [4.5].)

where n is the injected carrier concentration and A and B are constants dependent on the crystal doping level. Similar expressions apply to the electron and hole mobilities [4.2, 4.3].

Figure 4.1 is a plot [4.2] of the ambipolar mobility μ versus injected carrier density in n-type silicon at 300°K at the crystal doping level $N_B = 10^{15}$ donors/cm³. Figure 4.2 is a plot of diffusion constants D_n, D_p, and the ambipolar D [4.4]. Figure 4.3 represents the ambipolar and the electron and hole mobilities versus carrier densities up to 10^{20} cm⁻³, as computed by Burtscher *et al.* [4.5].

The decrease in mobility and diffusion constants with increasing carrier density affects appreciably the p-n-p-n device forward drop at high injection levels.

4.4
Lifetime at high injection levels

The recombination of carriers in semiconductors at low and moderately high injected carrier densities is adequately described by the Shockley–Read–Hall theory of carrier recombination through recombination centers. At very high injection levels, other processes may become important: (1) the direct radiative band-to-band recombination which involves two carriers (a hole and an electron) without the intermediary of a trap, and (2) the Auger recombination which can be visualized as an inverse of the impact ionization which causes the avalanche breakdown. In impact ionization, an electron or hole, when accelerated by a very high field, gives its energy to one or more carriers.

In the Auger recombination process the energy of the electron-hole recombination is imparted also to a third carrier which becomes more energetic. It is then a three-carrier process, since it involves two electrons and one hole, or two holes and one electron.

At high injection levels the number of excess electrons n equals the number of excess holes p, so that the charge neutrality is preserved. Under this condition, the rate of recombination may be represented by the expression [4.6 and 4.7].

$$U = -an - bn^2 - cn^3 \qquad (4.5)$$

At low carrier concentrations (4.5) reduces to the familiar expression

$$U = -an$$

where a is the reciprocal of the low injection-level lifetime τ. For $\tau = 5 \; \mu sec$, $a = 2 \times 10^5$. The constant b has a value of 2×10^{-15} cm^3/sec and c is approximately equal to 5×10^{-30} [4.6, 4.7] at room temperature and increases with temperature. According to [4.8–4.10], c is equal to $2 \times 10^{-31} \; cm^6/sec$. The second term due to the direct radiative transistion is usually negligibly small so that (4.5) reduces to

$$U = -an - cn^3 \qquad (4.6)$$

The high injection-level effective carrier lifetime is defined by

$$U = \frac{-n}{\tau_{eff}} \qquad (4.7)$$

Using (4.6) and (4.7), we obtain

$$\tau_{eff} = \frac{1}{a + cn^2} \qquad (4.8)$$

When the injected excess carrier concentration $n = 10^{18} \; cm^{-3}$ and $\tau = 1/a = 5 \; \mu sec$, then $\tau_{eff} = 2.5 \; \mu sec$. At this high injection level the lifetime is reduced from $5 \times 10^{-6} \; sec$, to a value of 1.4×10^{-6} sec, for c equal to $2 \times 10^{-31} \; cm^6/sec$.

At low injection levels but *high doping concentrations* ($> 10^{17}$ cm^{-3}) Auger recombination depends on the silicon doping levels; effective carrier lifetime still follows (4.8). However, the constant c is equal to $2.9 \times 10^{-31} \; cm^6/sec$ [4.11].

4.5
High–low junctions at high current densities

At very high injected current densities the voltage drop across a p^+-p or n^+-n junction may become significant [4.12]. In this type of a junction the intervening space-charge region is determined by the same rules as in the case of pn junction, except that its width is much smaller and no depletion layer exists, so it cannot support any reverse potential. The majority carrier current through such a forward-biased junction is similar to the minority carrier current of a p-n junction and is determined from

$$J = J_0 \exp \frac{qV}{KT} \qquad (4.9)$$

where J_0 is a constant and V the applied potential. Consequently, the voltage drop

$$V = \frac{KT}{q} \ln(J/J_0) \qquad (4.10)$$

The high–low junction is relatively impermeable to the passage of minority carriers in either direction. There is little impedance to the flow of majority carriers at low current densities, however.

4.6
Current gain fall-off

Because of the presence of conductivity modulation in the base region of a transistor at high current densities, the current gain falls off with increasing current density. This effect was first analyzed by Webster and is discussed fully in any standard handbook on transistor theory such as [4.13].

An approximate expression for the common-emitter current gain at high-level conditions is given by Clark [4.14] as

$$h_{FE} \simeq \frac{K_1}{J_s J_c W^2} \qquad (4.11)$$

where J_c is the collector current; J_s, the emitter saturation current; W, the base width; and K_1, a constant.

References

4.1 S. M. Sze. *Physics of Semiconductor Devices*. New York: John Wiley, 1969.

4.2 M. Lietz. Analytical expressions for carrier distribution in a wide thyristor base under high injection. *Electron. Letts.*, 8: 275–284, 1965.

4.3 D. M. Caughey and R. E. Thomas. Carrier mobilities in silicon empirically related to doping and field. *Proc. IEEE, 55*: 2192–2193, 1967.

4.4 N. R. Howard and G. N. Johnson. p^+i-n^+ silicon diodes at high forward current densities. *Solid State Electron., 8*: 275–284, 1965.

4.5 J. Burtscher, F. Dannhäuser, and J. Krause. Die Rekombination in Thyristoren and Gleichrichtern auz Silizium: Ihr Einfluss auf die Durchlasskennlinie und das Freiwerdezeitverhalten. *Solid State Electron., 18*: 35–63, 1975.

4.6 E. A. Bobrova, V. S. Vavilov, and G. N. Galkin. Interband impact recombination in silicon and germanium. *Sov. Phys. Solid State, 13*: 2982, 1972.

4.7 L. M. Blinov, E. A. Bobrova, V. S. Vavilov, and G. N. Galkin. Recombination of equilibrium carriers in silicon in the case of high photoexcitation levels. *Sov. Phys. Solid State, 9*: 2537–2542, 1968.

4.8 J. Krause. Auger recombination in the middle region of a forward conducting silicon rectifier and thyristor. *Solid State Electron., 17*: 427–429, 1974.

4.9 N. G. Nilsson and K. G. Svantenson. The spectrum and decay of the recombination radiation from strongly excited silicon. *Solid State Commun., 11*: 155–159, 1972.

4.10 J. P. Woerdmann. Some optical and electrical properties of a laser-generated free carrier plasma in Si. *Phil. Res. Rep. Suppl., 7*: 1–80, 1971.

4.11 J. D. Beck and R. Conradt. Auger recombination in silicon. *Solid State Commun., 13*: 93–95, 1975.

4.12 D. P. Kennedy. Mathematical Simulation of the Effects of Ionizing Radiation on Semiconductors. Contract No. F19628-67-C0116; final report February 1, 1970. Air Force Cambridge Research Laboratories.

4.13 A. B. Philips. *Transistor Engineering*. New York: McGraw-Hill, 1962.

4.14 E. E. Clark. High current density beta diminution. *IEEE Trans. Electron Devices, ED-17* (9): 661-666, 1970.

The gate-triggered SCR turn-on transient

Summary

Soon after the application of the forward bias to the p gate of an SCR in the off-state, the cathode starts emitting electrons in an area in the vicinity of the gate contact. These electrons are collected by the center junction, are swept into the n base, and slightly forward bias the p emitter (the anode junction). Holes begin to flow in the direction of the collecting center junction; when they reach the p base they increase further the electron injection. This process continues until eventually the holding current density J_h of the device is exceeded and the device is latched in. The time necessary to reach 10 percent of the final anode current—counted from the beginning of the gate current pulse—is designated as the delay time. The time interval between the 10 and 90 percent of I_A is called the rise time t_r. Beyond the 0.9 I_A point, the current increases usually with much smaller rate: the current spreads laterally from a comparatively small area until the whole cathode area conducts. The full current is reached after some time t_s. These three time intervals are evaluated on the basis of the thyristor charge-control model.

5.1
Introduction

The analysis of the turn-on transient initiated by a gate-current pulse is greatly complicated by the dependence of the many device parameters—such as lifetimes, transit times, and current gains—on the current density varying over very wide range during the turn-on process.

After the application of a trigger pulse to the gate, a certain time interval elapses before the SCR starts conducting and the current

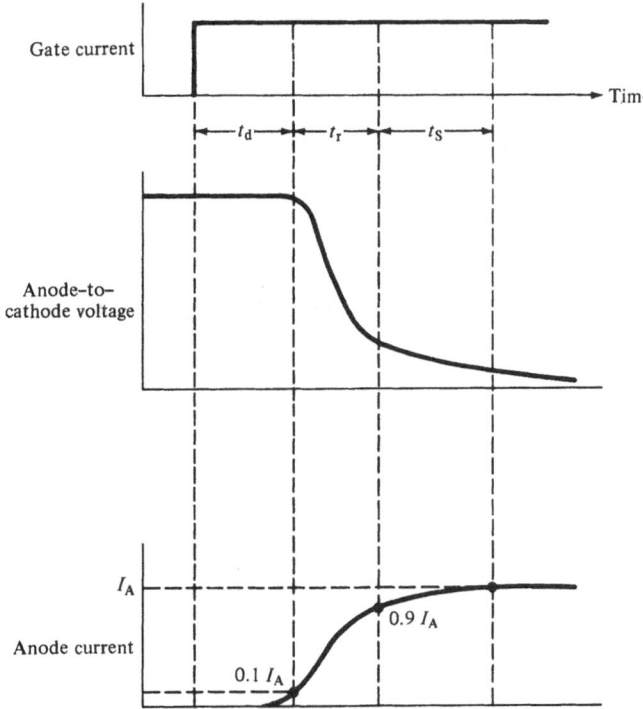

Figure 5.1   Typical turn-on waveforms.

reaches its full value determined by the circuit impedance. While the current is increasing, the voltage drop across the device goes down from the high off-state to the low on-state value. Figure 5.1 shows the current as a function of time and in relation to a step-function gate-current pulse.

The delay time t_d is defined as the time interval necessary to reach a 10-percent level of the final anode current. The rise time t_r, also defined arbitrarily, is the time required to reach 90 percent of the maximum anode current. In the meantime, the current spreads from the initial turn-on area until the entire device is conducting. The time interval between the 90-percent point and the final current value point is designated by t_s.

The analysis of the various time delays profits greatly from the charge control model of an SCR [5.1–5.3] which leads to interpretations in terms of device design parameters. For this reason and also because of its general usefulness in the thyristor studies, we devote some space to the introduction to the charge-control concept.

47

5.2
Charge-control model of a bipolar transistor

5.2.1
Active region

The concept of the charge control was introduced originally by Beaufoy and Sparkes [5.4–5.6] as a mathematical–physical tool in the analysis of switching transistors. In this model, the base, the emitter, and the collector currents are all related to the excess electric charge in the transistor base. When the transistor's emitter-base junction is forward biased, *majority* carriers in excess of those existing in equilibrium enter the device base—holes for the n-p-n and electrons for the p-n-p transistor. The excess charges are provided by the applied base current and redistribute themselves in a negligibly small time equal to the semiconductor relaxation time. The forward-biased emitter injects the minority carriers into the base. Since we assume that charge neutrality prevails in the base, the number of injected majority carriers and injected minority carriers are almost equal to each other. We recall that the current continuity expression in the p base of an n-p-n transistor is

$$\frac{\partial J}{\partial x} = q\frac{\partial n}{\partial t} + q\frac{n}{\tau} \tag{5.1}$$

where J is the current density, q is the electronic charge, and τ is the electron lifetime. The density of electrons is a function of time and position x, i.e., $n(x, t)$. The left- and the right-hand side of Equation (5.1) can be multiplied by dx and integrated between the limits of $x = 0$ and $x = W =$ base width. The integration gives the expression for the difference between the current entering the base at $x = 0$, i.e., the emitter current, and the current leaving the base at $x = W$, i.e., the collector current. Neglecting the collector leakage current, this difference is the base current J_B

$$J_B = J_E - J_C = \frac{dq_N}{dt} + \frac{q_N}{\tau} \tag{5.2}$$

In (5.2) q_N is the total excess charge per unit area in the base. No approximations or restrictions as to the specific charge distributions are implied in Equation (5.2). Therefore, this expression holds for the diffused as well as for the uniform base transistors. The interpretation of (5.2) is that the current entering the base is used in part for storing the charge and in part for recombination. The same thought can be expressed by stating that the excess charge stored, q_N, is altered by the supply rate, J_B, and by the loss rate q_N/τ.

The important feature of this approach is that *through the integration of the continuity equation, the carrier density as a variable is replaced by the total charge in the base.*

In the charge-control theory, the relations between the charges and the base, collector, and emitter currents in the steady state are determined from

$$\tau = \frac{q_N}{J_B} \qquad \tau_C = \frac{q_N}{J_C} \qquad \tau_E = \frac{q_N}{J_E} \tag{5.3}$$

where τ, τ_C, and τ_E are the base, the collector, and the emitter time constants, respectively.[1] Similar time constants can be defined likewise for the inverse transistor.

Equation (5.2) relating the base current to the base charge in all accuracy, should be somewhat modified to take into account the variation of charges stored in the transistor emitter depletion region. For the first-order analysis these charges may be neglected in many cases.

For the cases when the emitter efficiency differs appreciably from unity, it is necessary to take into account the recombination of the emitter charge q_E so that the total base current expression (5.2) is changed to

$$J_B = \frac{dq_N}{dt} + \frac{q_N}{\tau} + \frac{q_E}{\tau_E} \tag{5.4}$$

with τ_E being the minority carrier lifetime in the emitter.

5.2.2
Saturation mode

When a transistor is switched from the active into the saturation mode, both the collector and the emitter junctions become forward biased. While the device operates in the active region, only the emitter injects minority carriers into the base; in the saturated mode, on the other hand, not only the emitter but also the collector inject. As a result, the *total base charge* q_B consists of the charge due the emitter injection q_N, which we call the "normal" charge, and an additional charge due to the collector injection which we designate by q_S; therefore

$$q_B = q_N + q_S \tag{5.5}$$

In most cases there is an appreciable charge present in the collector body as well, especially when the collector is lightly doped and there

---

[1] τ_C is interpreted as the base transit time and τ is the minority carrier lifetime in the base.

is not only injection of minority carriers from the collector to the base, but from the base into the collector. If the charge in the collector body is q_{CS}, Equation (5.5) should be changed to

$$q_B + q_{CS} = q_N + q_S \tag{5.6}$$

In order to remove the transistor out of saturation, it is therefore necessasy to remove all the charges which were stored due to the presence of a forward-biased collector, i.e., $q_S = q_B + q_{CS} - q_N$.

For the device in saturation, the relation (5.4) should be replaced by

$$J_B = \frac{dq_B}{dt} + \frac{q_B}{\tau} + \frac{q_E}{\tau_E} \tag{5.7}$$

5.3
Charge-control model of a p-n-p-n structure [5.1]

5.3.1
Base charges and currents

A p-n-p-n structure can be represented by a two-transistor equivalent circuit as per Figure 5.2, and for each of the two transistors it is possible to write the charge-control equations. In particular, when the device is in full conduction and all three junctions are forward biased, the total base charge for each transistor can be expressed by Equation (5.5) since both transistors are in saturation. Figure 5.3 shows an approximate excess-charge distribution in the p-n-p-n device for a situation when the current through the device is above the holding current.

From the equivalent circuit of Figure 5.2 we may observe that the base current J_{CN} of the p-n-p transistor is the same as the collector

Figure 5.2   SCR equivalent circuit.

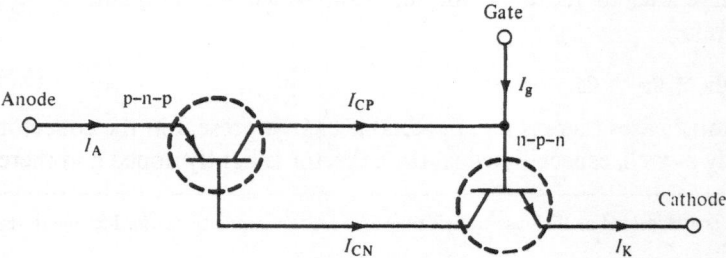

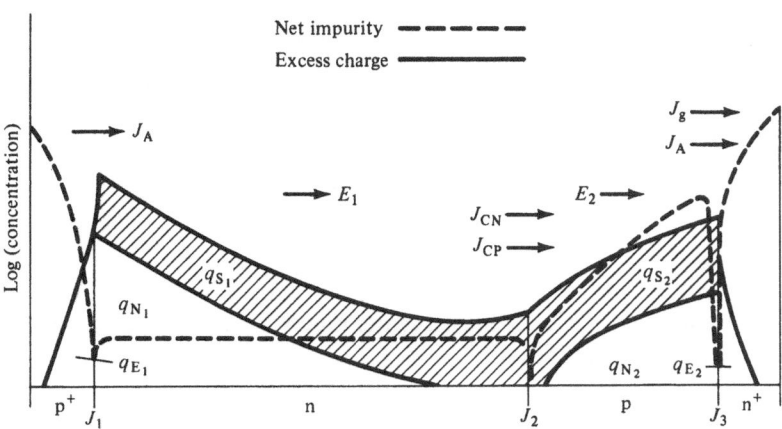

Figure 5.3 Forward-conduction charge model of a p-n-p-n device; J_1, J_2, J_3 are device junctions. (After Davies and Petruzella [5.1].)

current of the n-p-n transistor. Similarly, the p-n-p collector current J_{CP}, when added to the gate current J_g, is the base current of the n-p-n transistor. We can write, therefore [5.1]

$$J_{CN} = \frac{dq_{B_1}}{dt} + \frac{q_{B_1}}{\tau_1} + \frac{q_{E_1}}{\tau_{E_1}} = \text{n-base current} \qquad (5.8)$$

with τ_1 and τ_{E_1} being minority-carrier lifetimes in the n base and p emitter, respectively; and q_{B_1} and q_{E_1} are n-base and p-emitter charges. The majority charge which enters the base goes into storage—recombination in the base and recombination in the emitter. For the base current of the n-p-n transistor we have

$$\text{p-base current} = J_{CP} + J_g = \frac{dq_{B_2}}{dt} + \frac{q_{B_2}}{\tau_2} + \frac{q_{E_2}}{\tau_{E_2}} \qquad (5.9)$$

where q_{B_2} and q_{E_2} are p-base and n-emitter charges, respectively; τ_2 and τ_{E_2} are the p-base and n-emitter minority-carrier lifetimes, respectively. From the charge-control model we have also, according to Equation (5.3),

$$J_{CP} = \frac{q_{N_1}}{\tau_{C_1}} = \frac{q_{B_1}}{\tau_{C_1}} - \frac{q_{S_1}}{\tau_{C_1}} \qquad (5.10)$$

$$J_{CN} = \frac{q_{N_2}}{\tau_{C_2}} = \frac{q_{B_2}}{\tau_{C_2}} - \frac{q_{S_2}}{\tau_{C_2}} \qquad (5.11)$$

51

where q_{S_1} and q_{S_2} are saturation charges in the n- and p-base, respectively, and τ_{C_1} and τ_{C_2} are the carrier transit times in the n-base and p-base, respectively. Since the anode current (Figure 5.2)

$$J_A = J_{CP} + J_{CN} \tag{5.12}$$

we obtain, by adding (5.10) and (5.11),

$$J_A = \frac{q_{B_1} - q_{S_1}}{\tau_{C_1}} + \frac{q_{B_2} - q_{S_2}}{\tau_{C_2}} \tag{5.13}$$

For the *reverse- or zero*-biased center junction, i.e., for zero-saturation charge this simplifies to

$$J_A = \frac{q_{B_1}}{\tau_{C_1}} + \frac{q_{B_2}}{\tau_{C_2}} \tag{5.14}$$

5.3.2
Transit times

When the transport of the minority carriers takes place only by diffusion, the transit time through the base may be obtained from

$$\tau_C = \frac{W^2}{2D} \tag{5.15}$$

where W is the base width and D is the diffusion constant of the minority carriers.

On the other hand, if the high-field conditions prevail in the base and the diffusion can be neglected, the transit time may be calculated from

$$\tau_C = \frac{W}{\mu E} \tag{5.16}$$

where μ is the mobility of the minority carriers and E is the electric field. Otherwise, the above expressions may serve only for rough transit-time estimates.

5.4
Delay time

Following the application of the p-gate current pulse, the electrons injected by the n emitter cross the p base, arrive at the collecting center junction J_2, and get into the n base, charging it negatively; the p-n-p transistor's base-emitter junction becomes thus slightly forward biased so that the p emitter (the anode) starts injecting holes into the n base. These holes in turn are transported through the wide n base, get collected by J_2, and go to the p base, increasing consequently the n-emitter forward bias. This process continues until the anode current reaches its holding value I_h, the device turn on is established, and the current continues growing with some definite rate until the end of the delay time interval.

Gentry et al. [5.2] and Davies and Petruzella [5.1] approached the problem of the current build-up from the charge-control point of view. They have assumed that during the delay time both transistors were out of saturation, that the emitter-charge terms in Equations (5.8) and (5.9) could be neglected, and that the transit times through both bases were independent of the current density. By combining Equations (5.8), (5.9), (5.10), and (5.11), one obtains

$$\frac{dq_{B_1}}{dt} = \frac{q_{B_2}}{\tau_{C_2}} - \frac{q_{B_1}}{\tau_1} \tag{5.17}$$

$$\frac{dq_{B_2}}{dt} = \frac{q_{B_1}}{\tau_{C_1}} - \frac{q_{B_2}}{\tau_2} + J_g \tag{5.18}$$

with

$$J_A = \frac{q_{B_1}}{\tau_{C_1}} + \frac{q_{B_2}}{\tau_{C_2}} \tag{5.19}$$

Solving this system of differential equations, Davies and Petruzella [5.1] obtained an expression for the anode current density of the form

$$J_A = C_1 J_g \left[C_2 \exp\left(\frac{t}{\tau_s}\right) + C_3 \exp\left(\frac{t}{\tau_f}\right) + C_4 \right] \tag{5.20}$$

The constants C_1, C_2, C_3, and C_4, as well as τ_s and τ_f, are rather complicated algebraic functions of lifetimes and transit times in both device bases.

The assumption of the transit time independence from the current density, especially in the ungated n base, really does not apply to most commercial thyristors. The authors conclude that in an actual device the rate of current increase will be somewhat faster than exponential if the effects of the current on the transit time are taken into account. The results of this analysis apply to devices with narrow emitter stripes.

Bergman [5.3] makes the assumption that for thyristors with the gate terminal located on the base of the more sensitive transistor, and for relatively large gate currents, the initial rise in current is made up largely of the collector current of the more sensitive transistor. He further assumes that the less sensitive transistor, which for low-current levels has a slower rise time, does not make a contribution to the turn-on process until after the 10 percent of the final anode current has been exceeded.

53

The rough estimate of the current to rise to 10 percent of the final anode current and for a step-function gate signal J_g, the anode current density can be expressed [5.3] by

$$J_A = \frac{\tau_2}{\tau_{C_2}} J_g \left[1 - \exp\left(\frac{t}{\tau_2}\right) \right] \tag{5.21}$$

where τ_2 and τ_{C_2} are the electron lifetime and transit time in the p base (gate contact on the p base). Since $\tau_2/\tau_{C_2} \simeq h_{FE}$, the common emitter current gain, one obtains from (5.21) the delay time

$$t_{d_1} = \tau_{C_2} \left[h_{FE} \ln\left(1 - \frac{K}{10 h_{FE}} \right)^{-1} \right] \qquad K = \frac{I_A}{I_g} \tag{5.22}$$

The delay time expression is valid only for $K < 100$, i.e., for relatively large currents. The expression contains h_{FE}, which varies with current; however, the term in the square brackets does not vary as much as h_{FE} does. For more accuracy, it is necessary to add the time t_{d_2} needed to charge the cathode junction capacitance

$$t_{d_2} = \int_0^{V_g} \frac{C(V)dv}{J_g} \tag{5.23}$$

Where $C(V)$ is the cathode junction capacitance per unit area and V_g is the gate bias.

These results show that the delay time is shortened by the application of high gate currents and by the existence of the short transit time of the more sensitive transistor to which the gate contact is made.

5.5
Rise time with a resistive load

For a very crude approximation of the sum of the rise time and the delay time, Sze [5.7] neglects entirely the recombination in both bases so that from (5.8), (5.9), (5.10), and (5.11) we obtain

$$\frac{dq_{B_1}}{dt} = \frac{q_{B_2}}{\tau_{C_2}} \tag{5.24}$$

$$\frac{dq_{B_2}}{dt} = J_g + \frac{q_{B_1}}{\tau_{C_1}} \tag{5.25}$$

Differentiating the first equation and substituting for dq_{B_2}/dt from the second, the following expression is obtained:

$$\frac{d^2 q_{B_1}}{dt^2} - \frac{q_{B_1}}{\tau_{C_1}\tau_{C_2}} = \frac{J_g}{\tau_{C_2}} \tag{5.26}$$

The solution is of the form:

$$J_A \propto J_g \left\{ \exp \left[\frac{t}{(\tau_{C_1} \tau_{C_2})^{1/2}} \right] - 1 \right\} \qquad (5.27)$$

Consequently, the rise time from the 0- to 90-percent level is proportional to the time constant:

$$t_d + t_r \propto (\tau_{C_1} \tau_{C_2})^{1/2} \qquad (5.28)$$

Equation (5.27) also shows that the anode current rises faster with larger gate currents.

Bergman [5.3] solves charge-control equations without neglecting the recombination term q_B/τ and for the case when neither of the transistors is saturated and the collecting junction J_2 is slightly reverse or zero biased.

Bergman avoids the direct solution of the system of the charge-control equations. In place of it, starting with the charge equations, he derives transfer functions for each transistor and a loop transfer function $G(p)$ of the SCR equivalent circuit. The poles of the loop transfer function p_1 and p_2 are found by equating $G(p)$ to unity. The response to the gate signal (assumed to be the delta function) consists of the sum of two exponentials—one rising and one falling rapidly, with time constants being reciprocals of p_1 and p_2. For currents larger than the holding current, the negative exponential decays rapidly so that the rise in current is essentially exponential:

$$I_A \approx I_0 \exp(p_1 t) \qquad (5.29)$$

I_0 is a constant and p_1 is the positive root of the equation $G(p) = 1$. For the p- and n-emitter efficiencies assumed to be equal to unity, the rise time (10 to 90 percent) derived by Bergman is given approximately by

$$t_r \simeq 2 \left(\frac{\tau_{C_1} \tau_{C_2}}{\alpha_N + \alpha_P - 1} \right)^{1/2} \qquad (5.30)$$

where α_N and α_P are common base-current gains, and τ_{C_1} and τ_{C_2} are minority carrier-transit times. The rise time depends, therefore, upon the geometric mean of the carrier-transit times of the bases and is inversely proportional to $(\alpha_N + \alpha_P - 1)^{1/2}$. Since the alphas and transit times are current and voltage dependent, Formula (5.30) can serve only as a guide for the rise-time optimization.

All the present-day rise-time theories suffer in accuracy from the assumption of the nonforward-biased center junction J_2.

5.6
Rise time with inductive load

The above considerations regarding the rise time with a purely resistive load assumed that the collecting center junction was kept out of saturation so that the resulting rate of current rise was only device limited.

When the inductance of the thyristor external load circuit is not negligible the rate of current rise is dependent on the external inductance L. The rate of charge increase in the p-n-p-n structure becomes higher than that necessary to support the current [5.1]. This way there is a build up of an extra charge (the saturation charge) which heavily forward biases the center junction. As a result, the voltage drop across the device is appreciably reduced. The analysis of the current rise under these conditions, which require the solution of charge-control equations with the inclusion of stored charges, becomes intractable, except for some special situations discussed in the work of Davies and Petruzella [5.1].

The sum of the delay time and the rise time, $t_d + t_r = t_{gt}$, is normally given by thyristor manufacturers. Figure 5.4 shows gate-controlled turn-on time versus gate trigger current curve for the 20-A S 6200 silicon-controlled rectifier [5.8]. The effect of the gate current on the turn-on time t_{gt} is very clearly visible.

Figure 5.4 Gate-controlled turn-on time versus gate-trigger current for SCR S6200 [5.8].

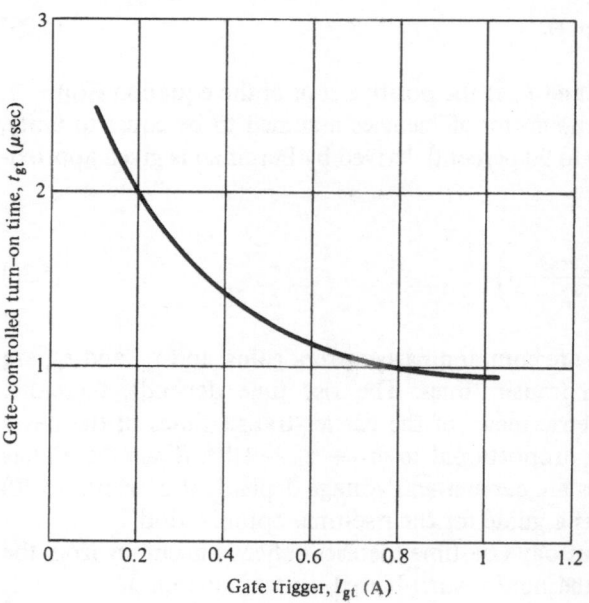

5.7
Propagation of the on state

Most of the thyristor types used in the power circuits do not have narrow emitters; these are encountered only in the fastest devices with interdigitated geometries. For wide emitters, it is necessary to take into account the time required by the turn-on region to propagate along the cathode; the steady state is reached only when the entire cathode emits electrons.

In large p-n-p-n structures, the gate current pulse turns only a small part of the device immediately adjacent to the gate electrode. After that, the on-state spreads to the rest of the cathode emitter until, eventually, the whole device area is turned on. For large-size cathodes, the propagation time is usually considerably larger than the current rise time and has, therefore, an important effect on the device dynamic behavior. The propagation times range from a few microseconds to tens of microseconds depending on the device lateral dimensions. Longini and Melngailis [5.9], Dodson and Longini [5.10], and Grekhov et al. [5.11] assume that the on-state propagation is due to the lateral diffusion of carriers from the active region into the off region. Dodson and Longini [5.10] have constructed special SCRs to permit the direct observation of the lateral spread of the turn-on within the structure and have found out that the spreading velocity of the on state and the load current at high current densities are related approximately by the expression $v_s \propto J^{1/n}$, where J is the anode current density and n is a constant.

Ruhl [5.12] attributes the propagation of the on state to the presence of an electric field, and neglects entirely the diffusion effects. Somos and Piccone [5.13] demonstrated the validity of the model for at least some range of current densities.

At first, when the current is already turned on in the cathode region close to the gate, a large current is flowing through a small area from cathode to anode. That part of the p base where large current is crossing is obviously at a higher positive bias than the rest of the base. As a result, according to Ruhl [5.12] a lateral field is established that drives the majority carriers (holes in the p base) out of the edge of the turn-on region toward the off region. As soon, however, as the base resistivity becomes conductivity modulated, the potential difference between these two points becomes negligibly small. Consequently, the electric field is effective only over a small distance from the edge of the active region. The value of the lateral field is typically about 5 V/cm and the length of the region where the field is effective may range between 150 and 400 μm. The model assumes that the spreading velocity is a linear function of the hole current in the p base and that the lateral field in the transition region is constant. Under these conditions, the spreading velocity is of the form

$$v_s = A \ln J + B \tag{5.31}$$

where J is the current density and A and B are constants. The field theory predicts that emitter shorts (called shorting dots or holes), when placed anywhere in the middle of the emitter, will slow down the propagation velocity, since they shunt the lateral conduction current out of the p base.

Recently, Matsuzawa [5.14] experimentally studied the spreading velocity of the on state in fast, gold-doped thyristors by measuring the on-state voltage across the thyristor. The thyristor under test had its gate biased with several hundred milliamperes DC, and turned on a small area next to the gate contact. At this stage, when the thyristor was subjected to a square wave load current pulse, the on-state voltage across the thyristor first increased and then decreased with passing of time and became constant after some determined time t_s. When the voltage across the thyristor has reached its final constant value, the entire thyristor area had been turned on. From the knowledge of the device geometry and of the spreading time t_s, it was possible to compute the *average* spreading velocity.

5.7.1
Current density effect

Matsuzawa's experimental results show that the average spreading velocity is proportional to nth root of the current density, in agreement with the results of Dodson and Longini [5.10]:

$$v_s = J^{1/n} \tag{5.32}$$

This relation held all the time whenever the resistivity, the base width, and other design parameters of the thyristor were changed. The value of n was approximately equal to 3. At the very low current densities on the order of 10–20 A/cm^2, the relationship (5.32) does not hold. In this range the spreading velocity is higher than predicted by the Equation (5.32) because only a part of the device is ever turned on. In the range from about 40 to 1000 A/cm^2, the spreading velocity varied from about 2.5×10^3 to about 9×10^3 cm/sec.

5.7.2
Effect of gold-diffusion temperature

In order to achieve shorter turn-off times it is customary to use gold-doped silicon. While the gold doping is beneficial from the turn-off point of view, it is deleterious from the spreading-velocity point of view. Matsuzawa's results [5.14] show that the spreading velocity can be expressed as

$$v_s \propto \tau^{1/2} \qquad (5.33)$$

where τ is the minority carrier lifetime in the n-base region. This confirms the theoretical results of Longini and Melngailis [5.9]. The spreading velocity decreases as the gold-diffusion temperature increases and lifetime decreases. For the SCR structure studied and a constant current density, v_s varied typically from 3×10^3 to 20×10^3 cm/sec as the lifetime τ increases from 1 to 30 μsec. The author attributes the lifetime effect on v_s to the variation of current gains α_1 and α_2.

5.7.3
Base-width effect

The results of Matsuzawa [5.14] demonstrate that the spreading velocity is inversely proportional to the n-base or p-base width as

$$v_s \propto \frac{1}{W} \qquad (5.34)$$

This agrees with the data of Dodson and Longini [5.10] and Somos and Piccone [5.13]. It is in conflict, however, with Ruhl's [5.12] theory; he reported a dependence primarily on the p-base width and only slight dependence on the n-base width. In the Matsuzawa [5.14] experiments, both p- and n-base widths have nearly the same effect on the spreading velocity. According to Matsuzawa, this tendency seems due to the current gains of the n-p-n and p-n-p transistors decreasing with increased base width. Combining relations (5.33) and (5.34) and assuming constant diffusivity D, we see that

$$v_s \propto \frac{\tau^{1/2}}{W} \propto \frac{L}{W} \qquad (5.35)$$

where L is the minority-carrier diffusion length. The quantity L/W in this last expression is related closely to the transistor-transport factor:

$$\beta_T = \text{sech}\, \frac{W}{L} \qquad (5.36)$$

High-voltage devices require large n-base widths and, consequently, exhibit smaller spreading velocities.

5.7.4
Shorting-holes (dots) effect

The inclusion of shorting dots in the emitter region of an SCR increases the dv/dt as well as the thermal capability of the device. It has been found experimentally that the spreading velocity decreases as the shorting-dot density increases, due to the reduction in the lateral flow rate [5.14]. This agrees essentially with Ruhl's data [5.12].

5.7.5
Gate-current effect

Experimental results of Dodson and Longini [5.10] indicate that for a given gate geometry the amplitude and width of the gate signal has little influence on the spreading velocity. Once the spreading of the on state has been started, the gate loses its control.

References

5.1   R. L. Davies and J. Petruzella. p-n-p-n charge dynamics. *Proc. IEEE, 55*: 1318–1330, 1967.

5.2   F. E. Gentry, F. W. Gutzwiller, N. Holonyak, and E. E. Von Zastrow. *Semiconductor Controlled Rectifiers.* Englewood Cliffs, N.J.: Prentice-Hall, 1964.

5.3   G. D. Bergman. The gate triggered turn-on process in thyristors. *Solid State Electron, 8*: 757–765, 1965.

5.4   R. Beaufoy and J. J. Sparks. The junction transistor as a charge-controlled device. *ATEJ, 13*: 310–327, 1957.

5.5   J. J. Sparks and A. Beaufoy. The junction transistor as a charge controlled device, *Proc. IRE, 45*: 1740–1742, 1957.

5.6   J. J. Sparks. A study of the charge control parameters of transistors. *Proc. IRE, 48*: 1696–1705, 1960.

5.7   S. M. Sze. *Physics of Semiconductor Devices.* New York: Wiley-Interscience,1969.

5.8   RCA Solid State Data Book Series. Thyristors, rectifiers. Somerville, N.J., 1975.

5.9   R. L. Longini and J. Melngailis. Gated turn-on of four layer switch. *IEEE Trans. Electron Devices, ED-10*: 178–185, 1963.

5.10 W. H. Dodson and R. L. Longini. Probed determination of turn-on spread of large area thyristors. *IEEE Trans. Electron Devices, ED-1e*: 478, 1966.

5.11 V. Grekhov, M. E. Levinshtein, and A. I. Uverov. Simple model for the propagation of the on-state along a p-n-p-n structure. *Sov. Phys.—Semicond.*, *5* (6): 978–981, 1971.

5.12 H. J. Ruhl, Jr. Spreading velocity in the active area boundary in a thyristor. *IEEE Trans. Electron Devices*, *ED-17* (9): 672–680, 1970.

5.13 I. Somos and D. E. Piccone. Behaviour of thyristors under transient conditions. *Proc. IEEE*, *55* (8): 680–687, 1967.

5.14 Takeo Matsuzawa. Spreading velocity of the on-state in high speed thyristor. *Trans. IEEE Japan*, *98-c*: 16–21, 1973.

Also of interest

J. Cornu and A. Jaecklin. Processes at turn-on of thyristors. *Solid States Electron.*, *18*: 683–689, 1975.

J. McGhee. A transient model for a three terminal p-n-p-n switch and its use in predicting the gate turn-on process. *Int. J. Electron.*, *35* (1): 73–79, 1973.

6

Nongated, undesirable thyristor triggering

Summary

A thyristor can be turned on not only by the application of the gate current, but also by such mechanisms as the thermal effect, carrier generation by radiation, or the application of rapidly rising and/or very high forward potential. The thermal effect and the dv/dt effects are undesirable in thyristors. Light irradiation, if so desired, may be utilized in light-activated switches.

The best technique to improve thyristor capability from the operating temperature and dv/dt point of view is to use the so-called "shorted" emitter which bypasses the thermal and displacement currents and makes the device less sensitive to these effects.

6.1
Introduction

The gate triggering of a thyristor represents the most common method of the intentional device turn-on. The device can be turned on, however, by other mechanisms which are not usually used for intentional triggering. Unintentional switching may take place due to thermal effects, exposure to light or other radiation, or by the application of rapidly rising or very high forward potential.

6.2
Thermal turn on

The center junction of a p-n-p-n structure is, in effect, a common collector for both the n-p-n and the p-n-p transistors of the two-transistor circuit analog (Figures 1.7 and 6.1). The device is assumed to be in the forward-blocking state, i.e., a positive potential is applied

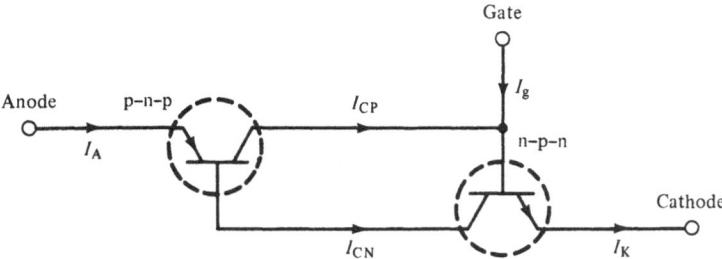

Figure 6.1 SCR equivalent circuit.

to the anode, but only a small current flows in the circuit due to the thermal carrier generation in various parts of the device.

The electron and hole DC currents collected by the center junction are

$$I_{CN} = \alpha_N I_K + I_{SN} \qquad (6.1)$$

$$I_{CP} = \alpha_P I_A + I_{SP} \qquad (6.2)$$

where α_N, α_P are common base current gains of the n-p-n and p-n-p transistors, respectively; I_{SN} and I_{SP} are electron- and hole-saturation currents, respectively; I_K is the cathode current; and I_A is the anode current. These are DC currents; the displacement (capacitance) currents are not included in the equations. The total saturation current is

$$I_S = I_{SN} + I_{SP} \qquad (6.3)$$

Since no gate current is applied

$$I_A = I_K \qquad (6.4)$$

From (6.1), (6.2), (6.3), and (6.4) we have

$$I_A = \frac{I_S}{1 - \alpha_N - \alpha_P} \qquad (6.5)$$

It can be easily shown following the procedure of the Appendix of Chapter 1 that

$$\frac{dI_A}{dI_S} = \frac{1}{1 - \alpha_2 - \alpha_1} \qquad (6.6)$$

where α_1 and α_2 are the p-n-p and the n-p-n transistor small-signal alphas.

The saturation current I_S is generated mainly in the depletion region of the reverse-biased J_2 junction. A surface leakage may be often much higher, at room temperature at least, than the bulk-saturation current, but the situation may be just opposite at elevated

63

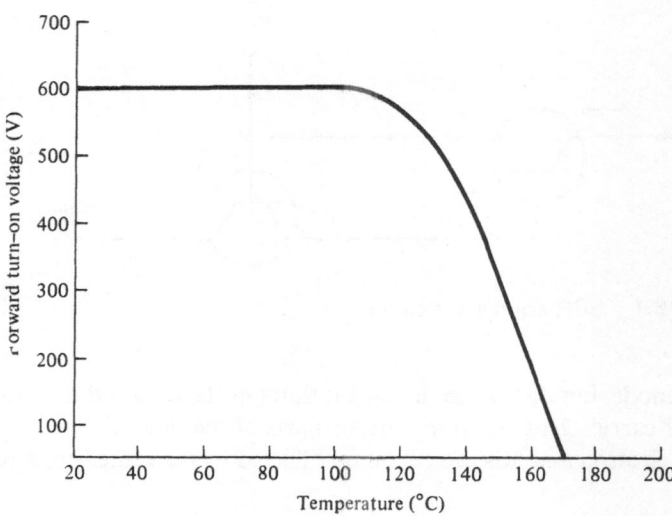

Figure 6.2 Breakover voltage versus temperature.

temperatures. The saturation current is exponentially dependent on the junction absolute temperature since, as the temperature increases, the number of electron holes generated in the device increases. Current gains are strongly dependent on current density, and if the saturation current reaches sufficiently high value because of the temperature rise, the sum of current gains may become equal to unity, i.e.,

$$\alpha_1 + \alpha_2 = 1 \qquad (6.7)$$

and the device will turn on. In addition, the lifetime of the minority carriers is higher at higher temperatures, which still further accelerates the onset of the turn on.

Because of the possible undesirable turn on, thyristors cannot operate at temperatures much above 125°C. This becomes obvious by the examination of the rapid fall of the breakover voltage with temperature (Figure 6.2). By the use of the shorted emitter it is possible to increase the operating temperature range and avoid the premature device firing. The shorted emitter is discussed with some detail in connection with the dv/dt effect.

6.3
Light triggering

When light impinges on a semiconductor, hole–electron pairs are generated. When an SCR is in the forward-blocking condition, the effect of exposing it to the light or other carrier-generating radiation will increase the magnitude of the generation-recombination current and will trigger the device in a way similar to that of the high-saturation current. Conventional devices are normally shielded from light by opaque enclosures.

The photoeffect is successfully used in special photo-activated devices such as light-activated silicon-controlled switch (LASCS).

6.4
Voltage triggering

There are two possible modes of voltage triggering: by the application of sufficiently high anode voltage or by having a rapidly rising anode voltage wavefront, commonly referred to as the dv/dt triggering.

If a sufficiently high voltage is applied to the device, the condition for the turn-on, as derived in Chapter 1, may be fulfilled, i.e.,

$$M_p \alpha_1 + M_n \alpha_2 = 1 \tag{6.8}$$

Consequently, all the device junctions become forward biased and a large current may flow through the device.

In the expression (6.8) the avalanche-multiplication factors M_N and M_P, as well as the small-signal alphas, are functions of the potential applied to the device. In Chapter 3 this case was fully reviewed, and the acceptable voltage limits to avoid triggering were determined.

While this mode of triggering is undesirable for a thyristor, it constitutes the basic mechanism of Shockley diode triggering. The dv/dt effect is also normally an undesirable effect and takes place when a rapidly rising, sufficiently high positive voltage is applied to the anode of an SCR [6.1–6.8]. The dv/dt turn-on may occur, for instance, when the device is operated from a voltage source and a transient potential is superimposed on the anode voltage [6.8].

Voltage transients which occur as a result of disturbance on the AC line caused by various sources such as load switching, solenoid closure, etc. may generate spike voltages which may either be high enough to turn the thyristor on if their duration is long enough, or exceed the critical rate of rise of off-state voltage capability (dv/dt). Commercial devices are, therefore, characterized by the so-called static dv/dt capability which sets the acceptable limit to the voltage rate of rise. Usually the device is additionally characterized by another dv/dt which forms a part of thyristor's turn-off time specification and really is a turn-off time condition rather than a specification per se. It is defined as the maximum allowable rate of application of the off-state blocking voltage while the device is regaining its rated off-state blocking voltage, following the device's turn-off time t_q under stated circuit and temperature conditions. This specification is reviewed shortly in Chapter 9.

6.4.1
dv/dt triggering

Figure 6.3 shows a two-transistor analog circuit with the inclusion of all three junctions' capacitances. A positive potential is applied to the anode, and the device remains in the forward-blocking state. Consequently, the outer junctions J_1 and J_3 are forward-biased and the center junction J_2 is reverse-biased. As a result, the capacitances C_1 and C_3 are very large compared to the capacitance C_2. C_1, C_2, and C_3 are connected in series, but since C_1 and C_2 have small reactances, only C_2 may be taken into account. The transient voltage will appear mainly across C_2.

The displacement (capacitance) current flowing through the device will therefore be

$$I_{dis}(t) = C_2 \frac{dv}{dt} + v \frac{dc_2}{dt} \tag{6.9}$$

The second term on the right-hand side of the equation usually can be neglected so that

$$I_{dis}(t) = C_2 \frac{dv}{dt} \tag{6.10}$$

For the sake of simplicity, it is assumed that the J_2 junction is a one-sided step junction and its depletion-layer capacitance C_2 per unit area can be expressed by

$$C_2 = \left(\frac{qK\varepsilon_0 N}{2(V_O + V_A)} \right)^{1/2} \tag{6.11}$$

where q is the electronic charge, ε is the silicon permitivity, N is the impurity concentration in the n base, V_O and V_A are the built-in and

Figure 6.3 Two-transistor circuit analog with junction capacitances shown.

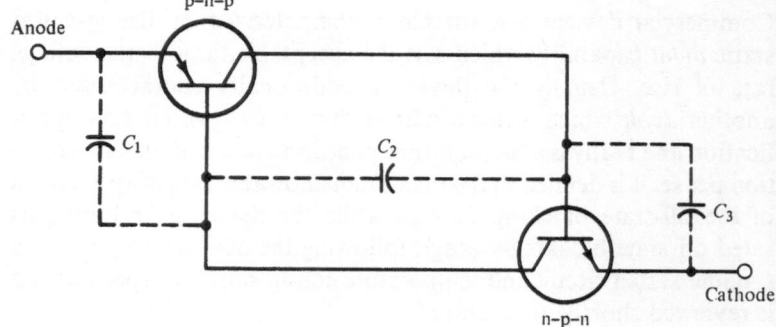

the applied potentials, respectively, and C_2 is a function of the applied voltage.

The electron and hole currents collected by the reverse-biased junction J_2 are

$$I_{CN} = \alpha_N I_K \qquad (6.12a)$$

$$I_{CP} = \alpha_P I_A \qquad (6.12b)$$

The anode current is

$$I_A(t) = \alpha_N I_A + \alpha_P I_K + I_{dis} \qquad (6.12c)$$

if the saturation current is neglected and I_{dis} represents the displacement current. Without the gate drive

$$I_A = I_K \qquad (6.13)$$

From Equations (6.12a), (6.12b), (6.12c), and (6.13),

$$I_A = \frac{I_{dis}}{1 - \alpha_N - \alpha_P} \qquad (6.14)$$

It can be easily shown following the procedure of the Appendix of Chapter 1 that

$$\frac{dI_A}{dI_{dis}} = \frac{1}{1 - \alpha_1 - \alpha_2} \qquad (6.15)$$

where α_1 and α_2 are the small-signal alphas of the n-p-n and p-n-p transistor sections, respectively, and are current-density dependent.

The condition for the turn-on will be the same as for the gate triggering, i.e., the sum of alphas should be at least equal to unity. If the displacement current is of sufficient magnitude, the current-dependent alphas will become large enough to fullfill this requirement and the device will turn on.

A further insight into the dv/dt effect can be gained from the understanding of the charge movement in the device upon the application of a transient voltage. In the forward-blocking condition, the depletion region of the center junction J_2 supports the applied anode potential. When this potential is suddenly increased, the depletion region must widen in order to support the increase. Therefore, the majority carriers on either side of J_2 should be removed in order to deplete more donors on the n side and more acceptors on the p side of J_2 if a two-sided junction is considered. The excess electrons start moving toward the positively biased anode while the holes move to the negative cathode. These excess majority carriers cause the J_1 and J_2 junctions to be more forward biased temporarily so that more minority carriers are injected from both the emitters (junctions J_1 and J_3). The injected minority carriers

67

are transferred by diffusion and by field toward the collecting junction J_2, and after a short period of time the charge neutrality is re-established. If the majority carrier flow, i.e., the current through the device, is sufficiently high to make the sum of alphas equal to unity, the device turns on.

The U.S. industry standard (dv/dt) capability is determined by using a linear ramp or an exponential waveform test. Static dv/dt capability is an inverse function of device junction temperature as well as a complex function of the transient waveform shape.

6.4.2
Methods of dv/dt capability improvement

The dv/dt turn on is nondestructive; therefore, in some applications it may have no harmful effects. A heater control is of this nature. In other applications, however, such as in an inverter circuit, the dv/dt effect may cause a complete malfunction. One way to overcome this problem is to use protective circuits called snubbers which consist of a network hooked between the cathode and anode. A network which may consist, for example, of a series-connected capacitor and a resistor, presents a very low impedance to a fast-varying voltage transient but shows a very high impedance to low-frequency potentials or to DC current flow.

Reverse-biasing of the gate with respect to the cathode may increase the dv/dt capability since it permits the removal of the majority carriers from the device p base [6.9]. This is effective mainly in small-area or interdigitated devices, with low sheet resistance of the p base. In place of—or in addition to—these circuit methods, it is necessary to design the power thyristor for good dv/dt performance. One way to do it is to minimize the magnitude of the displacement current by minimizing the junction J_2 capacitance. Lowering the efficiency of both emitters enhances also the dv/dt capability. All these measures are usually in conflict with other device specifications such as dissipation. The best way to minimize the displacement current effects is to use a device with a so-called "shorted" emitter [6.7, 6.8, 6.10].

6.4.2.1 The shorted emitter. Ideally a thyristor having the highest resistance to the rapid-voltage rate of rise would be one in which the alphas of both transistor sections are low when the applied voltage is high and the current low, and high at low voltage and high current. A shorted emitter provides an emitter which has zero efficiency at low currents but a high efficiency at high current levels. When the emitter is shorted the displacement current does not flow out through the emitter, but through the n-emitter short (Figure 6.4).

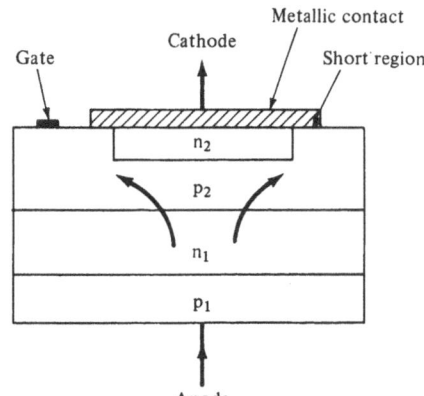

Figure 6.4 Schematic representation of the shunted emitter.

The same holds for the thermally generated current; consequently the device can be operated over a wider range of temperatures than a device without the short.

One way to obtain an emitter short is to extend the cathode metallization over the p base as per Figure 6.4. Another approach is to provide the emitter with openings (holes) filled with metal by plating or evaporation (Figure 6.5). A third technique consists of the combination of both.

6.4.2.2 Shorted emitter current–voltage relationships. The current–voltage relationships can be analyzed on the basis of a two-transistor analog as per Figure 6.6 with R_{shunt} added representing the parallel

Figure 6.5 Multiple shorting holes.

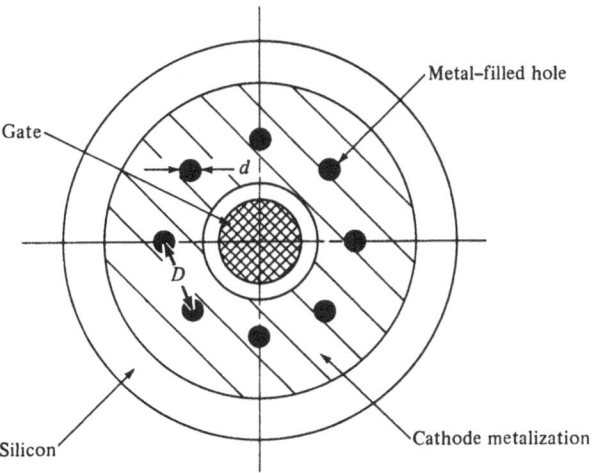

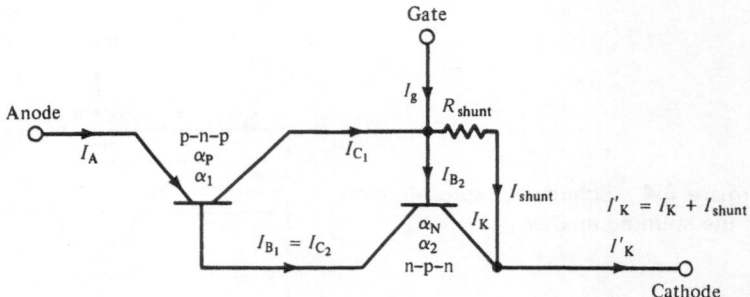

Figure 6.6 Two-transistor analog of a p-n-p-n switch with shunted emitter.

resistance existing between the gate contact and the emitter short. This is the lateral sheet resistance of the p base.

With the device in the forward-blocking state, the outer junctions J_1 and J_3 are forward-biased and J_2, the center-collecting junction, is reverse biased. In the shorted device, the n emitter (junction J_3) will be ineffective at low current levels since all the holes injected from the anode side will be flowing along the p base to the shorted emitter sites so that no feedback mechanism can be established. At sufficiently high current levels, the laterally flowing hole current (Figure 6.4) will forward bias the emitter–base junction sufficiently, and significant injection will take place. The voltage drop V_3 across the J_3 junction should reach a value of at least 0.5 V before this happens and the device may be turned on. The turn-on conditions will be the same as for a device without any shunt if we replace α_N, the DC current gain of the n-p-n section of the device by an effective gain α_{shunt}, which was already discussed in Chapter 1. For an approximate computation it is quite satisfactory to use the DC alphas [6.7]. An exact analysis would require the use of small-signal alphas for the determination of the turn-on conditions.

The α_{shunt} is defined by

$$\alpha_{shunt} = \frac{I_K \alpha_N}{I_K + I_{shunt}} \tag{6.16}$$

The value of the shunting current is

$$I_{shunt} = \frac{V_3}{R_{shunt}} \tag{6.17}$$

where R_{shunt} is the lateral p-base resistance between the short and the gate. The current flowing from the n emitter is

$$I_K = I_S\left[\exp\left(\frac{qV_3}{mkT}\right) - 1\right] \tag{6.18}$$

where the symbols have the following meaning: k is the Boltzmann constant, T is the absolute temperature, q is the electron charge, I_S is the saturation current, and m is a constant with a value between 1 and 2. The expressions (6.16) and (6.18) show that for $V_3 > 0.5$ V, $\alpha_{\text{shunt}} \simeq \alpha_N$, since $I_{\text{shunt}} \ll I_K$.

A sufficiently high lateral voltage V_3 will develop in the p base for a given current density J crossing the center junction if the resistivity of the p base is high and the path it follows is long and narrow. The exact determination of this voltage is rather involved; it is sufficient to know, however, that the maximum possible voltage drop is

$$V_{3\text{max}} = \frac{\rho L^2 J}{2t} \tag{6.19}$$

where ρ is the p-base average resistivity, L is the transverse current path along the p base from the edge to the center of the emitter, and t is the p-base width. This expression was derived [6.7] for a rectangular-shaped emitter. The short is more effective when the length L is small, as in the case of integrated emitters and the multiple-hole emitters.

6.4.2.3 Multiple-hole shorted emitter. A relationship between the diameter of the shorting holes d and the separation between holes D and their effectiveness in reducing the emitter efficiency of p-n-p-n devices have been discussed [6.8, 6.10]. This relationship was derived on the assumption that the voltage drop V_{dis} caused by any displacement current does not exceed 0.5 V. When V_{dis} exceeds this value, the emitter starts injecting very significantly. The relationship is

$$16V_{\text{dis}} \frac{A}{C_2} \left(\frac{dv}{dt}\right)^{-1} \sigma t = d^2 + D^2 \left[2 \ln\left(\frac{D}{d}\right) - 1\right] \tag{6.20}$$

In this expression C_2 is the capacitance of the center junction J_2, σ is the average conductivity of the p base having thickness t, and A is the junction J_2 area. Graphs to indicate how the calculations could be used in the design of particular devices are given in [6.8], where the experimental results compare favorably with the theoretically computed values. The experimentation showed that dv/dt and also high-temperature performances can be improved very significantly by the multiple-hole shorted emitter if a sufficiently large number of holes is used. As the number of holes was gradually increased, the dv/dt performance gradually improved and eventually reached its maximum value. If larger holes are used, a smaller number of them is required.

By using the multiple-hole emitter it was possible [6.10] to produce thyristors with breakdown voltages up to 5000 V, capable of withstanding voltage rise rates of several hundred volts per microsecond.

References

6.1 F. E. Gentry. Recent advances in p-n-p-n devices. IEEE Conf. Paper 63–430, January 1963.

6.2 A. K. Jonscher. Notes on the theory of four-layer semiconductor switches. *Solid State Electron., 2*: 143–148, 1961.

6.3 V. A. Kuz'min. Theory of the *dv/dt* effect in thyristors. In *Physics of the p-n Junctions and Semiconductor Devices*. (S. M. Ryokin and Y. V. Shmartsev, eds.). London: Consultants Bureau, 1971.

6.4 I. Somos. Switching characteristic of silicon power-controlled rectifiers. *AIEE Trans., 83*: 861–871, 1964.

6.5 F. E. Gentry, F. W. Gutzwiller, N. Holonyak, and E. E. Von Zastrow. *Semiconductor Controlled Rectifiers*. Englewood Cliffs, N.J.: Prentice-Hall, 1969.

6.6 T. C. McNulty. A review of thyristor characteristics and applications. Thyristor, RCA Application Note, AN-4242.

6.7 F. E. Gentry, R. E. Scace, and J. K. Flowers. Bidirectional triode p-n-p-n switches. *Proc. IEEE, 53*: 355–369, 1965.

6.8 P. S. Raderecht. A review of the "shorted emitter" principle as applied to p-n-p-n silicon controlled rectifiers. *Int. J. Electron., 31* (6): 541–564, 1971.

6.9 C. D. Root. Negative current bias and resistance bias for preventing turn-on by *dv/dt* of controlled rectifiers. *Proc. IEEE, 51*: 1672, 1963.

6.10 C. K. Chu. Geometry of thyristor cathode shunts. *IEEE Trans. Electron Devices, ED-17* (9): 687–690, 1970.

7

Thyristor voltage drop in the on-state

Summary

When a thyristor is in the forward-conducting state, all device junctions are forward biased and the device presents a minimum resistance to the forward current flow. At moderate and high injection levels, two of the three forward-biased junctions usually exhibit ·nearly equal but opposite potentials. Consequently, one junction contributes to the total device voltage drop. Since the junction voltage is a logarithmic function of the current, it varies only very slowly and its value is typically 0.8 to 1.0 V. for moderate or high current levels.

All regions of the device between the junctions, as well as metal contacts, contribute to the voltage drop across the device, but the most significant drop usually occurs in the long n base. This voltage drop is a sensitive function of the minority carrier lifetime in the n base, base width, and current density. At very high current densities the voltage drop consists basically of three components: the junction voltage drop, the n-base voltage drop, and the ohmic contact drop. At the extremely high currents, because of the injected carriers' effect on mobility and lifetime, the voltage drop in the long base may also show an ohmic behavior.

7.1
Introduction

After the initial transient occurring upon the application of a forward potential to the device gate, the thyristor is in a steady-state forward-conducting state. A DC current flows from the external circuit to the anode of the device, traverses the device, and leaves through the cathode. In order to minimize the power losses and junction temperatures, it is important to reduce the thyristor's internal impedance. This requirement is not different than the one expected from any other switch. Any mechanical current breaker is expected, as a matter of course, to show only an insignificant voltage drop across its contacts.

When a p-n-p-n structure is in the steady-state condition, the current flow is in the direction from the anode to the cathode, and lateral currents may be neglected. Therefore, a one-dimensional model for the device can be well justified, and the six equations describing the steady state can be written in one dimension:

Transport equations

$$J_n = q\mu_n nE + qD_n\frac{dn}{dx} \tag{7.1}$$

$$J_p = q\mu_p pE - qD_p\frac{dp}{dx} \tag{7.2}$$

$$J_T = J_n + J_p \tag{7.3}$$

Continuity equations

$$\frac{dJ_n}{dx} = qR \tag{7.4}$$

$$\frac{dJ_p}{dx} = -qR \tag{7.5}$$

Poisson's equation

$$\frac{d^2V}{dx^2} = -\frac{q(p - n + N)}{K\varepsilon_0} \tag{7.6}$$

n and p are electron and hole densities, respectively; N is the doping concentration; J_n and J_p are electron and hole currents; μ_n and μ_p are electron and hole mobilities; D_n and D_p are electron and hole diffusion constants; K is the dielectric constant; and R is the recombination rate. All these quantities are functions of position x and/or of carrier concentration. Because of this dependence, it is impossible to solve the above equations analytically and obtain a closed-form solution without making many simplifying assumptions such as current independent lifetimes, mobilities, etc. Jonscher [7.1] considered only low-level conditions; Kuz'min [7.2] carrier density independent mobilities and 100-percent-efficient emitters. Moll et al. [7.3] studied low injection levels only. Herlet and Raithel [7.4] have shown that under very high injection-level conditions, the forward drop in the low-doped base of a p-n-p-n structure is approaching the voltage drop in a silicon p-i-n rectifier. Such voltage drop was studied by Hall [7.5], Fletcher [7.6], Howard and Johnson [7.7], Herlet [7.8], Shields [7.9], Spenke [7.10], Kao and Muss [7.11], Choo [7,12, 7.13], and Burtscher et al. [7.14].

Most recent p-n-p-n voltage drop studies were made *numerically* by Kokosa [7.15] and Cornu and Lietz [7.16], and analytically by Otsuka [7.17]. It is beyond the scope of this work to go into details and discussion of all these studies, so we limit ourselves to reporting here the results of some of them only.

7.2
Herlet's closed-form p-i-n diode analysis

Herlet [7.8] extends Hall's theory of the p-i-n rectifier by considering, in addition to the diffusion current in the i region, the diffusion currents in the p and n regions. All regions are assumed to be uniformly doped. At small current densities Herlet's general solution simplifies to Hall's [7.5] and Kuz'min's [7.2] solutions which predict an exponential dependence of the base voltage drop for $(W/L) \gg 1$ (W is the base width and L is the minority carriers' diffusion length). For the high current densities a quadratic dependence on the current density is found. An approximate solution for the current–voltage relationship is given at high current densities by

$$J_T = A \tanh^2 \frac{W}{2L} \frac{V^2}{V_m} \qquad (7.7)$$

where V is the voltage drop, J_T is the device current density, and A and V_m are factors depending in a very complex way on mobility, base lifetime, and width. Herlet does not consider the variation of the mobilities and lifetimes with the injection level.

7.3
Kokosa's numerical analysis

Kokosa [7.15] solves numerically the transport, continuity, and Poisson's equations only for the high–injection-level conditions with injected carrier densities of more than 10^{16} cm^{-3} and emitter concentrations on the order of 10^{19} cm^{-3}. The assumed injected carrier densities are larger than the net impurity density *in both bases*. This determines the lower limit of current densities for which the solution is valid. The dopings of all regions are assumed to be uniform and the recombination in the space charge regions of the junctions is neglected. Rather than using constant, average mobility, Kokosa takes into account the carrier–carrier scattering so that appropriate mobilities are used at each point. However, lifetimes were considered current independent. The results show that the base

75

voltage drop V is proportional to $J_T^{1/2}$, similar to Herlet's results, and that within a narrow current density range when

$$\frac{W}{L_p} \ll 1 \qquad V = \frac{2kT}{q}\left(\frac{W}{2L_p}\right)^2 \tag{7.8}$$

as predicted by Hall [7.5] and when

$$\frac{W}{L_p} \gg 1 \qquad V = \frac{\pi kT}{2q}\exp\left(\frac{W}{2L_p}\right) \tag{7.9}$$

as predicted by Kuz'min [7.2]. Kokosa's results can be represented [7.18], for at least some current range, by an expression for the n-base voltage drop as

$$V = c_1\sqrt{J_T}\exp\left(\frac{W}{2L_p}\right) \tag{7.10}$$

The value of the constant c_1 according to Matsuzawa [7.19] is 2.36×10^{-2} for currents greater than 400 A/cm^2.

In the limit when the current densities are on the order of many thousands of amperes, as occur in current surges, Kosoka predicts an ohmic behavior in the n base.

Another conclusion of [7.15] is that when the n-emitter concentration is larger than 10^{19} cm^{-3}, then the n-base forward drop will depend on the p-emitter concentration and will be inversely proportional to $(p^+)^{1/4}$, p^+ being the anode emitter concentration.

7.4
Numerical analysis of Cornu and Lietz

The results of this analysis are based on the numerical solution of the transport, continuity, and Poisson equations for the one-dimensional thyristor with wide n bases, greater than several diffusion lengths. The doping mobility and lifetime were varied from point to point. The *uniformity* of the doping was *not assumed*, although it was assumed in Kokosa's analysis. Dependence of lifetime on current density was not taken into account. Furthermore, the authors have not used the usual depletion-layer approximation of the space-charge region. The space-charge density was modified to take into account the mobile carriers traversing the space-charge region at high current-density conditions. Cornu and Lietz conclude that for high currents and base widths large compared to the diffusion length, the dependency of the mobility on the carrier concentration leads to a base *voltage drop that is significantly lower than that predicted* by the analytical theory of Herlet [7.8]. Another conclusion is that the effect of the doping width and lifetime of the highly doped n and p regions in the p-i-n rectifier considered by Herlet have much less influence than predicted by Herlet's theory. Cornu and Lietz [7.16] finally conclude that the analytical theories usually hold well enough for base widths up to twice the ambipolar diffusion length.

7.5
Otsuka's forward-drop analysis

This section is based mostly on the results of Otsuka's [7.17] closed-form analysis which, like other analytical approaches, makes several simplifying assumptions and approximations. Its advantage, however, is that due to the comparatively simple analytical expressions, it makes salient those physical p-n-p-n device parameters that affect mostly the forward-voltage drop.

The following notation is used (Figure 7.1): V_0 is the voltage drop across the p_0^+-p high-low junction; V_1, V_2, and V_3 are the voltage drops across corresponding p-n junctions; and $V(n_2)$, $V(p_2)$, $V(n_1)$, $V(p_1)$, and $V(p_0)$ are the voltage drops in the various semiconductor

Figure 7.1 Thyristor schematic structure. J_0, J_1, J_2, J_3—junctions; g, t, w, d, g—thicknesses of various layers; V_0, V_1, V_2, V_3—voltages across junctions; $V(p_0)$, $V(p_1)$, $V(n_1)$, $V(p_2)$, $V(n_3)$—voltage drops in respective regions. (After Otsuka [7.17].)

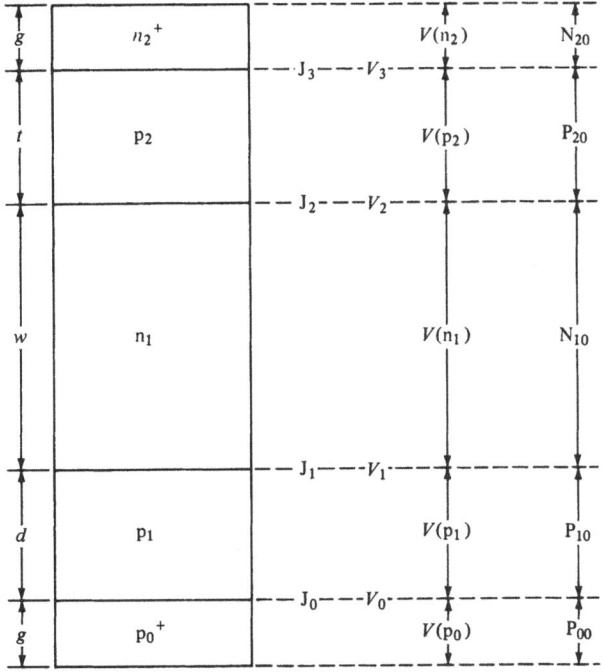

layers. $J_T R$ is the ohmic drop in the metal–semiconductor and metal-metal interfaces for a forward-current J_T. In order to package the device and to be able to connect it to an outside circuit, metal contacts are used and several metal–semiconductor and metal-to-metal interfaces of finite-series resistance always will be present.

The total voltage drop across the device is then

$$V_T = V_0 + V_1 + V_2 + V_3 + V_2(n_2) + V(p_2) + V(n_1)$$
$$+ V(p_1) + V(p_0) + J_T R \qquad (7.11)$$

It is convenient, for the sake of simplification, to view the device at three different current-density levels which correspond to the low-, moderate-, and high-injection levels. The following assumptions are made which permit the derivation of analytical expressions [7.17]: (1) all semiconductor regions are uniformly doped; and (2) the impurity concentration in the anode emitter p_0^+ is P_{00} and equal to the concentration N_{20} in the n_2^+ emitter. The concentrations in the p_1 and p_2 region are the same, so that $P_{10} = P_{20}$; (3) the concentration in all regions are much higher than the intrinsic concentration $n_i = 1.4 \times 10^{10}$ carriers/cm³ for silicon, and the following relationships exist between the carrier concentrations in various regions when there is no external potential applied (thermal equilibrium):

$$P_{00} = N_{20} > P_{10} \qquad (7.12)$$
$$P_{10} = P_{20} > N_{10} \qquad (7.13)$$

(4) the diffusion length L_p of holes in the N_1 base of the device is smaller than the base width W; consequently

$$\frac{W}{L_p} > 1 \qquad (7.14)$$

This situation is typical for power thyristors because of the blocking voltage requirements; (5) all regions between junctions are assumed to be quasineutral at any injection level; (6) the analysis is one dimensional; all transverse currents are neglected in the on state; (7) t/L_n, g/L_p, and d/L_n are assumed to be less than unity. The assumption of g/L_p may be not always realistic, e.g., at very high doping levels; (8) all the relationships of Otsuka's analysis are greatly simplified if it is assumed that the ratio Z of the p-base sheet conductance to the n^+-emitter sheet conductance is much smaller than unity, i.e., $Z \ll 1$ (see Appendix). If Z is ever made to approach unity, then the n^+-emitter efficiency drops and so does the current gain α_2, so that the sum $\alpha_1 + \alpha_2$ may become less than unity and the device cannot be made to turn on; (9) carrier concentrations can be related to quasi-Fermi levels by the Boltzman relationships.

7.5.1
Low injection level

The low injection level occurs when the injected carrier density into the n base is lower than the equilibrium carrier density equal to the n-base doping density.

Power devices that require several hundred volts of blocking capability are fabricated with the n_1-base doping density N_{10} on the order of 10^{14} donors/cm^3; low-level injection will take place when the density of the injected carriers into the n_1-base is smaller than this value. The typical values of current are below 100 mA/cm^2.

The maximum allowable current density for the low injection level can be determined from the expression given in the Appendix. Since the current density is very small, the voltage drops in the uniformly doped regions are also negligibly small. The same is true for the voltage drops across the high–low junction p$^+$-p. Only the potential drops across the three p-n junctions J_1, J_2, and J_3 should be taken into account.

Since the device is in the on state, i.e., in the forward-conduction mode, all three junctions are forward biased and, consequently, the overall device impedance is small. Because the potentials across J_1 and J_3 are of opposite directions, the total voltage drop is

$$V_T = V_1 - V_2 + V_3 \tag{7.15}$$

It can be shown [7.17] that for a low injection level

$$V_T = \frac{kT}{q} \ln K_1 J_T \tag{7.16}$$

where k is the Boltzmann coefficient, T is the temperature in degrees Kelvin, q is the electronic charge, J_T is the current density in the device, and K_1 is a constant dependent on device parameters (see Appendix).

The voltage drop at low current densities is on the order of 0.8 V.

7.5.2
Moderate injection level

By definition, the moderate injection level takes place when the injected excess carrier density exceeds the doping level (the majority carrier density) in the n_1 base, but not, however, in the p_1 and p_2 regions. The maximum allowable current density for the moderate injection level can be determined from the relationship of the Appendix. The typical values of moderate levels are 10–50 A/cm^2.

Due to the quasineutrality assumption, the hole and electron densities in the n_1 region during the on state are nearly equal. Since, in the simplified device model the equilibrium concentrations P_{10} and P_{20} were assumed to be the same, the *total carrier concentrations* at the J_1 and J_2 *junction edges are equal.* This leads to the conclusion that the voltage drops V_1 and V_2 across these two junctions must also be equal, but opposing each other.

Since the current density is still comparatively small, the voltage drop across the p^+-p, high–low junction, may be entirely neglected. Under these conditions the sum of the junction voltage drops is limited to only one voltage drop, V_3. Its simplified form is

$$V_j = V_3 = \frac{kT}{q} \ln K_2 J \qquad (7.17)$$

where K_2 is a constant dependent on device parameters (see Appendix).

The voltage drop in the p_1 region (Figure 7.1) is almost ohmic and is normally small. If the average resistivity of the p_1 region is ~ 1.0 Ωcm, then the voltage drop for p_1 layer width of $20\,\mu$m will be about 0.02 V at 10 A/cm^2.

A significant voltage drop occurs in the n_1 base, which is conductivity modulated at moderate current densities. The voltage drop in the n_1 base becomes current independent and is given [7.17] by

$$V(n_1) = \frac{kT}{q} \frac{b}{b+1} \left(\frac{W}{L_1}\right)^2 \qquad (7.18)$$

W is the n_1-base width; b is the ratio of the electron mobility to the hole mobility (equal to about 2.7 at moderate current densities); and L_1 is the ambipolar diffusion length:

$$L_1 = L_p \left(\frac{b}{b+1}\right)^{1/2} \qquad (7.19)$$

Typical values $V(n_1)$ for lifetimes on the order of a few microseconds are about 0.1 V. Since all other voltage drops can be neglected, the total voltage drop at a moderate-injection level will be

$$V_T = V_j + V(n_1) \qquad (7.20)$$

7.5.3
High injection level

This is the case of the greatest interest for the high-current, high-voltage power devices. At high injection levels the injected carrier densities are greater than the doping densities in all device regions with the exception of the highly doped n^+ and p^+ layers. The expression for the maximum limiting current density for this case is given in the Appendix. Typical values are from 50 to 5000 A/cm².

All regions, with the exception only of the n^+ and p^+ layers, are conductivity modulated. In order to preserve the charge neutrality in the p_1, n_1, and p_2 regions, the concentrations of electrons and holes in each of them must be equal. Because of the continuity of the current flow, the hole and electron concentrations at the both sides of junctions J_1 and J_2 must also be equal. Therefore, since it was assumed that $P_{00} = N_{20}$, the hole and electron concentration distributions become symmetrical and have a form [7.17] as per Figure 7.2. This implies that the hole and electron concentrations and their gradients are equal, i.e.,

$$p(x) = n(x) \tag{7.21}$$

$$\frac{dp}{dx} = \frac{dn}{dx} \tag{7.22}$$

The device resembles very much a p-i-n diode with the i region having an effective thickness W_T equal to the sum of t, W, and d of the p_1, n_1, and p_2 regions, respectively.

Figure 7.2 Distributions of carrier concentration in high-level operation. (After Otsuka [7.17].)

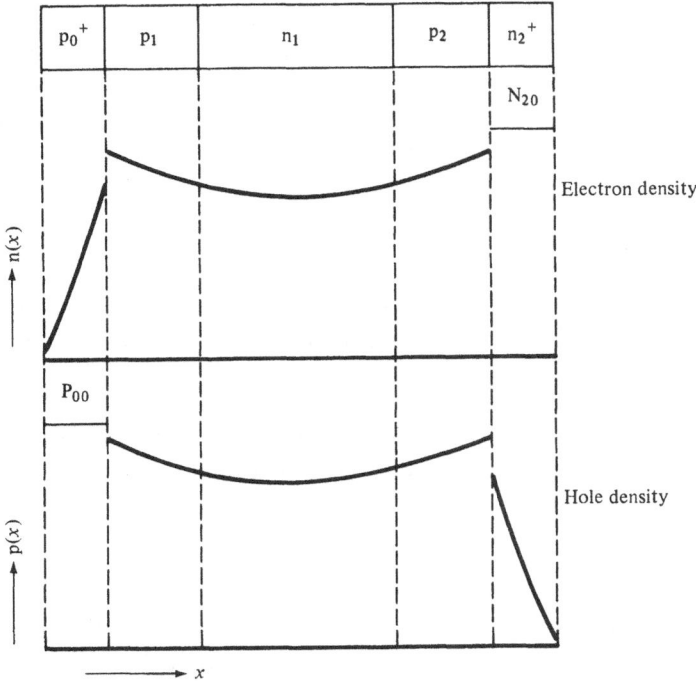

81

At high injection levels the hole and electron mobilities are much smaller than those at the low carrier densities due to the carrier–carrier scattering mechanism, and they become equal to each other $\mu_n = \mu_p = \mu$. The same is true of the hole and electron diffusion constants which at the very high injection levels fall much below 10 cm^2/sec and become almost equal to each other and to the ambipolar diffusion constant: $D_p = D_n = D$. Adding the two transport equations (7.1) and (7.2) and utilizing the equality of the mobilities and diffusivities, we obtain an expression for the total current density

$$J_T = J_p + J_n = 2q\mu n(x)E(x) \tag{7.23}$$

This implies that at the high injection levels the current is carried entirely by the field. Since the mobilities and hole- and electron-carrier concentrations are equal, the hole-current component is equal to the electron-current component, so that

$$J_n = J_p = \frac{J_T}{2} \tag{7.24}$$

At lower levels this equality is not true, and the electron current will dominate over the hole current because of much higher electron mobility.

The total junction potential drop at the high injection level is

$$V_j = V_0 + V_1 - V_2 + V_3 \tag{7.25}$$

But, because of the symmetry, $V_1 = V_2$, and (7.25) reduces to $V_j = V_0 + V_3$. This voltage drop is given by

$$V_j = \frac{kT}{q}\ln(K_3 J_T) \tag{7.26}$$

where K_3 is a constant dependent on device parameters and its expression, as well as the expression for V_0, can be found in the Appendix.

The total junction potential V_j is a logarithmic function of the current density and therefore varies slowly with it. Its typical value is on the order of 1 V even at very high densities; a one-hundred-fold variation in J_T will add only an increment of about 50 mV at room temperature.

The voltage drop in the highly doped n$^+$ and p$^+$ regions is purely ohmic and usually not very significant. It adds typically another 50 mV or so to the total voltage drop at 1000 A/cm^2. On the other hand, the voltage drop in the wide effective base of W_T length is usually very important. Its magnitude may be determined from the following expression [7.17]:

$$V_{W_T} = \frac{W_T^2}{2\mu\tau^{1/2}} \frac{\sinh\left(\dfrac{W_T}{L_2}\right)}{\left[\cosh\left(\dfrac{W_T}{L_2}\right) - 1\right](2q\, g N_{20})^{1/2}} J_T^{1/2} \qquad (7.27)$$

where g is the n_2^+ region thickness, μ is the mobility at high current densities, τ is the hole lifetime in the n_1-base, and L_2 is the ambipolar length at high current densities, when $D_n = D_p$. It is of interest to note that this voltage drop follows the so called "square-law" relationship [7.20, 7.21] for double injection in ideal solids.

The doping density in the n_1 base is negligibly small as compared to the injected carrier density, so that the n_1-base region is conducting current almost entirely by the injected carriers and only very little by the equilibrium carriers. As a result, at high-injection levels the voltage drop in the n_1 base is independent of the n_1-base resistivity. The n-base voltage drop increases rapidly with the base width and the ratio W_T/L. It is inversely proportional to the square root of lifetime and the square root of the emitter conductance (the concentrations of both emitters were assumed to be equal).

The total voltage drop of the device is

$$V_T = V_j + V_{W_T} + J_T R \qquad (7.28)$$

The square-law relationship of the voltage drop in the n base approaches eventually simple Ohm's law behaviour at extremely high carrier densities (about 10^{18} carrier/cm³) due to the Auger recombination effects on lifetime [7.22]. If the contact ohmic drop and n-base ohmic drop are combined at the extremely high current densities, the total voltage drop in the device will become

$$V_T = V_3 + J_T R_T \qquad (7.29)$$

where R_T is the total ohmic resistance per unit area and J_T is the current density.

7.5.4
Voltage drop of a nonuniformly doped thyristor

Most of the present day devices are fabricated by impurity diffusion into silicon, and only the n_1 base may be considered to be doped uniformly. All other regions are graded. The voltage-drop expressions in the n base are still applicable with some further sacrifice of accuracy if the average concentrations are used in place of the uniform densities.

Any part of the n_2^+-emitter region and, similarly, of the p_0^+-emitter region, which is doped above 10^{19} atoms/cm^3, contributes very little to the emitter efficiency. It is so because, at impurity concentrations above 10^{19}, the lifetime of the minority carriers is very small due to Auger recombination [7.23 and 7.24] and the energy gap of silicon is narrower than that at low doping levels [7.25] and the intrinsic carrier concentration is increased [7.26]. Consequently, regions with dopings much above 10^{19} cm^{-3} will not inject at all into the base. As a result, it is quite satisfactory to consider as an active emitter only that part of the n_2^+ region which is below the 10^{19} cm^{-3} concentration.

7.5.5
Experimental data

Otsuka [7.17] found a good agreement between the theory at all injection levels and experimental data taken on SCR samples with alloyed emitters, diffused p regions, and uniformly doped n base ($\sim 10^{14}$ cm^{-3}). W was equal to 170 μm and $W_T = 380$ μm. In his

Figure 7.3 Forward drop versus forward current of RCA S260 SCR. Lifetime in n base = 4 μsec. (After Assour and Bender [7.26].)

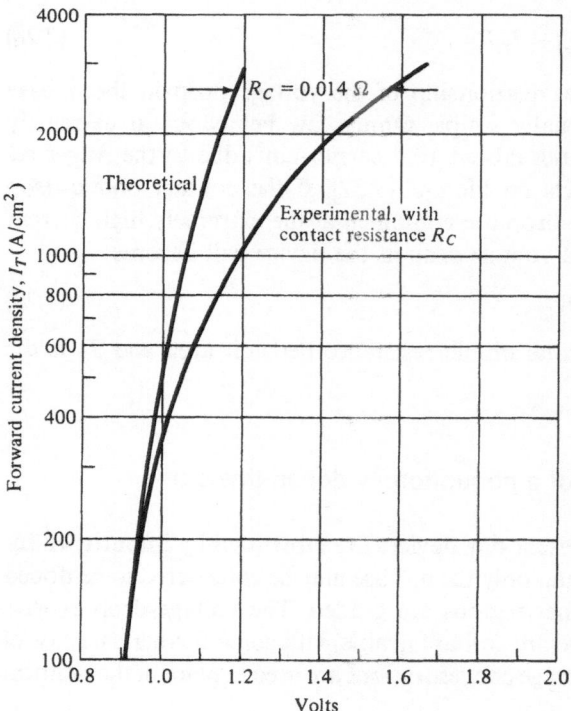

calculation, the following diffusion constants were used: $D_p = 13$ cm^2/sec and $D_n = 30$ cm^2/sec for the low and moderate current levels. For high injection levels $D_n = D_p = D = 7$ cm^2/sec. The lifetimes in the n base varied from 3.5 to 25 μsec, and $\mu = 200$ cm^2/sec V.

Assour and Bender [7.27] made an extensive experimental study of the forward drop at the *high current* densities. Their experimental results for a variety of diffused SCR structures have shown consistent and reasonable agreement with theoretical calculations based on Otsuka's formulations. In most cases, the difference in voltage drop between the calculated values of $V_j + V_{W_T}$ [Equations (7.26), (7.27)] and experimental data has been related to the ohmic voltage drop in the metal contacts and lead wires. The investigated devices had an n-base width $W \simeq 110$ μm), n-base concentration at about 10^{14} cm^{-3}, and lifetimes of holes in the base were varied between 4.0 and 0.2 μsec. The results of the data obtained on several samples of one of the types investigated are shown in Figure 7.3. The high injection-level diffusion constants were assumed to be $D_n = D_p = D = 7$ cm^2/sec. The contact resistance R_c was equal to 0.014Ω.

APPENDIX

1
Low injection level

The maximum current density:

$$J_T < \frac{q N_{10} D_p}{L_p} \qquad \text{for } z \ll 1$$

$$z = \frac{\mu_p}{\mu_n} \cdot \frac{t}{g} \cdot \frac{P_{20}}{N_{20}}$$

$$K_1 = \frac{t P_{20}}{q D_n n_i^2}$$

2
Moderate injection level

The maximum current density:

$$J_T < \frac{q D_p b P_{20}}{t} \qquad \text{for } z \ll 1$$

$$K_2 = \frac{P_{20} t}{q D_n n_i^2}$$

85

3
High injection level

Maximum current density:

$$J_T < \frac{qDN_{20}}{g}$$

$$K_3 = \frac{qN_{20}}{2_q Dn_i^2}$$

where D is the ambipolar diffusion constant at high-injection level.

4
High–low junction voltage drop

$$V_0 = \frac{kT}{2q} \ln\left(\frac{gJ_T P_{00}}{2qDP_{10}^2}\right)$$

References

7.1 A. K. Jonscher. PNPN switching diodes. *J. Electron. Control*, 3: 573–586, 1957.

7.2 V. A. Kuz'min. Volt–ampere characteristics of p-n-p-n type semiconductor devices in the "on" condition. *Radio Engrg. Electron. Phys.*, 8: 150–156, 1963.

7.3 J. L. Moll, M. Tanenbaum, J. M. Goldey, and N. Holonyak, Jr. P-N-P-N transistor switches. *Proc. IRE*, 44: 1174–1182, 1956.

7.4 A. Herlet and K. Raithel. Forward characteristics of thyristors. *Solid State Electron.*, 9: 1089–1105, 1966.

7.5 R. N. Hall. Power rectifiers and transistors. *Proc. IRE*, 40: 1512–1518, 1956.

7.6 N. H. Fletcher. The high current limit for semiconductor junction devices. *Proc. IRE*, 45: 862–872, 1957.

7.7 N. R. Howard and G. W. Johnson. P^+IN^+ silicon diodes at high forward current densities. *Solid State Electron.*, 8: 275–284, 1967.

7.8 A. Herlet. The forward characteristic of silicon power rectifiers at high current densities. *Solid State Electron.*, 11: 717–742, 1968.

7.9 J. Shields. The forward characteristics of p^+-n-n^+ diodes in theory and experiment. *Proc. IEE, Part B*, 106: 342, 1959.

7.10 E. Spenke. Notes on the theory of the forward characteristic of power rectifiers. *Solid State Electron.*, 11: 1119–1130, 1968.

7.11 Y. C. Kao and D. R. Muss. Analytical design theory for high voltage PIN rectifiers. *Solid State Electron.*, 13: 825–841, 1970.

7.12 S. C. Choo. Theory of a forward biased diffused junction P-L-N rectifier. *IEEE Trans. Electron Devices, ED-20*: 418–426, 1973.

7.13 S. C. Choo. Analytical approximation for an abupt PN junction and high-level condition. *Solid State Electron., 16*: 793–799, 1973.

7.14 J. Burtscher, F. Danhauser, and J. Krausse. Die Rekombination in Thyristoren und Gleichrichtern aus Silizium: Ihr Einfluss auf die Durchlasskennlinre und das Freiwerdezeitverhalten. *Solid State Electron., 18*: 35–63, 1975.

7.15 R. A. Kokosa. The potential and carrier distribution of a PNPN device in the on-state. *Proc. IEEE, 55* (8): 1389–1400, 1967.

7.16 J. Cornu and M. Lietz. Numerical investigation of the thyristor forward characteristic. *IEEE Trans. Electron Devices, ED-19* (8): 975–981, 1972.

7.17 M. Otsuka. The forward characteristics of a thyristor. *Proc. IEEE, 55* (8): 1400–1408, 1967.

7.18 W. E. Newell. A design tradeoff relationship between thyristor ratings. *PESC*: 294–304, 1974.

7.19 T. Matsuzawa. Pulse current capability of high-speed thyristors. *Electrical Engineering in Japan, 94* (1), 1974. [Translated from *Denki Gakkai Ronbunshi, 94C*, (1): 8–17, 1974.]

7.20 A. Rose. Comparative anatomy of models for double injection of electron and holes into solids. *J. Appl. Phys., 35* (9): 2664–2678, 1964.

7.21 M. A. Lampert and R. B. Schilling. Current injection in solids: The regional approximation method. *Semiconductors and Semimetals*, Vol. 6. New York: Academic Press, 1970.

7.22 A. Blicher. The effect of the Auger recombination on the forward drop. RCA, Somerville, N.J., private communication, 1973.

7.23 J. D. Beck and R. Conradt. Auger recombination in Si. *Solid-State Commun., 13*: 93–95, 1973.

7.24 W. W. Sheng. The effect of Auger recombination on the emitter injection efficiency of bipolar transistors. *IEEE Trans., ED-22*: 25–27, 1975.

7.25 H. J. J. DeMan. The influence of heavy doping on the emitter of a bipolar transistor. *IEEE Trans., ED-18* (10): 833–835, 1971.

7.26 R. P. Mertens, H. J. DeMan, and J. J. VanOverstraeten. Calculation of the emitter efficiency of bipolar transistors. *IEEE Trans. Electron Devices, ED-20* (9): 772–778, 1973.

7.27 J. M. Assour and J. R. Bender. The forward characteristic of an SCR under high injection. RCA, private communication, 1974.

8

SCR turn-off transient

Summary

When a reverse bias is applied to the SCR anode in order to turn off the device, it is possible to distinguish four separate phases of the ensuing transient: two storage times and two fall times. The device recovers first at the n-emitter junction because there are only very few electrons injected from the wide n base into the narrow p base. The opposite is true for the n base, since the p base injects heavily into it and replenishes holes as fast as they are collected by the reverse-biased anode. The n-emitter junction breaks down as soon as the junction recovers its depletion region if the applied external potential is high enough. The first storage time and fall time are comparatively short. The second storage time is required to remove the excess charge from the long n base where the holes are continuously replenished by injection by the p base. Finally, all the charge disappears almost at the end of the second fall time.

Fast turn offs can be achieved for devices which have low minority-carrier lifetimes (gold doping) and are turned off with high reverse currents. The reverse biasing of the device gate may speed up primarily the first phase of the turn-off transient.

8.1
Introduction

If a forward voltage is reapplied prematurely to an unidirectional thyristor (SCR) after the forward anode current ceased to flow, the device will go into the conduction state again. Because of the stored charge present, it is necessary to wait for a definite interval of time after current cessation before the reapplication of the forward potential if the device is expected to block the reapplied voltage. The turn-off time is the time necessary for the removal of the excess charges from all the parts of the p-n-p-n device.

88

In many practical applications the forward current is removed from the SCR by the reversal of the current flow in the outside circuit; the decreasing anode current passes through zero and goes negative. It has its maximum value just after the reversal and decays until all the excess charges are removed and a depletion layer has fully developed across the reverse-blocking junction. Only then is the SCR ready for the reapplication of the forward potential.

Generally speaking, the device can be turned off not only by the application of a reverse potential but also by opening the external circuit or by applying a reverse bias to the device gate. The most rapid turn-off is achieved when the anode current is reversed with simultaneous reversal of the gate current. We consider first the case when the anode current only is reversed; consequently, the two outer junctions, which were forward biased during the on state, become eventually reverse biased. The effect of the additional application of the reverse gate current is shown, then, to decrease the turn-off time.

The device behavior can be analyzed one-dimensionally by the charge-control method using the superposition principle [8.1–8.4]. The results of the one-dimensional analysis are applicable obviously with greater accuracy to structures with narrow emitters.

Figure 8.1 An SCR and an approximate carrier concentration at $t = 0$.

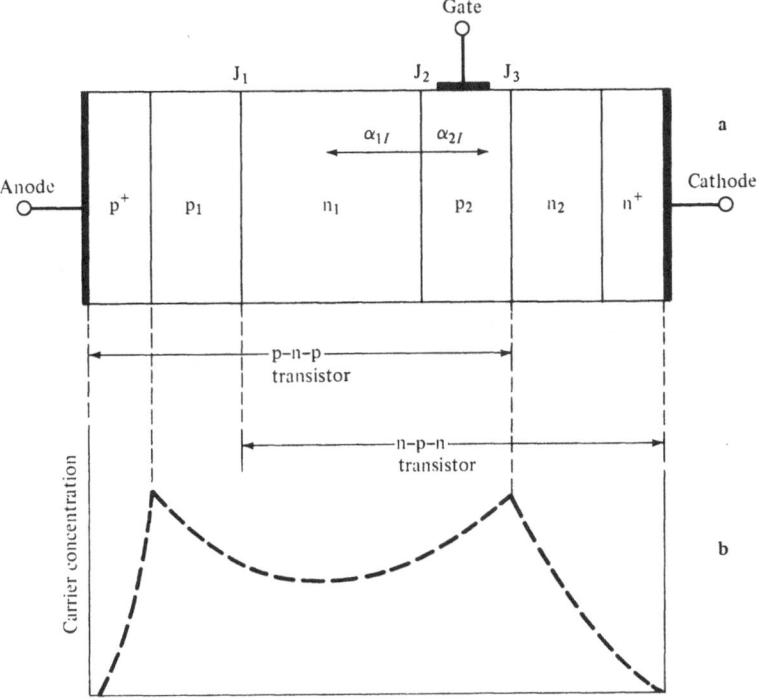

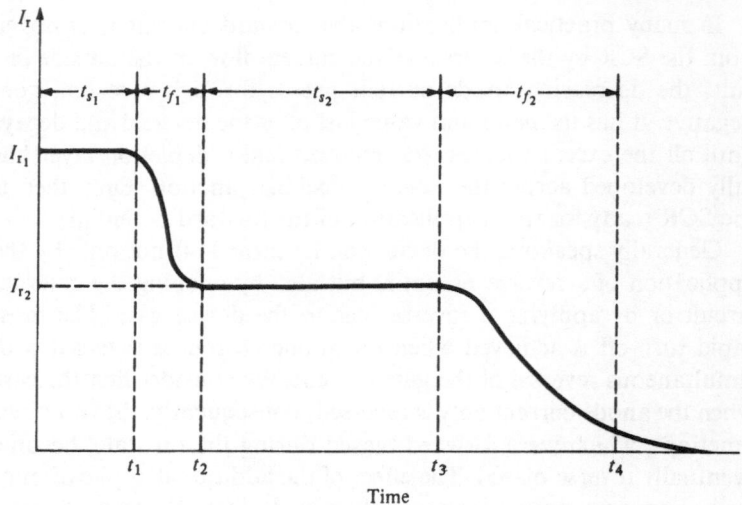

Figure 8.2 Four turn-off phases [8.3].

Figure 8.1 represents a p-n-p-n device just prior to the application of the reverse potential to the anode. For the reverse-bias turn off, the current waveform may be divided into four regions corresponding to the four turn-off phases (Figure 8.2).

8.2
Storage time t_{s_1}

The n base in a power device is usually uniformly doped with very low impurity concentration ($\sim 10^{14}$ atoms/cm³ or less). The p base, on the other hand, is doped nonuniformly and may have an average impurity concentration on the order of 5×10^{15} atoms/cm³. As a result, in the forward-blocking state when the center junction J_2 is forward biased and injecting, its emitting efficiency is good from the p base into the n base, but very poor in the opposite direction. For this reason, the turn off of the p-n-p-n device starts with the removal and decay of excess minority-carrier concentration (electrons) from the p base when the n-emitter efficiency is high. The reverse-biased cathode (positive) has to collect only the charges (electrons) present in the p base which are hardly replenished by the injection from the n base. It is not so in the n base, since as soon as the charges (holes) are collected by the reverse-biased anode (negative), more holes are injected efficiently by the p base into the n base. Consequently, the turn off starts with the n-emitter junction J_3.

During the t_1 (Figure 8.2) the current through the device is constant because there is a large charge stored in the p base which allows the current to flow when the bias is reversed. As long as the excess charges are not entirely removed from the p-base junction, J_3 will remain forward biased despite the external voltage reversal. Neglecting the forward voltage drop across the device, the constant current through the device during t_1 equals

$$I_{r_1} = \frac{V_r}{R} \tag{8.1}$$

where V_r is the reverse voltage pulse amplitude and R is the circuit resistance (ohmic load is assumed).

At the end of the time t_1, enough electrons are removed by the reverse (positive) potential applied to the cathode, the depletion region of the J_3 junction begins to widen, and J_3 starts supporting the reverse bias [8.1–8.3].

Using charge-storage relationships, it is possible to obtain the following expression for the first storage time (notation as per Table 8.1 and Figure 8.1):

$$t_{s_1} = \tau_2 \ln\left[\frac{\gamma_{1I} I_{r_1} + A_p I_F}{(1 - \alpha_{2I})I_{r_1}}\right] \tag{8.2}$$

$$\alpha_{2I} = \beta_{2I} \gamma_{2I} \tag{8.3}$$

where I_F is the on-state current, A_p is a constant, τ_2 is the lifetime in the p-base, γ_{1I} is the emitter efficiency of the p_2-n_1-p_1 transistor, α_{2I} is n_1-p_2-n_2 transistor DC gain, β_{2I} is the n_1-p_2-n_2 transistor transport factor, and γ_{2I} is the n_1-p_2-n_2 transistor emitter efficiency. For $\alpha_{2I} \ll 1$ and $\gamma_{1I} \approx 1$, this expression reduces to a simple relationship

$$t_{s_1} = \tau_2 \ln\left(1 + A_p \frac{I_F}{I_{r_1}}\right) \tag{8.4}$$

Thus, the first storage time can be made small if the lifetime in the p base is short and/or the reverse current is high.

8.3
Fall time t_{f_1}

There are two possible situations that should be considered: (1) the n-emitter junction recovers, i.e., its depletion region builds up, but the externally applied voltage is smaller than the junction-breakdown voltage; and (2) the applied voltage is large enough to cause the breakdown of the n-emitter junction J_3 (or the n emitter

Table 8.1
Notation used for the turn-off transient analysis[a]

Parameter	n_2-p_2-n_1 transistor	n_1-p_2-n_2 transistor	p_1-n_1-p_2 transistor	p_2-n_1-p_1 transistor
Common base DC current gain	$\alpha_N = \gamma_2 \beta_2$	$\alpha_{2I} = \gamma_{2I} \beta_{2I}$	$\alpha_P = \gamma_1 \beta_1$	$\alpha_{1I} = \gamma_{1I} \beta_{1I}$
Transport factor	β_2	β_{2I}	β_1	β_{1I}
Emitter efficiency	γ_2	γ_{2I}	γ_1	γ_{1I}
Base transit time	T_2	T_{2I}	T_1	T_{1I}
Minority-carrier lifetime in the base	τ_2	τ_2	τ_1	τ_1

[a]Letter I is used for inverse parameters (see Figure 8.1).

is shorted). In the first case it was shown by Sundresh [8.3] that the decaying current has the form

$$I(t) = \frac{K_1 \tau_2}{T_{2I}} I_{r_1} \exp\left(\frac{-t}{\tau_2}\right) \tag{8.5}$$

with

$$K_1 = \gamma_{2I}(1 - \beta_{2I}) \tag{8.6}$$

and the origin for time is assumed to be at the end of t_{s_1} ; T_{2I} is the minority carriers' (electrons) transit time in the p base in the inverse direction from the n base toward the n emitter. When there is no breakdown, the expression (8.5) permits the determination of the first fall time.

The second case in which breakdown occurs is usually more important from the practical point of view and corresponds to high-anode reverse-applied potentials. For this case [8.3]

$$t_{f_1} = \tau_2 \ln\left[\frac{\tau_2}{T_{2I}} \gamma_{2I}(1 - \beta_{2I}) \frac{I_{r_1} R}{I_{r_1} R - V_B}\right] \tag{8.7}$$

V_B is the n-emitter (junction J_3) breakdown voltage.

The requirements for the short fall time t_{f_1} are: small minority-carrier lifetime in the p base, low breakdown of the n-emitter junction, and low injection efficiency from the n base into the p base.

8.4
Storage time t_{s2}

As the voltage reaches the avalanche breakdown of the n-emitter junction J_3, the voltage drop across J_3 becomes constant and remains in series with the rest of the device which consists of the p-n-p portion still conducting fully and contributing a negligibly small voltage drop.

As soon as J_3 breaks down, a second storage time starts, during which the current remains essentially constant and can be determined from

$$I_{r_2} = \frac{(V_r - V_B)}{R} \tag{8.8}$$

The center junction J_2 is a good emitter of holes and an inefficient emitter of electrons so that the hole injection into the n base from J_2 will continue as long as J_2 is forward-biased. The n base recovers then more slowly than the p base because of this continuous replenishment of holes. Using the charge-control method, it is possible to obtain a simple expression for t_{s_2}. If we assume for simplicity that

93

the fall time τ_{f_1} is negligibly small and both emitters (n and p) have unity gamma, then for high-anode reverse-bias

$$t_{s_2} = \tau_1 \ln\left(C_1 + C_2 \frac{I_F}{I_{r_2}} \right) \tag{8.9}$$

where C_1 and C_2 are algebraic functions of the normal- and reverse-current gains [8.1]. Here again, large reverse current and short minority-carrier lifetime in the n base will reduce the duration of the storage time.

8.5
Fall time t_{f_2}

Assuming that the carrier flow during this phase is by diffusion only and that the charge distributions in both bases are linear, it is possible to obtain an analytical expression linking the currents to the charges in each base [8.3]. In power SCRs we have usually $\gamma_{2I} \ll \gamma_{1I}$, and the current during the second decay time may be expressed by

$$I(t) = I_{r_2} \tau_1 \gamma_{1I}(1 - \beta_{1I}) \frac{1}{T_{1I}} \exp\left(\frac{-t}{T_{1I}} \right) \tag{8.10}$$

where τ_1 is the lifetime in the n-base, β_{1I} is the p_2-n_1-p_1 transistor transport factor, and T_{1I} is the p_2-n_1-p_1 transistor base transit time. The fall time t_{f_2} can be determined from (8.10) by letting the current $I(t)$ drop down to the holding current level; for instance,

$$t_{f_2} = T_{1I} \ln \frac{I_{r_2} \tau_1 \gamma_{1I}(1 - \beta_{1I})}{I_h T_{1I}} \tag{8.11}$$

The second fall time can be decreased by short minority-carrier lifetime in the n base and short inverse transit time.

8.6
Effect of the gate current

To speed up the turn-off process, it is quite common to apply a reverse potential to the p gate of the device simultaneously with the application of the reverse potential to the anode. The gate permits the withdrawal of majority carriers from the p base and the reduction of the turn-off time.

During phase 1, the current flowing out the p-base region is increased by the gate current I_g so that (8.2) should be modified to

$$t_{s_1} = \tau_2 \ln\left[\frac{\gamma_{1I} I_{r_1} + A_p I_F + I_g}{(1 - \alpha_{2I}) I_{r_1} + I_g} \right] \tag{8.12}$$

During the phase 2, the n-emitter junction is in the avalanche-breakdown condition for anode voltages greater than V_B and the gate loses its effectiveness. The gate continues to be ineffective during the rest of the transient, i.e., during the second saturation and fall times.

8.7
Simplified approaches

By making several simplifying assumptions it is possible to obtain a very approximate expression for the total turn-off time. The assumptions are [8.2]: (1) the first storage and fall times are negligibly small compared to the second storage and fall times; (2) the injection efficiency of J_2 junction for holes into the n base is equal to unity and zero for injection of electrons into the p base; (3) the holes collected by J_1 are immediately replaced by the holes injected by J_2; and (4) the hole current injected into the n base at J_2 is equal to the hole current collected at J_1, and the difference between these two currents is zero (base current $= 0$).

The charge conservation equation (5.3) becomes, therefore,

$$\frac{dq_N}{dt} + \frac{q_N}{\tau_1} = 0 \tag{8.13}$$

where q_N is the excess charge in the n base at the time t and τ_1 is the hole lifetime in the n base. The solution of (8.13) is

$$q_N = \tau_1 \alpha_N J_F \exp\left(\frac{-t}{\tau_1}\right) \tag{8.14}$$

where $\tau_1 \alpha_N J_F$ is the charge at the time $t = 0$.

The hole current through the structure is proportional to the excess charge in the n_1 base; therefore

$$J = J_F \exp\left(\frac{-t}{\tau_1}\right) \tag{8.15}$$

The current will drop to the holding current level after the time

$$t_{off} \sim \tau_1 \ln\left(\frac{J_F}{J_h}\right) \tag{8.16}$$

Even after reaching the current level below J_h for a while the SCR will continue conducting some small current and, therefore, it may take some additional time before the forward potential can be re-applied to the anode. The total time necessary before the reapplication of the forward potential is the so-called circuit-commutated turn-off time t_q.

In Chapter 12, which discusses the commutation of triacs, the problem of charge removal from an SCR is handled in a manner different from that discussed in this chapter. The triac serves to control the AC current flow, and at the time when one part of the triac is turning off, a short-duration reverse current flows which has some finite slope di/dt and is not a step function. Under these conditions it is necessary to simplify the problem by assuming that the recovering SCR behaves as a diode rather than a multijunction device.

An accurate analysis of a turn-off transient with a sinusoidal recovery current was attempted for two extreme situations by Davies and Petruzella [8.4].

8.8
Experimental data

The availability in the literature of the direct experimental data is very limited. However, the information obtained by Baker et al. [8.1] confirmed the dependence of storage time on the forward- and reverse-current densities as per expressions (8.2) and (8.9). The same data show the direct proportionality of the saturation times to the minority-carrier lifetimes in respective device bases. The circuit-commutated turn-off-time t_q measurements[1] confirm essentially the validity of the above theoretical considerations in so far as the total switching time is concerned, since t_q behaves qualitatively in the same way as switching time $t_{s_1} + t_{s_2} + t_{f_1} + t_{f_2}$.

According to the EIA-NEMA standards [8.5] t_q is measured as follows: The thyristor under test is made to conduct the specified on-state current at the specified temperature. A reverse potential is then applied to the anode so that the current is reversed through the thyristor at the specified rate (di/dt). The reverse current recovers the stored charge from the anode and cathode junctions allowing the thyristor to support the *specified reverse-blocking voltage*. However, some additional time interval is required for the J_2 collector charge to recombine before the thyristor can block the reapplied forward voltage. The test is performed by applying an off-state voltage at the specified rate of (dv/dt) (Figure 8.3) after successively shorter waiting times until it is observed that the thyristor is unable to support the off-state voltage, without switching to the on state.

The thyristor current and voltage waveforms are illustrated in Figure 8.3 [8.5].

---

[1] See thyristor definitions, Chapter 1.

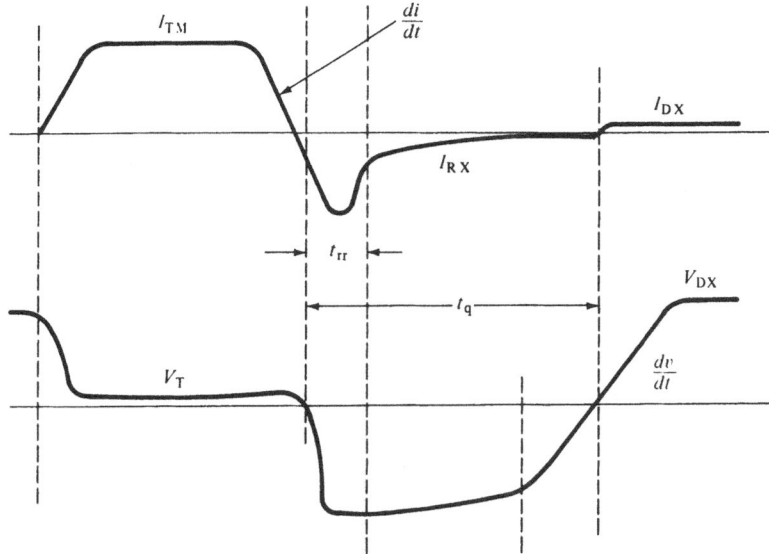

Figure 8.3 Thyristor current and voltage waveforms during circuit-commutated turn off as per EIA-NEMA standard [8.5]

References

8.1 A. N. Baker, J. M. Goldey, and I. M. Ross. Recovery time of PNPN diodes, 1959 *IRE Wescon Convention Record*, Part 3, pp. 43–48.

8.2 R. E. Gentry, F. W. Gutzwiller, N. Holonyak, Jr., E. E. Von Zastrow. *Semiconductor Controlled Rectifiers*. Englewood Cliffs, N.J.: Prentice-Hall, 1964.

8.3 T. S. Sundresh. Reverse transient in PNPN triodes. *IEEE Trans. Electron Devices, 14*: 400–402, 1967.

8.4 R. L. Davies and J. Petruzella. PNPN charge dynamics. *Proc. IEEE, 55* (8): 318–330, 1967.

8.5 EIA-NEMA Standard RS 397. Recommended standards for thyristors. Washington, D.C.: Electronic Industry Association, June 1972.

8.6 E. S. Yang. Turn-off characteristics of p-n-p-n devices. *Solid State Electron., 10*: 927–933, 1967.

Gate turn-off
thyristor (GTO)

Summary

A gate turn-off switch (GTO) is a four-layer p-n-p-n device similar in construction to an SCR. A GTO is triggered into conduction by applying a forward bias to its gate and *turned off by the application of a reverse-gate bias*. An SCR, on the other hand, is turned off primarily by reducing or reversing the anode current. A simplified two-dimensional model of the switching mechanism of a GTO is discussed. The model predicts the dependence of the storage time (plasma-pinching time) on the device turn-off gain, i.e., on the ratio of the anode current to the reverse-gate current. An expression for the maximum switchable current is also derived.

9.1
Introduction

In order to turn off an SCR it is necessary to reduce the main current flow below the holding current level. Usually this is done by reversing the anode and cathode polarities (Chapter 17). In a GTO it is not necessary to reverse the polarity of the main terminals and the turnoff is accomplished by the application of the *reverse-gate potential and current*. As soon as a negative potential is applied to the p-gate (Figure 9.1), holes from the anode begin to be partially removed by the gate. On the other hand, the electrons injected by the cathode emitter are pushed away from the negative gate potential toward the active center of the device. At some point in time the n-emitter regions closer to its edges stop to conduct entirely and the current flows in a narrow region far away from the gate. The current is "squeezed" (Figure 9.2) into a high–current-density filament [9.1].

98

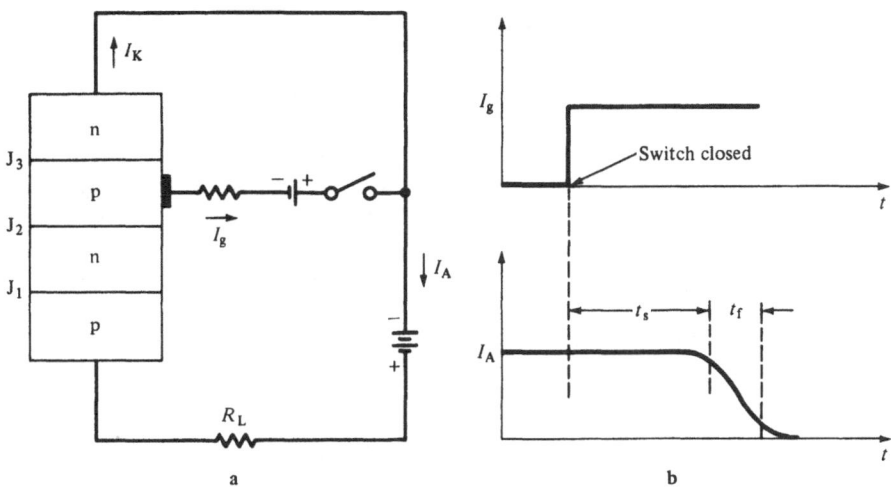

Figure 9.1 GTO biased for turn off. (After Wolley [9.1].)

The current in the filament consists of equal number of holes and electrons as required for charge neutrality at high injection levels. The charge in the filament consists therefore of the minority and majority carriers and it is often referred to as plasma. Since the two carrier densities are equal, it is sufficient to discuss just one of them.

In order to turn off the device it is necessary to remove all the excess carriers present in the bases during forward conduction. When this is accomplished, the blocking junction J_2 comes out of saturation, is no longer forward-biased, and starts blocking the applied forward potential. In the analysis [9.1] the time for the carrier removal by the gate and recombination after the plasma was squeezed to its minimum value is assumed to be very short as compared to the

Figure 9.2 Hole and electron flow in a partially turned-off GTO.

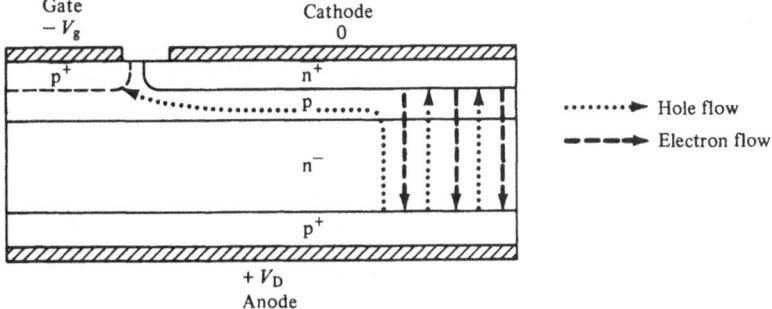

99

time necessary to focus the plasma to its minimum value. This time is the storage time t_s.

The quantitative analysis [9.1] is limited here to the time t_s required for the focusing or pinching of the plasma in a low-lifetime gold-doped device to its minimum value.

It is of interest to note that the reverse second-breakdown mechanism of transistors is akin to the mechanism described for the p-n-p-n gate turn-off.

9.2
Plasma-pinching mechanism

Figure 9.3 represents the n-p emitter–base junction with the plasma confined to the center portion of the p base, a short time after the junction was reverse-biased for the turn-off. One part of the emitter, closest to the gate, is already reverse biased, and stopped emitting because the biasing charge was already removed. The gate current $\frac{1}{2}I_g$ flows laterally through the p base toward the gate contact. The region between the plasma and the gate contact lacks enough charges to be conductivity modulated so it exhibits comparatively large sheet resistance. The center portion of the device is still turned on, i.e., all junctions J_1, J_2, and J_3 are forward biased. The anode current remains almost unchanged, and its value depends, at this point in time, on the external circuit impedance only because the device impedance itself is still low. Since the current remains more or less constant but the conduction area is reduced, the carrier concentration is highly increased.

The velocity with which the plasma is being pushed toward the center can be determined approximately from a simplified model

Figure 9.3 Plasma focussing in the p base. (After Wolley [9.1].)

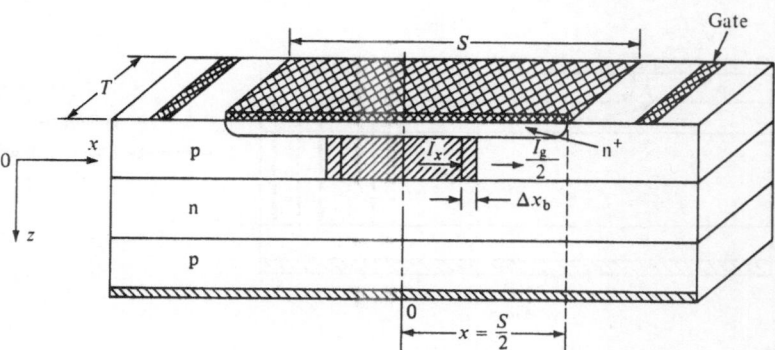

only [9.1]. An exact analysis has not been attempted so far, due to the complexity of the phenomenon. The simplified model avoids one of the many difficulties by assuming the gate current to be constant. This is rather a crude approximation since in actual devices a constant gate current is usually not observed. Constant current requires high-impedance gate circuit, which would lead to the emitter-base junction breakdown and would counteract the turning-off process. Nevertheless, the simplified analysis helps us understand the device behavior, and its results are confirmed essentially by the experimental data [9.1].

The model requires an introduction of a parameter L_{eff} which may be interpreted as an effective ambipolar diffusion length in the p base at the edge of the on region, in the presence of an electric field.

The net rate of electron removal at the right-hand boundary of the plasma region is assumed to be equal to the gate current minus the electron-diffusion current. It is sufficient to concentrate our attention on only one type of charge carrier, since—due to the charge-neutrality requirement—the opposite polarity charge is taken care of automatically. It is further assumed that the device has gate contacts on both sides of the cathode and that the active region is pinched uniformly from both sides. The electron-concentration distribution $n(z)$ in the p base along the z axis is assumed to have the form as in Figure 9.4 whereas the distribution $n(x)$ along the x axis is shown in Figure 9.5. The $n(x)$ concentration distribution has a constant value up to the point x_b on the x axis; beyond x_b it falls down exponentially, following the expression

$$n(x) = \exp\left(\frac{x_b - x}{L_{\text{eff}}}\right) n(x_b)$$

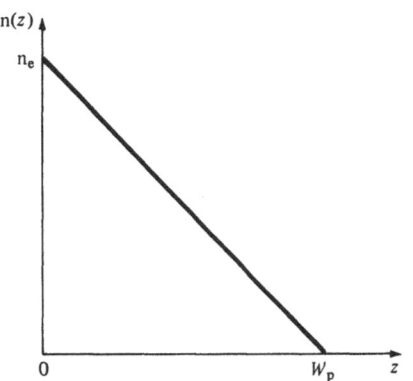

Figure 9.4 Electron concentration n as a function of distance z.

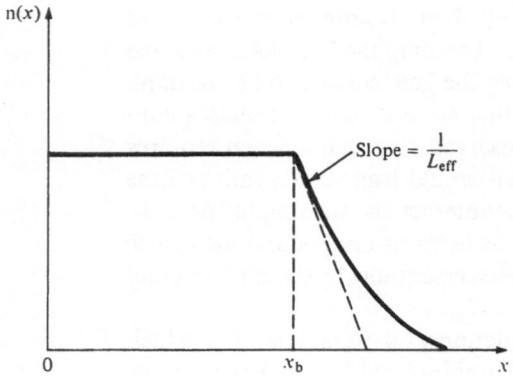

Figure 9.5 Electron concentration as a function of distance x. (After Wolley [9.1].)

9.3
Turn-off velocity

The turn-off velocity v is determined mainly by the pinching—also called the storage—time and is assumed constant. The pinching time is considered to be much larger than the fall time, so this last term is neglected in the quantitative considerations. The product $(TW_p dx)$ represents an elemental volume in the p base (Figure 9.3), where T is the emitter length and W_p is the p-base width.

The elemental volume contains an electronic charge

$$dq = \frac{-qn_e}{2}(TW_p)dx \tag{9.1}$$

where n_e is the injected electron density at the emitter-base junction. The distribution along the z axis is assumed to be linear as per Figure 9.4, so the average electron density[1] is $\frac{1}{2}n_e$.

The rate of change of the positive charge in the elemental volume is the sum of the gate current $\frac{1}{2}I_g$ at one side of the emitter and a lateral-diffusion current I_x that flows into the elemental volume from the conducting region:

$$\frac{dq}{dt} = \frac{I_g}{2} + I_x \tag{9.2}$$

with

$$I_x = -qD\frac{(TW_p)n_e}{2L_{eff}} \tag{9.3}$$

---

[1] Zero-charge density at J_2 is an approximation: so is the linearity assumption.

where D is a diffusion constant. Expression 9.3 is a diffusion equation in the x direction for a positive charge averaged in the p-base along the z axis. Substituting (9.1) and (9.3) into (9.2) and rearranging terms, we obtain

$$v = \frac{dx}{dt} = -\left[\frac{I_g}{2}\frac{2}{(qn_e)(TW_p)}\right] + \frac{D}{L_{eff}} \tag{9.4}$$

where v is the charge-propagation velocity during the pinching time t_s.

The cathode current during the turn off is given by

$$I_K = I_A - I_g \tag{9.5}$$

where [9.1]

$$I_K = -2qD(Tn_e)\frac{x_b + L_{eff}}{W_p} \tag{9.6}$$

The transit (diffusion) time through the p base is

$$t_p = \frac{W_p^2}{2D} \tag{9.7}$$

and the turn-off gain is defined as

$$\beta_{off} = \frac{I_A}{I_g} \tag{9.8}$$

Utilizing expressions (9.5), (9.6), (9.7), and (9.8), we modify (9.4) to

$$\frac{dx}{dt} = v = -\frac{x_b + L_{eff}}{t_p(\beta_{off} - 1)} + \frac{D}{L_{eff}} \tag{9.9}$$

9.4
Pinching (focusing) time and turn-off gain

The time required to focus the charges in the center part of the p base can be determined by the integration of the expression for dx/dt (9.9) between two limiting values of x. The upper limit is $\frac{1}{2}S$ (Figure 9.3) while L_{eff} is taken as the value of the lower limit. This is rather an arbitrary choice, but a reasonable one since L_{eff} value is on the order of one diffusion length.

The integration yields

$$t_s = (\beta_{off} - 1)t_p \ln\left(\frac{\dfrac{SL_{eff}}{W_p^2} + \dfrac{2L_{eff}^2}{W_p^2} + 1 - \beta_{off}}{\dfrac{4L_{eff}^2}{W_p^2} + 1 - \beta_{off}}\right) \tag{9.10}$$

103

From (9.10) we obtain a maximum possible turn-off gain

$$\beta_{\text{off}} \max = 1 + \frac{4L_{\text{eff}}}{W_p^2} \tag{9.11}$$

At this gain value the focusing time goes to infinity.

As an example: for $L_{\text{eff}} = 10 \ \mu\text{m}$ and $W_p = 20 \ \mu\text{m}$ the $\beta_{\text{off}} \max = 5$.

Other authors [9.2–9.4] have derived for the turn-off gain of the one-dimensional device.

$$\beta_{\text{off}} = \frac{\alpha_N}{(\alpha_N + \alpha_P) - 1} \tag{9.12}$$

where α_N and α_P are the p-n-p and n-p-n transistors' current gains, respectively. The expression (9.12) is derived in the Appendix. The gain given by (9.11) can be considered as useful while the active region is being reduced essentially to the one-dimensional device, while the expression for gain given by (9.12) can be used during the time interval when the one-dimensional device is being turned off. The smaller of the two is the limiting turn-off gain.

A more recent one-dimensional analysis [9.5] shows that at high injection levels a more appropriate expression for the turn-off gain is

$$\beta_{\text{off}} = \frac{\left(\dfrac{b+1}{b}\right)\alpha_N}{\left(\dfrac{b+1}{b}\right)\alpha_N + \alpha_P - 1} \tag{9.13}$$

Where b is the electron-to-hole mobility ratio and the α_N and α_P in (9.13) are the high injection level current gains which are much smaller than those at moderate current levels. Parameter b is a function of current density also.

9.5
Maximum anode and gate currents

When the plasma size is reduced to its minimum value L_{eff}, the resistance of the p base under the half-width of the emitter reaches its maximum value. The product of this unmodulated resistance and the gate current should be smaller than the reverse-breakdown voltage of the n emitter. For $L_{\text{eff}} \ll \frac{1}{2}S$ it was shown by Wolley [9.1] that this limits the gate turn-off current to its maximum value.

$$I_g(\text{max}) = \frac{4V_{BR}}{R_b} \tag{9.14}$$

where

$$R_b = \frac{\rho S}{W_p T} \tag{9.15}$$

V_{BR} is the emitter breakdown voltage and ρ is the average resistivity of the unmodulated gated base. From the definition of $\beta_{off} = I_A/I_g$, the maximum switchable anode current will be

$$I_A = \frac{4\beta_{off} V_{BR}}{R_b} \tag{9.16}$$

The storage time t_s can be reduced by increasing the gate turn-off current. If this current is increased, however, above the value given by (9.14), the emitter-avalanche breakdown will take place and the increased I_g will not be effective in charge-carrier-removal.

9.6
Plasma-pinching in the ungated n base

When the lifetime in the ungated n base is small because of gold-doping, and the transit time through the n base—due to drift—is smaller than the diffusion transit time through the p base, then the pinching of the plasma of minority carriers in the ungated base will follow that in the gated base.

If, however, the lifetime in the ungated n base is large and the transit time through the n base is longer than through the p base, then the pinching of the plasma in the ungated base will lag the focusing in the gated base. The plasma in the ungated base will follow, however, that of the gated base if high–turn-off gain, i.e., small-gate currents, are used since this will make the saturation time t_s larger than the transit time through the n base.

If the lag is present in the ungated base, a large number of minority carriers will be left in this base at the end of storage time and will have to be removed during the fall time. The gate turn-off current aids in removal of the carriers from the gated p base. In the ungated n base, however, the minority carriers must be transported for collection by drift and/or diffusion to the collecting junction J_2 or will have to recombine with the majority carriers. The fewer the number of carriers left in the injected base, the shorter, therefore, will be the fall time. Low lifetime and small β_{off} will be useful in reducing the fall time.

105

9.7
Theoretical model compared with experiment

Storage times were measured [9.1] for a gold-diffused GTO and plotted in (Figure 9.6) in function of the turn-off gain β_{off}. The anode current was 1 A. The device emitter stripe had a width of 600 μm and the p-base width was 27 μm. The ambipolar diffusion constant D was assumed to be about 20 cm^2/sec^1. Curves computed from [9.1] are plotted in Figure 9.6 for three values of L_{eff} (35, 50, and 65 μm). The experimental data agree with the theoretical predictions fairly well for $L_{\mathrm{eff}} = 50 \mu$m.

The experimental results [9.1] regarding the fall time versus turn-off gain do not show a satisfactory correlation of the curve shapes to the device parameters such as base widths or base resistance. If

Figure 9.6 Theoretical and measured values of t_f versus β_{off} for an experimental GTO with I_A = 1 A. (After Wolley [9.1].)

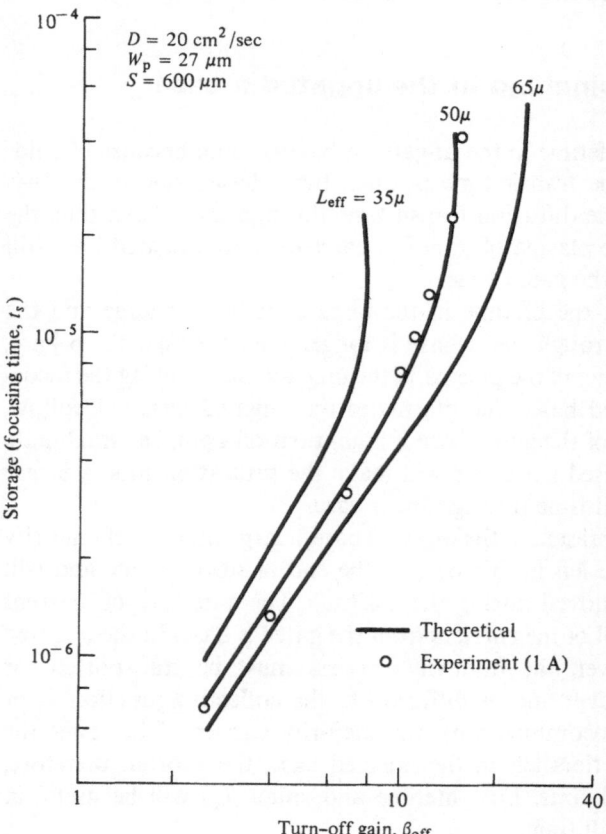

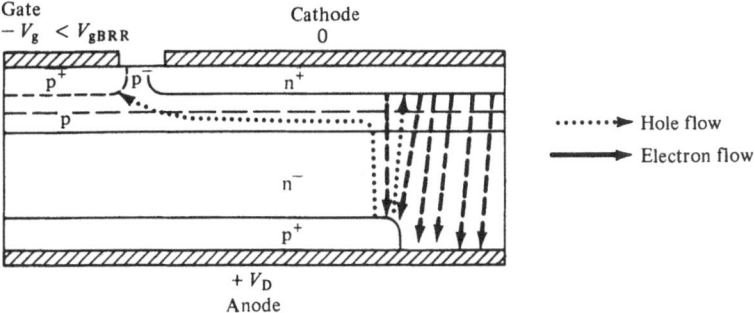

Figure 9.7 GTO with high emitter breakdown and low regeneration region (After Becke and Neilson [9.7].)

most of the carriers are stored in the gated base, the fall time should decrease with decreasing turn-off gains. The curves show that, in general, the fall time decreases with decreasing gain, but there are regions in which the fall time is relatively independent of gain for some devices.

At low turn-off gains, i.e., at high I_g currents, the breakdown of the n emitter could be used to explain this behavior. At low gate currents, i.e., high gains, this argument obviously cannot be used.

Experimental results confirmed that in agreement with Equation (9.16) the smaller the p-base resistance, the larger the current which could be controlled.

Some additional experimental data for devices with ratings of 10 A and 600 V were given [9.6].

Recently a structure has been developed [9.7] which tries to overcome the problems with excessive current focusing which may lead to the device destruction due to the local overheating under unfavorable conditions. The device p base consists of two layers: one of high resistivity and one of low resistivity (Figure 9.7). The high-resistivity region assures high breakdown voltage of the cathode-emitter base junction; the low resistivity, on the other hand, lowers the lateral resistance R_b of the gated-base region. In addition, there is a region away from the gate where the cathode and anode emitters do not overlap and where the final conducting filament will extinguish itself more easily since there is less regeneration [9.7] than in the region where the anode and cathode overlap. The device was gold-doped to minimize fall time for fast-switching applications. Using this device, it was possible to switch a power of 1.2 kW through a resistive load at a rate of at least 20 kHz. The silicon chip used had a size of 0.3 cm by 0.3 cm.

107

APPENDIX

Turn-Off Current Gain [9.2]

$$I_K = I_A - I_g \tag{9.5}$$

but also

$$I_K = \alpha_N I_K + \alpha_P I_A$$

Eliminating I_K yields

$$I_A = \frac{\alpha_N I_g}{\alpha_N + \alpha_P - 1}$$

$$\beta_{off} = \frac{I_A}{I_g} = \frac{\alpha_N}{\alpha_N + \alpha_P - 1} \tag{9.12}$$

References

9.1 E. Duane Wolley. Gate turn-off in pnpn devices. *IEEE Trans. Electron Devices, 13*: 590–597, 1966.

9.2 R. H. van Ligten and D. Navon. Base turn-off of pnpn switches. *1960 IRE Wescon Conv. Rec.*, Pt. 3, pp. 49–52.

9.3 D. R. Muss and C. Goldberg. Switching mechanism in the npnp silicon controlled rectifier. *IEEE Trans Electron Devices, ED-10*: 113–120, 1963.

9.4 J. M. Goldey, I. M. Mackintosh, and I. M. Ross. Turn-off gain in pnpn triodes. *Solid State Electron., 3*: 119–122, 1961.

9.5 I. A. Liniychuk and A. I. Palamazchuk. Gate turn-off of a pnpn device at high injection level. *Radio-Engineering and Elektronic Physics, 4*: 605–609, 1973.

9.6 T. C. New, W. D. Frobenius, T. J. Desmond, and D. R. Hamilton. High power gate controlled switch. *IEEE Trans. Electron Devices, ED-17* (9): 706–710, 1970.

9.7 H. W. Becke and J. M. Neilson. A new approach to the design of a gate turn-off thyristor. *IEEE Pesc. Record*, pp. 292–299, 1975.

Also of interest

M. Kurata. A new CAD-model of a gate turn-off thyristor. *IEEE Pesc. Record*, pp. 125–133, 1974.

Thyristor *di/dt* and current pulse capability

Summary

If the initial rate of the anode current rise, *di/dt*, exceeds a certain value, the thyristor may be either destroyed or permanently damaged due to the excessive junction temperature. The limiting *di/dt* value depends on the size of the initially turned-on area and on the velocity of the plasma spreading along the n emitter. It is, therefore, advantageous to increase the initially fired area. To achieve this, several methods—such as interdigitation, field-initiated turn on, amplifying gate, and emitter gate—are used.

10.1
Introduction

When a positive pulse is applied to the p gate of a thyristor, a current starts flowing between the gate and the n emitter. This way a lateral voltage drop is developed in the p base of such a direction that the emitter edge nearest to the gate becomes more forward-biased than the rest of the emitter. As a result, most of the injection occurs initially at this location.

Due to the diffusion and field effects, the turn-on area spreads along the emitter with a velocity usually smaller than about 10^4 cm/sec. Eventually, the entire emitter begins to inject uniformly.

If the initial rate of current rise *di/dt* exceeds a certain value [10.1, 10.2], the device may be either destroyed or permanently degraded due to the excessive localized overheating. This limiting *di/dt* value depends on the size of the initially turned-on area and on the velocity of the plasma spreading along the n emitter.

The *di/dt* limitation is not only of importance for intentional gate triggering, but also for the "false," undesirable triggering due to the *dv/dt* effect or due to the application of a voltage exceeding the forward-blocking capability (avalanche triggering).

While the turn on for the gated triggering starts at the emitter edge nearest to the gate, the turn on due to the voltage effects may start at any point of the n emitter due to the nonuniformities of the device. The localized area of the thyristor, which has the highest sum of alphas, will turn on first. Due to the very complex nature of these phenomena, the device *di/dt* capability can be determined on an experimental basis only.

In the cases where *di/dt* exceeds the device capability, reactors may be inserted in the circuit to lower the rate of the current rise. Among circuits that are likely to produce *di/dt* problems are those where capacitors are being discharged. However, even in the 60-Hz phase-control applications, situations can arise where *di/dt* is beyond the device capability. For instance, in three-phase SCR circuits with inductive loads, the rate of current commutation from one phase to another is inversely proportional to the system impedance. Where such thyristor circuits are directly connected to an AC system of low short-circuit impedance, the current rate of rise may be the limiting factor [10.3].

Figure 10.1 (a) Circuit for *di/dt* test; (b) *di/dt* test-circuit waveform.

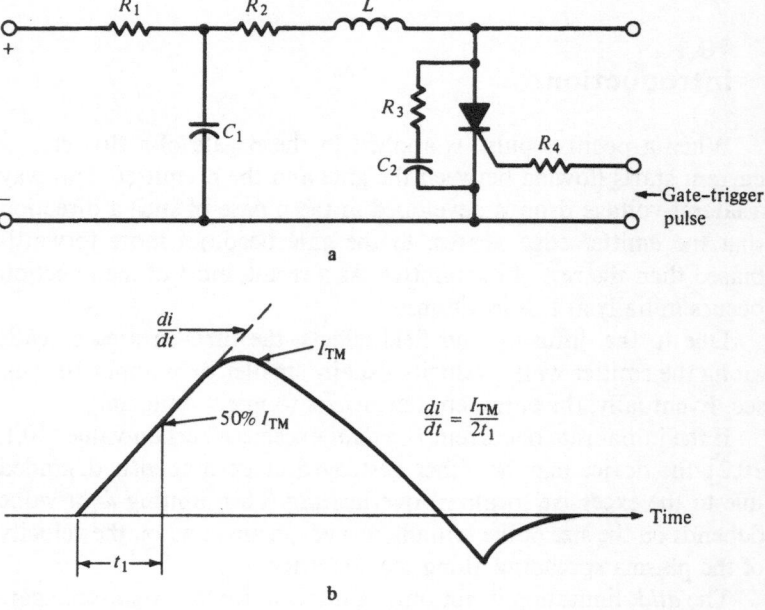

10.2
Test circuit for gate-triggered di/dt

The industry method of testing the di/dt capability consists of a power supply delivering current to an RLC circuit (Figure 10.1a). When the gate is pulse triggered, a current of the form shown in Figure 10.1b, starts flowing through the thyristor. If the peak value of this current is I_{TM}, the di/dt rate of rise can be determined approximately from the expression

$$\frac{di}{dt} = \frac{I_{TM}}{2t_1} \tag{10.1}$$

with t_1 defined by Figure 10.1b.

10.3
Initial turn-on region

As mentioned in Section 10.1, in the case of a gated turn on, the region that is the nearest to the control electrode fires first. This was confirmed by many experimentors who found that the burned-out regions were invariably located in the n-emitter regions adjacent to the gate. It is not so, however, when the turn on occurs due to the dv/dt or avalanche effects.

A series of experiments were performed by Ikeda and Araki [10.4] to determine the location of the initial turn-on region when a device is triggered by voltage effects. The basic circuit for these measurements consisted of a linear dv/dt source. The turn on could be achieved either by dv/dt effect or by the application to the anode of a potential exceeding the forward-blocking capability.

The device itself consisted of a silicon pellet 30 mm in diameter, with an effective area of about 4.4 cm^2. Multiple shorting holes 1 mm in diameter were distributed over the entire n-emitter area at intervals of 2.5 mm in order to increase the dv/dt capability. Proving electrodes were inserted through openings in the cathode metallization and the n emitter so that they were contacting the p base in several places underneath and around the n emitter. The turn on was determined by the measurement of the voltage drops between the cathode and the probing electrodes.

In order to restrict the on region to a small area, the load current was limited by an external resistance to a value only somewhat larger than the device holding current. Figure 10.2 shows a diagram of the

111

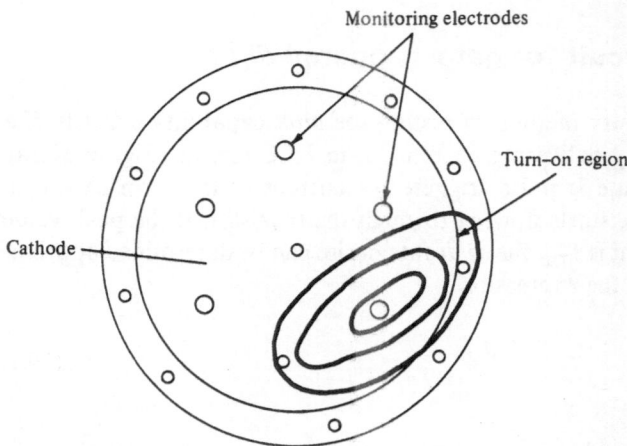

Figure 10.2 Turn-on region spreading. (After Ikeda and Araki [10.4].)

spreading of the turn-on region estimated from the device voltage-drop measurements. The initial area was found to be concentrated in a small region whose location was independent of the *dv/dt* value and was also independent of the load-current magnitude. It was assumed that at the area where the turn on occurred first, the emitter efficiency was the highest and the alpha sum became equal to unity before this occurred at any other area.

After the device was completely turned on, the load current was reduced below the holding level and the device turned off. The last region to turn off was located in almost the same area as the region which was initially turn on by the *dv/dt*.

When thyristors were turned on by the application of an anode voltage exceeding the maximum forward-blocking capability (avalanche firing), the area which turned on first was the same as that which occurred for the *dv/dt* mechanism.[1]

While performing these experiments the authors [10.4] observed that the plasma propagation velocity was anisotropic, i.e., the spreading velocity was not the same in all directions. They also found that in some thyristors some areas turned off in the course of spreading of the on region. This anomalous behavior seems to take place when the average current density over the cathode area is less than a few hundred milliamperes per square centimeter. In such a case, the current density in the initial turned-on region may become smaller than the holding-current density.

---

[1] Avalanche triggering is often destructive in devices not specifically so designed.

10.4
di/dt capability of the gated and nongated turn on

To determine the *di/dt* capability, experiments [10.4] were conducted on a thyristor fabricated on a silicon pellet 22 mm in diameter; the effective area was about 3.0 cm². The cathode had the same shorted emitter structure as described in Section 10.3.

For the gated *di/dt* capability test, gate-current pulses of 10-μsec duration and of 0.1-μsec rise time were used. The repetition rate was 5 pulses/sec. The samples were tested at 120, 400, and 800 mA of principal (load) current. The anode potentials of about 1000 V were within the device ratings. The measurements were made at the maximum allowable operating junction temperature of 125°C. The circuit and current waveform used were similar to those described in Section 10.2. The current amplitude, I_{TM}, in these tests was increased from zero to the value at which the device was destroyed.

For the *dv/dt* or avalanche triggering, the repetition rate of pulses was 5 pps at 125°C. In the *dv/dt* test the *dv/dt* was equal to 200 V/μsec.

Figure 10.3 Waveforms of *di/dt* test, fired by: (a) *dv/dt*;
(b) forward breakover voltage. (After Ikeda and Araki [10.4].)

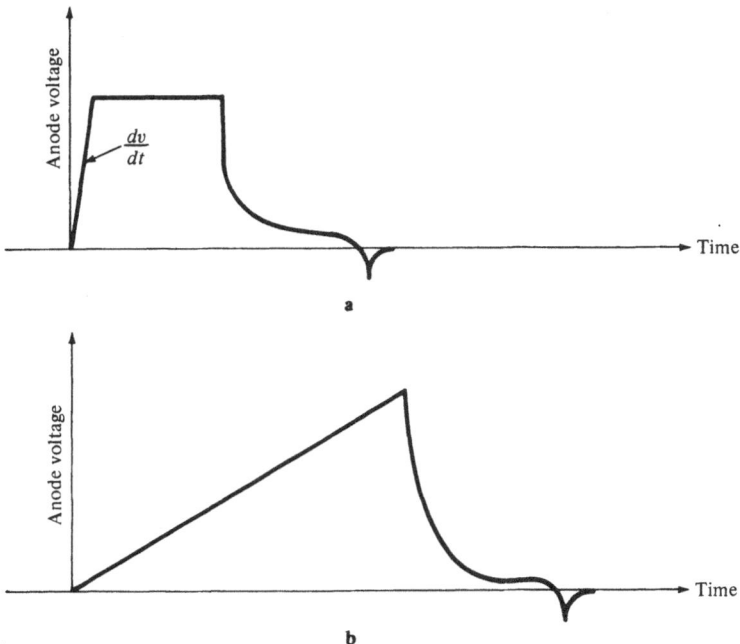

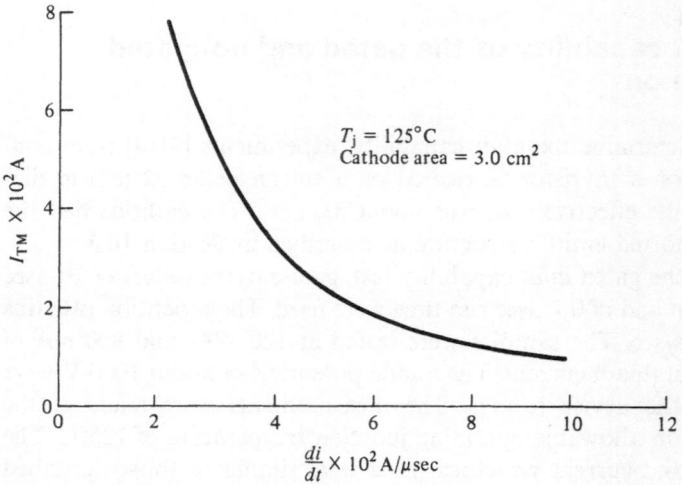

Figure 10.4 *di/dt* capability for various firing methods. (After Ikeda and Araki [10.4].)

For the avalanche firing test, the *dv/dt* of 20 V/μsec was used in order to avoid *dv/dt* triggering. The voltage waveforms for these tests are shown in Figure 10.3. The anode current waveform was monitored for the determination of the *di/dt* slope at the origin in a way similar, but a little more accurate, than described in Section 10.2 and Figure 10.1.

Figure 10.5 *di/dt* capability by the overdriving of a gate. (After Ikeda and Araki [10.4].)

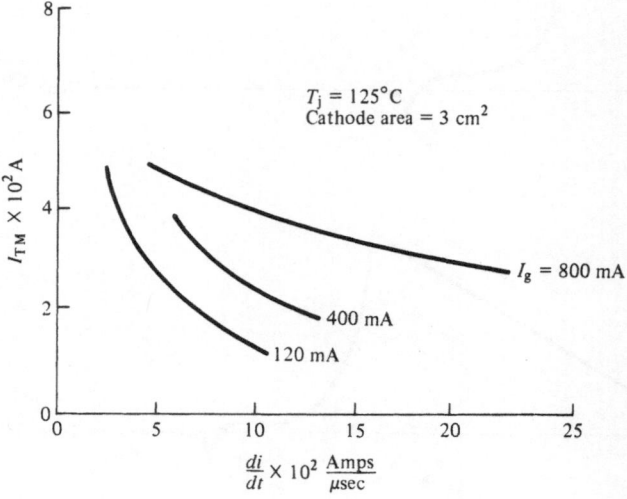

Figure 10.4 is a plot of *di/dt* versus peak current I_{TM}, and represents an average of several measurements. There was no significant difference among the three triggering methods; therefore, one averaged curve represents all three types of firing.

It should be pointed out, however, that for *dv/dt* rates different from 20 V/μsec, the *di/dt* values may differ for the avalanche turn on from those of the gated turn on.

Driving the gate hard by high-gate current increases the *di/dt* capability. Figure 10.5 shows *di/dt* versus I_{TM} for various gate-current amplitudes: the higher the gate current amplitude, the larger the *di/dt* for a given peak anode current I_{TM}.

10.5
Localized temperature rise for a short single pulse

The temperature rise of localized junction region for the case of a very short-duration pulse can be calculated [10.4] in the following manner. The load current is increased until the sample is destroyed, and the small burn-out area is assumed to be about equal to the initially turned-on area. The energy of a single pulse can be computed from the waveforms of the anode current and voltage. Because the time to failure is very short (on the order of very few microseconds), the thermal resistance to the heat conduction from the hot spot to the heat sink does not have to be considered for a rough evaluation. Only the heat capacity of the on region can be taken into account. With this simplification, the temperature rise of the hot spot will be

$$\Delta T = \frac{U}{C} \tag{10.2}$$

where ΔT is the temperature rise above the initial temperature, U is the dissipated energy, and C is the heat capacity of silicon of the turn-on region. The heat capacity

$$C = 4.2cp\, dA \tag{10.3}$$

where c is the specific heat of silicon (0.16 cal/g °C), ρ is the silicon density (2.3 g/cm³), d is the thickness of silicon pellet in centimeters, and A is the turn-on area (in square centimeters) as a function of time. For single, very short pulses on the order of few microseconds, it was found [10.4] that the device was destroyed when the hot spot reached a temperature between 1100 and 1300°C, i.e., below the silicon melting point (1412°C).

115

10.6
Temperature rise for long or recurrent pulses

When the single pulse or train of pulses is compared to the thermal time constant, it may be justified for crude computations to neglect the local heat capacitance effects [10.5]. Then the temperature rise will be dependent mainly on the thermal conductivity of silicon and, therefore, on the hot-spot thermal resistance.

For a circular spot, the spreading thermal resistance, when the spot diameter D is much smaller than silicon pellet thickness, is given for one-sided heat sink by

$$\theta = \frac{1}{2kD} \tag{10.4}$$

where k is silicon thermal conductivity. k is temperature-dependent; thus, at 300°K, $k = 1.5$ W/cm °C; at 500°K, $k = 0.75$ W/cm °C; and at 700°K, $k = 0.5$ W/cm °C. An estimated temperature rise of the hot spot is obtained from

$$\Delta T = P\theta \tag{10.5}$$

where P is the dissipated power. It was found [10.5] that below certain current density the heat generated in the hot spot decreases with increasing temperature and that above this current density the generated heat increases with the temperature. At the crossover point the amount of heat generated is independent of the temperature. This crossover point depends on device parameters and usually occurs at a lower current density with high-voltage devices, which have larger voltage drops than the low-voltage devices.

At the beginning of a pulse, the thyristor can operate above or below this crossover point depending on the *di/dt* and the conducting area magnitude. For many applications, such as inverters, the thyristors operate above the crossover point during the initial conduction period. Under these conditions a small change in the package temperature, e.g., from 100°C to 110°C, may lead to a thermal runaway, because above the critical point the amount of heat generated becomes greater than the amount of heat that can be dissipated by the package.

When the pulse repetition rate increases, the heat power generated increases also; therefore, in order to make the device stable it is necessary to lower the case temperature.

Since a thyristor operating at high case temperature will either stabilize or runaway thermally, one need only test a device for a length of time sufficient for it to reach thermal equilibrium (typically a few hundred seconds).

The repetitive pulses are normally more liable to cause device destruction than is a single pulse. The single pulse of very short duration will increase the junction temperature at one small spot, whereas repetitive pulses may increase the average temperature everywhere in the vicinity of the conducting area. The recurrent pulses may cause failure by comparatively slow temperature buildup, which leads to a thermal feedback mechanism and runaway condition.

Because of this feature, the nonrecurrent power ratings allow the devices to operate instantaneously at junction temperatures vastly exceeding the maximum junction steady-state temperatures. These ratings accommodate unusual circuit conditions.

10.7
Temperature rise and transient thermal impedance during turn-on spreading[2]

When discussing short, single-pulse behavior, it was assumed that the thermal resistance of the conducting region was not affecting the temperature rise and, therefore, only the thermal capacitance was considered. In the case of long or recurrent pulses, on the other hand, it was assumed, that the localized thermal resistance plays a role only.

Figure 10.6   Transient thermal resistance versus time. (After Ikeda *et al.* [10.6].)

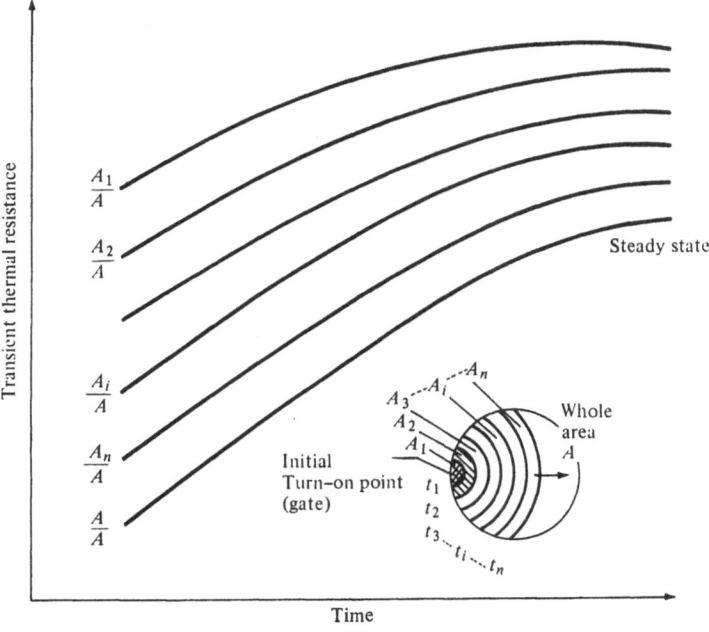

[2] This topic is discussed further in Chapter 16.

117

To obtain more accurate information about the maximum allowable temperature, calculations of the temperature rise should be based on the transient thermal impedance during the on-state spreading. The values of the thermal resistance and capacitance can be calculated from the geometry of the device, and the thermal impedance, in turn, can be determined by considering an analog electric RC network (see Chapter 16 and [10.6–10.10]. Ikeda *et al.* [10.6] have calculated the thermal resistance and capacitance for various ratios of active emitter areas to the total area, taking into account the heat spreading in silicon and in the heat sink. Figure 10.6 shows the transient thermal impedance versus time during the spreading of the on-state [10.6]. $A_1, A_2, \ldots, A_n$ of Figure 10.6 are the values of the turn-on area corresponding to increasing times $t_1, t_2, \ldots, t_n$. The total turn-on area at the steady state is A. The curves of Figure 10.5 are plotted with $(A_1/A), (A_2/A), \ldots, (A_n/A)$ as parameter. Using the values of the transient thermal impedance, it was possible to compute the instantaneous temperature rise of the initial turn-on region considered to have the highest temperature. Figure 10.7 shows the junction temperature rise of a 50-A stud-mounted thyristor computed for various values of sinusordal anode current with 20-μsec pulse base width.

Figure 10.7 Calculated temperature rise versus time. (After Akeda *et al.* [10.6].)

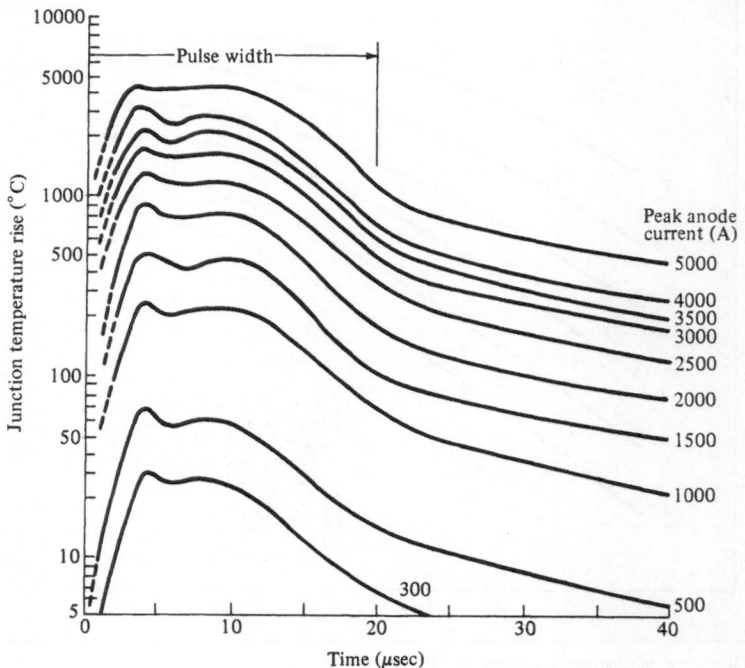

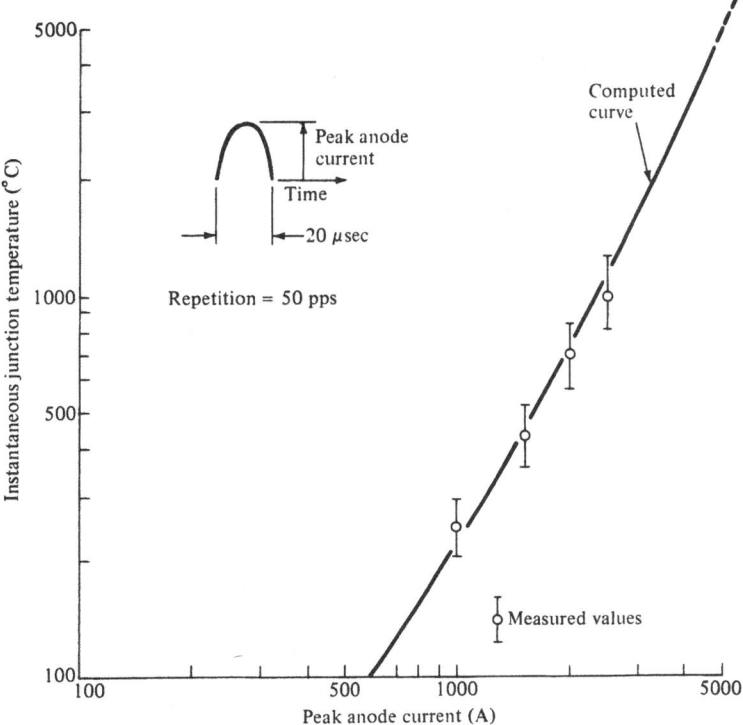

Figure 10.8 Junction temperature rise obtained by analytical and infrared methods. (After Ikeda *et al.* [10.6].)

Figure 10.8, on the other hand, shows the actual experimentally determined temperature rise using an infrared detector. The measured values agreed well with the computed values.

The same authors estimated the maximum allowable junction temperature based on time-to-failure, step-stress capability experiments. They found that the maximum allowable junction temperature for nonrepetitive current pulses was 1000°C for pulse widths ranging from 10 nsec to 100 μsec, 500°C for pulse widths from 500 μsec to 10 msec and 200–300°C for repetitive pulses of any width.

Runaway conditions could, however, exist for repetitive pulses even within the rated temperature limits, i.e., below 125°C.

10.8
Methods of increasing the initial turn-on area

The maximum allowable rate of current rise (di/dt) is higher for devices in which the initial turn-on area is larger. There are several methods of increasing the initial area, such as gate-cathode interdigitation, field initiated turn ons (FI), amplifying gates, and emitter

119

gates. Most of the methods utilize lateral n-emitter and p-base currents to induce lateral fields and utilize the principal current flow, rather than the gate current alone, to turn the device on.

10.8.1
Interdigitation

The interdigitation of the gate-cathode electrodes constitutes a logical approach to the increase of the originally turned-on area [10.11]. The gate fingers are surrounded by the n-emitter and a large periphery of the emitter can theoretically be turned on. The increase of the periphery provides an increased initial turned-on area and improves the *di/dt* capability and also the dynamic (instantaneous) voltage drop.

Thyristors with the interdigitated gate-cathode geometry may have gating characteristics [10.11] different from the conventional thyristor with centrally located gates. The minimum current required to trigger the device is directly proportional to the length of the emitter periphery. In order, then, to provide a sufficiently high triggering current, it may be useful to use a driving pilot device of small dimensions which is triggered by a small amount of current only. This auxiliary thyristor is normally integrated with the main device on the same pellet. This kind of arrangement is called an amplifying gate.

The conduction region, for a given pellet area, is much reduced with the interdigitated gate. But, in case of narrow pulses, the plasma will probably spread less than the emitter finger width. For wide-pulse applications, the noninterdigitated devices may be better suited from the total conducting area point of view.

10.8.2
Amplifying gate

Figure 10.9a shows schematically a centrally located amplifying gate used in a nonintegrated device [10.10, 10.12, 10.13, 10.14]. The application of a small current to the pilot device gate turns this device on rapidly because of its small lateral dimensions. The pilot current is much larger than the original triggering gate current and it provides a strong driving current to the main device. The larger the driving current, the larger will be the area of the initial conduction. Figure 10.9b represents the equivalent circuit of the amplifying gate SCR.

120

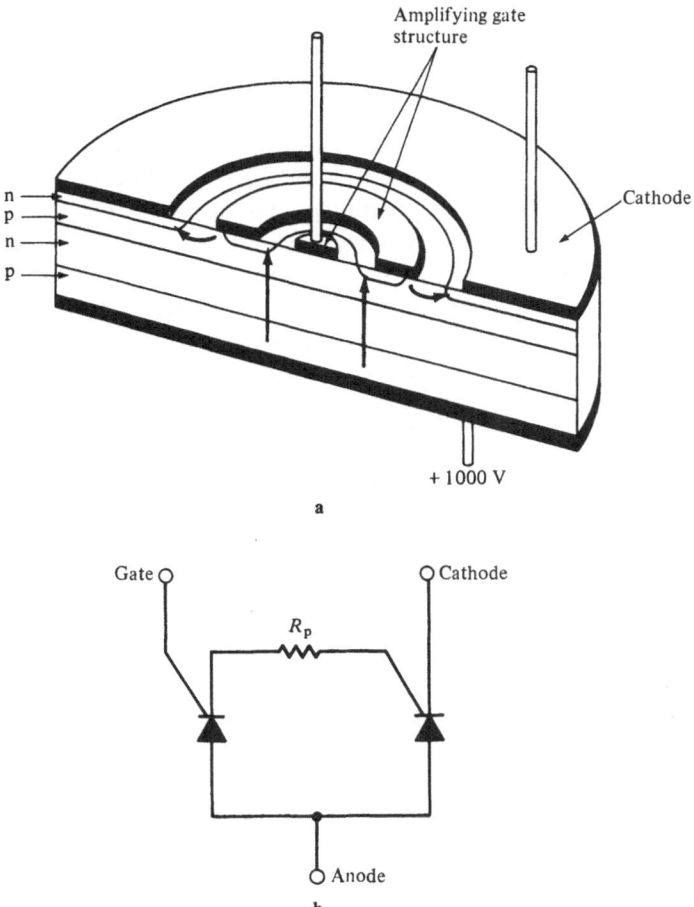

Figure 10.9 (a) Amplifying gate arrangement (after Voss [10.10]); (b) amplifying gate equivalent circuit.

For conventional amplifying gate center-fired thyristors, the use of a gate current larger than the minimum gate current to trigger the auxiliary thyristor is not beneficial since a large gate current tends to turn on the main thyristor at the same time as the auxiliary thyristor. For large levels of gate overdrive, center-fired thyristors approach performance levels similar to those obtained when the gate drive is just sufficient to trigger the pilot thyristor.

For interdigitated thyristors, similar variations in performance are expected. However, because the periphery of the interdigitated emitter is much longer, the point at which gate overdrive no longer improves the switching speed occurs at higher current levels.

121

10.8.3
Field-induced turn on

This approach is sometimes called a trigger-finger method because of the special n-emitter geometry. Figure 10.10 helps to explain the principle involved. The device's cathode metallization is kept some distance away from the n-emitter edge [10.5, 10.15, 10.16]. When the device is activated by the gate, only the region marked *a* next to the gate turns on first and a load current builds up through the turn-on region of the emitter. This load current passes laterally through the p base and the n-emitter layers toward the cathode contact. The lateral hole flow through the p base increases the lateral field in the p base and, therefore, the field between the turn-on portion and the immediately adjacent turned-off portion of the p base increases. The gating (base) current being supplied to the inactive region by the turned-on region of the emitter is considerably increased, and the off region in the immediate vicinity of the on region turns on faster. At the same time, the lateral current flowing in the n emitter increases the emission of the n-emitter part at the point marked *b*. This is because the site *b* of the emitter becomes more negatively biased due to the lateral current flow in the emitter. Therefore, point *b* injects more heavily, and the device turns on at this point. As a result, there will be two turn-on regions: region *a* that spreads toward *b*, and *b* which spreads both toward *a* and in the opposite direction. The whole noncontacted emitter area becomes rapidly turned on. The

Figure 10.10 Field-initiated gate. (After Somos and Piccone [10.5].)

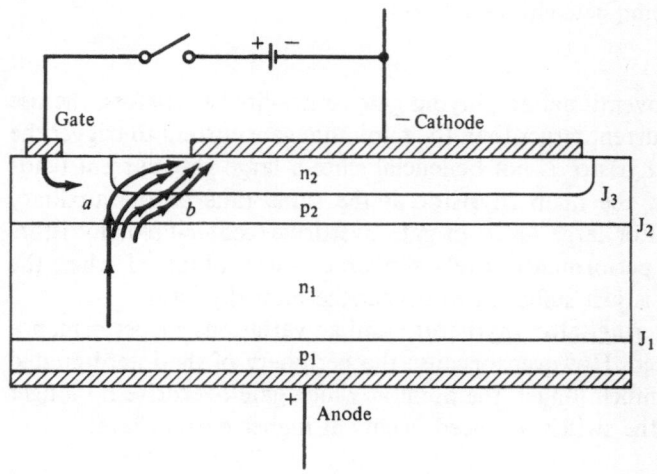

described sequence of events will take place if the load current has a chance to grow sufficiently high. In the initial period, the current is not only limited by the load resistance but also by the lateral resistance of the uncontacted n-emitter region. If too high, this resistance may prevent firing.

The lateral current present in the n emitter increases the injected current density as the distance from point a to point b increases. The lateral base current, on the other hand, has an opposite effect and tends to crowd the current at the emitter edge. Proper selection of the relative resistances of the noncontacted n-emitter region and the p region is, therefore, of great importance.

One rather small disadvantage of the FI method is that, after the end of the transient, the noncontacted emitter region contributes only insignificantly to the current injection; therefore, the effective emitter area is reduced. As reported by Gerlach [10.15], thyristors with n-emitter diameter of 14 mm and a loss of 10 percent of the effective area could withstand energy dissipations five to seven times higher than the devices without the FI action. The FI-gated SCR can be considered from the equivalent-current point of view as consisting of two SCRs integrated on one chip. The main device is triggered by the application of a positive gate pulse to the gate of the driving device, as shown by the equivalent circuit of Figure 10.11 in which R_n is the sheet resistance of the trigger finger (the noncontacted emitter region) while R_p is the sheet resistance of the p base under the finger.

The field-initiated gate structure can turn on a device even with a comparatively small driving current because a small amount of gate drive is necessary to initiate conduction in the pilot structure. To achieve high plasma-spreading velocity, however, a stiff, i.e., fast rising, large gate current will still be required.

Figure 10.11   Equivalent circuit of the field-initiated gate SCR.

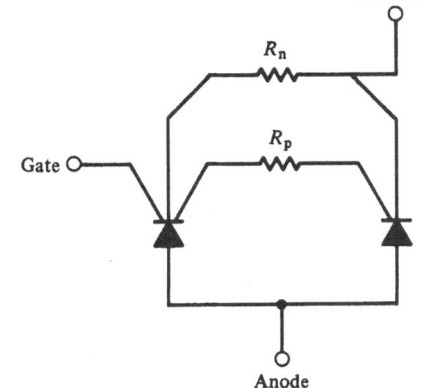

123

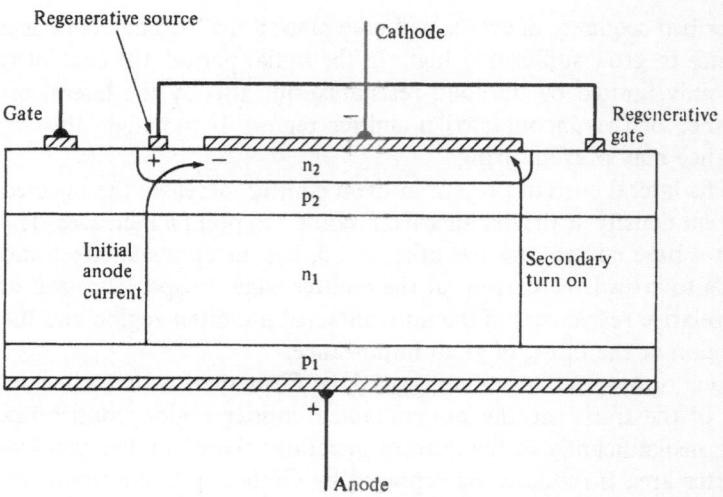

Figure 10.12   Regenerative gate thyristor. (After D. I. Gray [10.17].)

10.8.4
Regenerative gate thyristors

The principle of the field-induced turn on is carried still further by regenerative gate thyristors [10.17]. The resistance of the emitter finger is made large enough so that a significant voltage drop develops in the uncontacted n-emitter region. A potential positive with respect to the equipotential part of the emitter is picked off as in Figure 10.12 and fed to one or more additional regenerative gate contacts on the p_2-base. The initiating gate signal starts the localized anode flow at one end of the n-emitter while the regenerative gate almost immediately starts a secondary turn on at the other end. If more than two gates are arranged around the emitter, several intermediate spots will turn on simultaneously. Therefore, in principle this approach assures more uniform and rapid current turn on.

10.9
Emitter gate

The principle of the emitter gate [10.5] is illustrated by Figure 10.12. The gate consists of a metallic contact placed in the vicinity of the cathode electrode. Although not shown in Figure 10.13, the highly conducting layer of the n emitter is usually removed (e.g., etched chemically away) between the gate and the cathode so that the path between the gate and the cathode has some significant resistance.

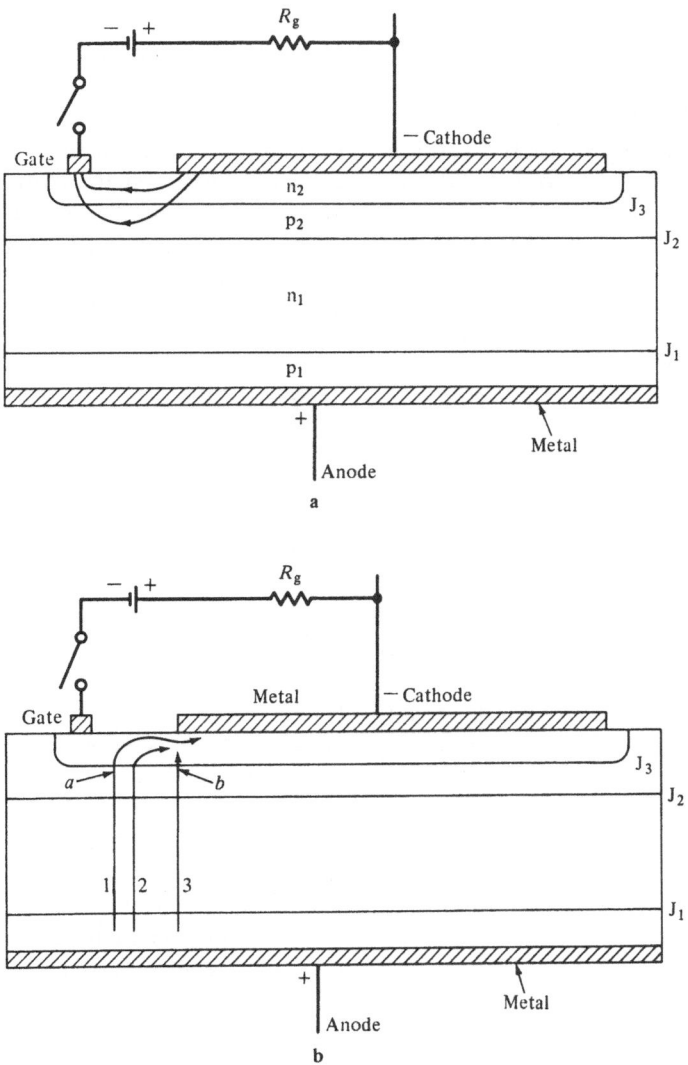

Figure 10.13 Emitter gate. (a) Gate-current path; (b) load-current path. (After Somos and Piccone [10.5].)

This type of a gate can operate with a negative or with a positive potential. When a negative potential in respect to the cathode is applied to the emitter gate, a turn-on action similar to the FI action takes place. The gate current that flows laterally from the cathode to the gate follows the paths in the emitter and in the p-base (Figure 10.13a). A forward bias across the n^+-p junction develops under the gate contact so that the thyristor turns on at point a and the principal (load) current starts flowing following paths 1 and 2 of Figure 10.13b.

125

The current of path 2 develops a lateral voltage drop and initiates a second triggering at point *b*, and as a result the principal current starts following path 3.

A different situation develops when the emitter gate is positively biased in respect to the cathode. The gate current flows in the opposite direction to that of the negatively biased gate and, therefore, region *b* turns on first. In this mode of operation the initially turned-on area depends on the magnitude of the gate current. To attain high *di/dt* capability, high gate current, about ten times higher than the device switching current, should be used.

10.10
Emitter shorts versus the turn-on time

The emitter shorts (shunts) are primarily used to improve thyristor *dv/dt* capability and to allow the device to operate at higher temperatures, when saturation currents are large. The same shunts affect, however, other device parameters such as the forward-voltage drop and the turn-on time.

It has been reported by Chu [10.18] that the *dv/dt* capability depends not only on the area of the shorting holes but also on their geometrical pattern. A close hexagonally packed pattern appears to give very good *dv/dt* characteristics.

When emitter shorts are employed, the n emitter bias is due to the lateral-voltage drop in the p base. The bias is zero at the short site itself and increases in the base in the direction from the short toward the device gate. At the point where the bias exceeds the threshold potential, appreciable injection takes place, and this region of the device turns on. If the short is only a very short distance away from the gate, a larger current will be required to obtain a bias potential higher than the threshold voltage. Close-spaced shorting holes will be very effective in the increase of the *dv/dt* capability. They will have a negative effect, however, on the total emitter injecting area. Also, since the region surrounding a short is biased below the threshold level, the spreading of the plasma will be hampered in that region. This leads to a requirement that the shunts be placed so that they contribute the most to the *dv/dt*-improved performance but have the least effect on plasma spreading. One way to achieve a compromise is to skip the one row of shorts closest to the centrally located gate.

10.11
Beam-fired thyristor

A properly designed thyristor capable of supporting large and fast-rising currents should be capable of a rapid turn on of the initial area—a fast spreading of the plasma to the rest of the device without

126

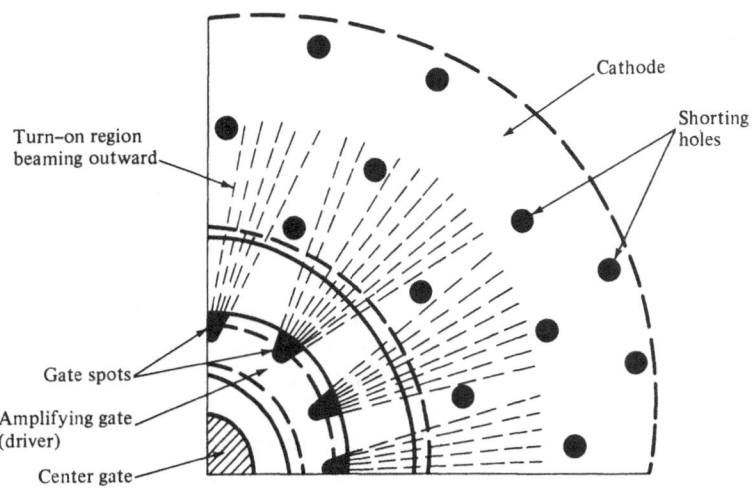

Figure 10.14 Beam-fired thyristor. (After New and Cooper [10.18].)

affecting deleteriously other device characteristics. The goal of a beam-fired thyristor [10.19] is to achieve this result for very large current devices (100–800 A).

The device is provided with a centrally located amplifying gate (driver) (Figure 10.14). The cathode has several shorting holes. The first row of shunts is placed as far out as consistent with the maximum permissible distance for control of dv/dt and leakage. When the driving device is turned on, the anode current flows to the gate spots, which are arranged in a circle. Each gate spot faces a channel where a beam of turn-on current can travel outward between the shunts. Consequently, the anode current through the driver thyristor must enter into the multiple-gate areas with a current density much higher than in a current uniformly distributed along the entire driver perimeter. This concentrated secondary gate current helps to turn on rapidly a large n-emitter area and enhances the plasma-spreading velocity. The open region ahead of the beam can accommodate a large initial turn-on area and permits a shorter spreading path. The overall effect accomplishes, then, several desirable features necessary for high-current switching.

The device described by New and Cooper [10.19] was capable of a critical rate of rise off-state voltage (dv/dt) greater than 1000 V/μsec, and $di/dt > 1000$ A/μsec for nonrepetitive rating and 200 A/μsec for repetitive rating. The device forward-blocking voltage was 700 V, the steady-state on-current was 110 A.

10.12
Field-controlled thyristor

Conventional thyristors and transistors are at the present state of the art unable to meet the performance and/or cost requirements for operation at frequencies on the order of 20 to 30 kHz with currents of tens of amperes and breakdown voltages on the order of 1000 V. A device which shows promise of meeting these requirements is the recently described [10.20] field-controlled rectifier, which in the on condition behaves essentially like a p-i-n rectifier. When it is forward biased (Figure 10.15), the high-resistivity n^- region is heavily conductivity modulated and, as a result, the forward voltage drop is reasonably low. A third electrode, the p^+ grid, is added to the diode, forming a p^+-n^- junction with the n^-(*i*) region of the device. An application of a reverse bias between the grid and the cathode makes it possible to turn the device off and to leave it in a forward-blocking state since the anode–cathode current is diverted to the grid. Simultaneously, the grid bias depletes part of the n^- region from the mobile carriers and, for sufficiently high voltage, when both depleted sides meet under the cathode (Figure 10.15) a potential barrier from the cathode is established. The device remains in the forward-blocking state until the grid bias is removed so that the depletion region can collapse and the current can again start flowing. Because the field-controlled thyristor is a nonregenerative device, it does not latch on. The current gain at turn off is only equal to one and the grid voltage required to turn the device off may be rather high—a few hundred volts. The device is immune to static *dv/dt* and can operate at higher temperatures than the conventional thyristor or GTO since it is not activated by triggering. Since the device is turned

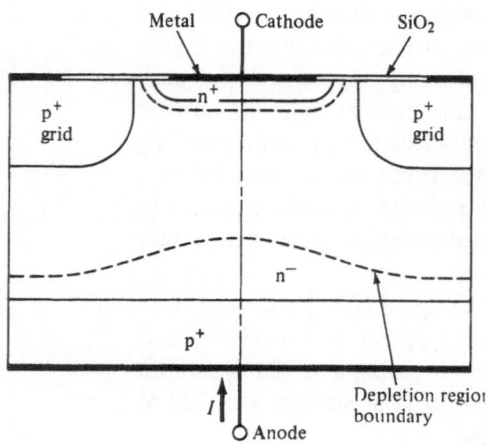

Figure 10.15 Cross section of the field-controlled thyristor (not to scale). (After Houston *et al.* [10.20].)

on by simply removing the grid bias and there is no plasma spreading, a higher di/dt rating is possible than with a conventional thyristor. The turn off of the device is not delayed by the presence of regeneration as in a conventional p-n-p-n structure. As a result, the switching times may be made short enough for operation at high frequencies.

The field-controlled thyristor developed by Houston et al. [10.20] had a voltage drop of 1.45 V at 10 A (1300A/cm^2 current density). An anode voltage of 650 V could be blocked with 10 μA of leakage current when a grid of -150 V was applied. At turn off, the current fell from 14 A to 2 A in 0.3 μsec.

References

10.1 N. Mapham. Overcoming turn-on effects in silicon controlled rectifiers. *Electronics*, *35*: 50–51, 1962.

10.2 N. Mapham. The rating of silicon controlled rectifiers when switching into high currents. *AIEE Trans. (Commun. Electron.)*, *83*: 515–519, 1964.

10.3 Recommended standards for thyristors. *EIA-NEMA Standard* RS 397. Washington, D.C.: Electronic Industries Association, June, 1972.

10.4 S. Ikeda and T. Araki. The di/dt capability of thyristors. *Proc. IEEE*, *55* (8): 1301–1305, 1967.

10.5 I. Somos and D. E. Piccone. Behaviour of thyristors under transient conditions. *Proc. IEEE*, *55* (8): 1306–1311, 1967.

10.6 S. Ikeda, S. Tsada, and Y. Waki. The current pulse ratings of thyristors. *IEEE Trans. Electron Devices*, *ED-17* (9): 690–693, 1970.

10.7 N. L. Potter. An electrical analogue for heat flow problems in semiconductors. *Electron. Eng.*, *31*: 454–457, 1959.

10.8 I. Lundstrom. Temperature rise in thyristors during turn-on. *Int. J. Electron.*, *23* (1): 59–82, 1967.

10.9 F. E. Gentry, F. W. Gutzwiller, N. Holonyak, Jr., and E. E. Von Zastrow. *Semiconductor Controlled Rectifiers*. Englewood Cliffs, N.J.: Prentice-Hall, 1964.

10.10 F. E. Gentry and J. Moyson. The amplifying gate thyristor. Int. E.D.M., October, 1968, Washington, D.C.

10.11 J. B. Brewster, Y. C. Kao, and J. Urish. Fast switching thyristors. *IEEE 8th Conf. Record*, Milwaukee, Wis., IAS, 1973, pp. 245–249.

10.12 P. Voss. The turn-on of thyristors with internal gate current amplifying. *IEEE 9th Conference Record, IAS*, 1974, Pittsburgh, Pa., pp. 467–476.

10.13 A. Silard and V. Marinescu. Computed-aided experimental investigation of the correct turn-on in thyristors with amplifying gate. *Electron. Letts.*, *11* (17): 419–420, 1975.

10.14 G.E. SCR Manual, Syracuse, N.Y.: General Electric, 5th ed., 1972.

10.15 W. Gerlach. Thiristor mit Querfeld-Emitter. *Z. Angew. Phys.*, Heft 5: 396–400, 1965.

10.16 W. H. Dodson and R. L. Longini. Skip turn-on of thyristors. *IEEE Trans. Electron. Devices, ED-13*: 598–604, 1966.

10.17 D. I. Gray. This SCR is not for burning. *Electronics, 30*: 96–100, 1968.

10.18 C. K. Chu. Geometry of thyristor cathode shunts. *IEEE Trans. Electron Devices, ED-17* (9): 687–690, 1970.

10.19 T. C. New and D. E. Cooper. Turn-on characteristics of beam fired thyristors. October, 1973, pp. 259–265.

10.20 D. E. Houston, S. Krishna, D. Piccone, R. J. Finke, and Y. S. Sun. Field-controlled thyristor (FCT)—A new electronic component. International Electronic Devices Meeting, Washington, D.C., 1975.

Bidirectional
p-n-p-n switches

Summary

The p-n-p-n bidirectional switches conduct current in the positive or negative direction and are, therefore, useful in AC applications. The devices that belong to this class of switches are the p-n-p-n diode switch and triac (acronym for triode, AC switch). A triac can be triggered by either the positive or negative pulses applied to the gate which acts, depending on the switching mode, as a conventional gate, a junction gate, or as a remote gate. There are four possible triggering modes of a triac. Two in the first quadrant of the *I–V* characteristics and two in the third quadrant of the *I–V* characteristics. Since each of these switching modes depends on a different physical basis, the triggering current is different for each mode. Table 11.1 summarizes the state of various emitter and base layers and junctions of a triac during the turn-on transient.

11.1
Introduction

With the p-n-p-n bidirectional switches, the load current can flow in either direction. Two of these types of devices are the bidirectional p-n-p-n diode switch and bidirectional p-n-p-n SCR switch, the triac (triode, AC switch; see Figure 11.1). The triac can switch the current in either direction by the application of low-voltage, low-current pulse, of either polarity, between a gate terminal and one of the two load current terminals MT_1 or MT_2. These devices are very useful in a broad range of applications such as light dimming, motor speed control, temperature control, etc. For triggering, triacs depend largely on the lateral voltage drops developed by carrier injection during the switching period. These voltage drops determine the current distribution across the forward-

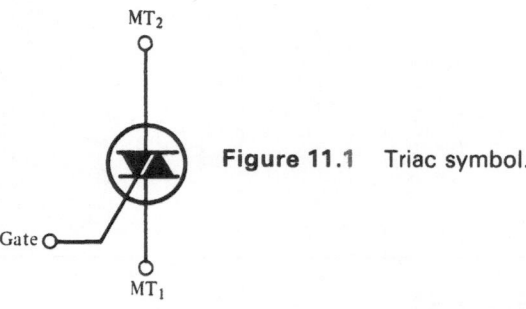

Figure 11.1   Triac symbol.

biased junctions during the turn-on transient. There is one gate only in the triac for switching both current polarities. This gate is considerably more complicated in its structure than the conventional thyristor gate. The triac gate is behaving, depending on the applied gate potential polarity, either as a *conventional gate*, a *junction gate*, or as a *remote gate* (see Sections 11.2, 11.3, and 11.4 below). The gate-trigger polarity is referenced always to the main terminal MT_1 (Figures 11.2 and 11.4) [11.1]. The potential difference between the two terminals is such that gate current flows in the direction indicated by the dotted arrow (Figure 11.2). The polarity symbol at the main terminal MT_2 is also referenced to the main terminal MT_1.

For the various operating modes, the polarity of the voltage on the main terminal MT_1 is given by the quadrant in which the triac operates (quadrant I or III). The polarity of the gate signal used to trigger the device is given by the proper symbol next to the operating quadrant. For the I(+) operating mode, for instance, the main terminal MT_2 and the gate are both positive with respect to the main terminal MT_1.

The initial gate-current flow is shown by the dotted arrows; the principal (load) current flows through the p-n-p-n structure, as shown by the solid arrows. Because the direction of the principal current influences the gate-trigger current and because the operation depends in each switching mode on different physical basis, the magnitude of the current required to trigger the triac differs for each of the four modes shown in Figure 11.2.

The I(+) mode (gate and MT_2 positive with respect to MT_1), which is equivalent to SCR operation from the gating point of view, is the most sensitive. Therefore, the smallest gate current is required to trigger the triac in this mode. The I(−) and III(−) are usually somewhat less sensitive. They may require about twice as much current for triggering as the I(+) mode. The least sensitive is the III(+) mode. In this mode the triggering current may be four times as high as that required to trigger the device in mode I(+) (see Table 11.1).

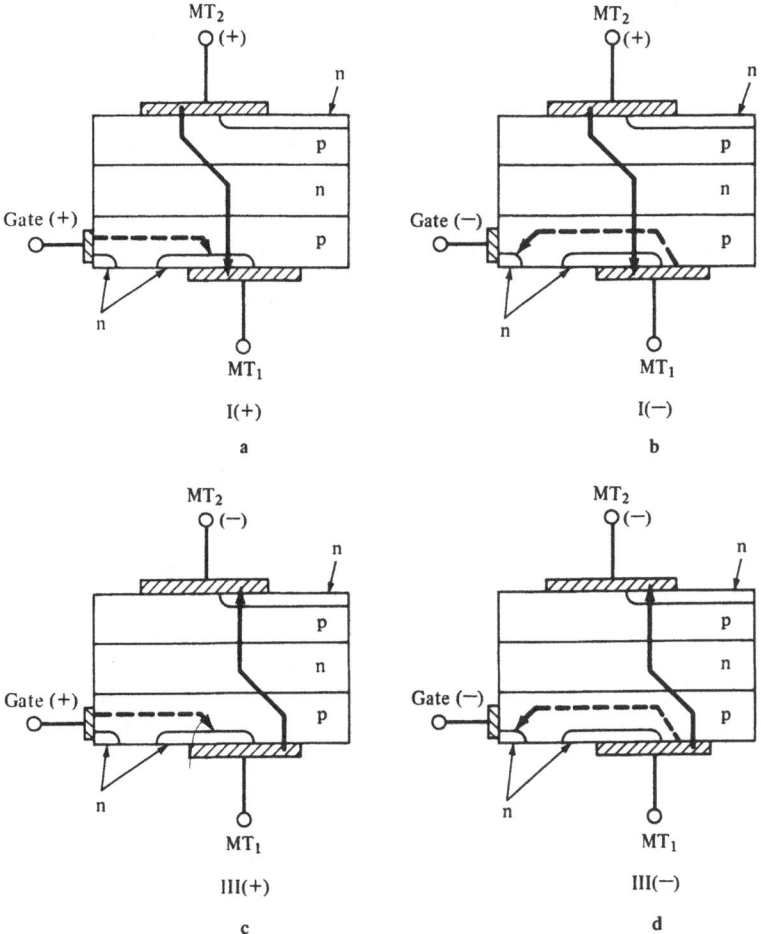

Figure 11.2 Current flow in the four triggering modes of a triac: (a) mode I(+); (b) mode I(−); (c) mode III(+); (d) mode III(−). (After T. C. McNulty [11.1].)

Like other parameters of semiconductor devices, the magnitude of the gate-trigger current varies with junction temperature. At temperatures above room temperature, the saturation current (leakage) increases and the gate becomes more sensitive in all operating modes. The opposite is true below room temperature.

11.2
Bidirectional p-n-p-n diode switch

A four-layer Shockley diode (an SCR without gate) can conduct current in one direction only. On the other hand, the bidirectional diode can conduct in both directions [11.2]. The device consists

133

basically of two p-n-p-n sections in an antiparallel connection (Figure 11.3). It can be thought of as two four-layer Shockley diodes integrated on one silicon chip.

When an AC potential is applied to the MT_1 and MT_2 terminals, current will flow for both voltage alternations (first and third quadrant of Figure 11.4). A diode of this type can be triggered, e.g. by the avalanche breakdown, by the dv/dt effect or by increasing the device temperature and, consequently, its saturation current to a level sufficient for $\alpha_1 + \alpha_2$ to become equal to unity.

When terminal MT_2 is negatively biased with respect to terminal MT_1, junctions J_2 and J_4 are forward biased while the J_1 and J_3 junctions are reverse biased. At some sufficiently high external potential, the J_1 junction will go into avalanche breakdown so that it will present only a small resistance to the current flow. This will initiate the triggering of the device in the following way: the J_3 junction is reverse biased; therefore, the n_3 emitter cannot emit into the p_2 base and the current starts flowing from the positively biased left-hand terminal MT_1 across the right-hand portion of the device. The current must overcome the lateral resistance of the p_2 base and develops a voltage drop from left to right. The J_2 junction becomes, therefore, more forward biased as the current increases and the p_2 base starts injecting more holes into the n_2 region. The holes diffuse across n_2 and are collected by the reverse-biased J_1. The collected holes reach the p_1 region and raise its potential in respect to the terminal MT_2 and, therefore, with respect to the n_1 emitter. Consequently, the n_1 region starts emitting electrons.

At low levels, the hole current follows the path of the smallest resistance along the J_4 junction, and a lateral voltage drop develops

Figure 11.3 Bidirectional p-n-p-n diode switch.

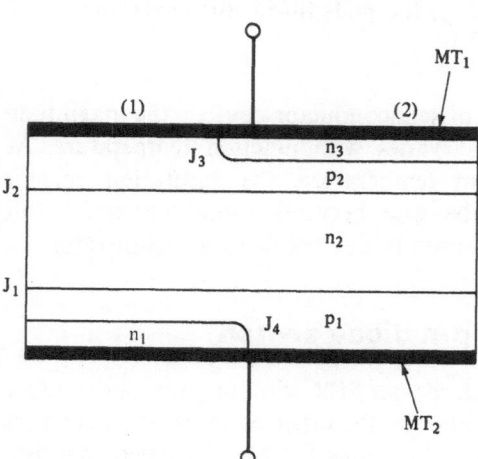

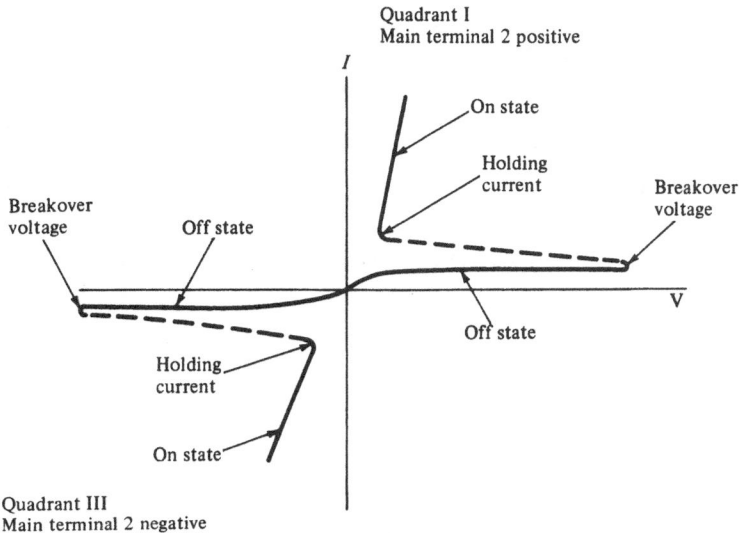

Figure 11.4   Principal voltage–current characteristic of a triac.

in the p_1 region, biasing one part of the J_4 junction in the forward direction so that eventually a large current can flow now through J_4. When a sufficiently high current level is reached, the p_2-n_2-p_1-n_1 section (1) will turn on.

If the polarity of the two electrodes MT_1 and MT_2 is reversed now, the right-hand section of the device, i.e., the p_1-n_2-p_2-n_3 section (2) will turn on by the same mechanism as the right-hand section of the device.

Referring now to Figure 11.3 we notice that the n emitters of the two sections overlap. This helps to trigger the device at a lower current level. The overlap makes the lateral path for current flow longer and increases the lateral voltage drop for the same current level. The higher the voltage drop, the larger the forward bias will be, and the n emitter will start injecting sooner.

11.3
Current distribution across a forward-biased junction bounded by resistive layers (current-crowding effect)

From the discussion of the bidirectional p-n-p-n diode, it is evident that it is important to know and to be able to control the current distribution as a function of the lateral position before the device is fully turned on. This concerns both the bilateral diode and the triac. The analysis of the current distribution [11.2, 11.5] can be

135

simplified by ignoring the surface effects, the conductivity modulation, and the variation of current gain with current density. Although the absolute values may be affected by these assumptions, the general current and potential behavior, which is important for understanding triac operation, remains essentially unmodified. Figure 11.5 [11.2] helps in the understanding of the theoretical approach. The

Figure 11.5 Model for current distribution. (After Gentry *et al.* [11.2].)

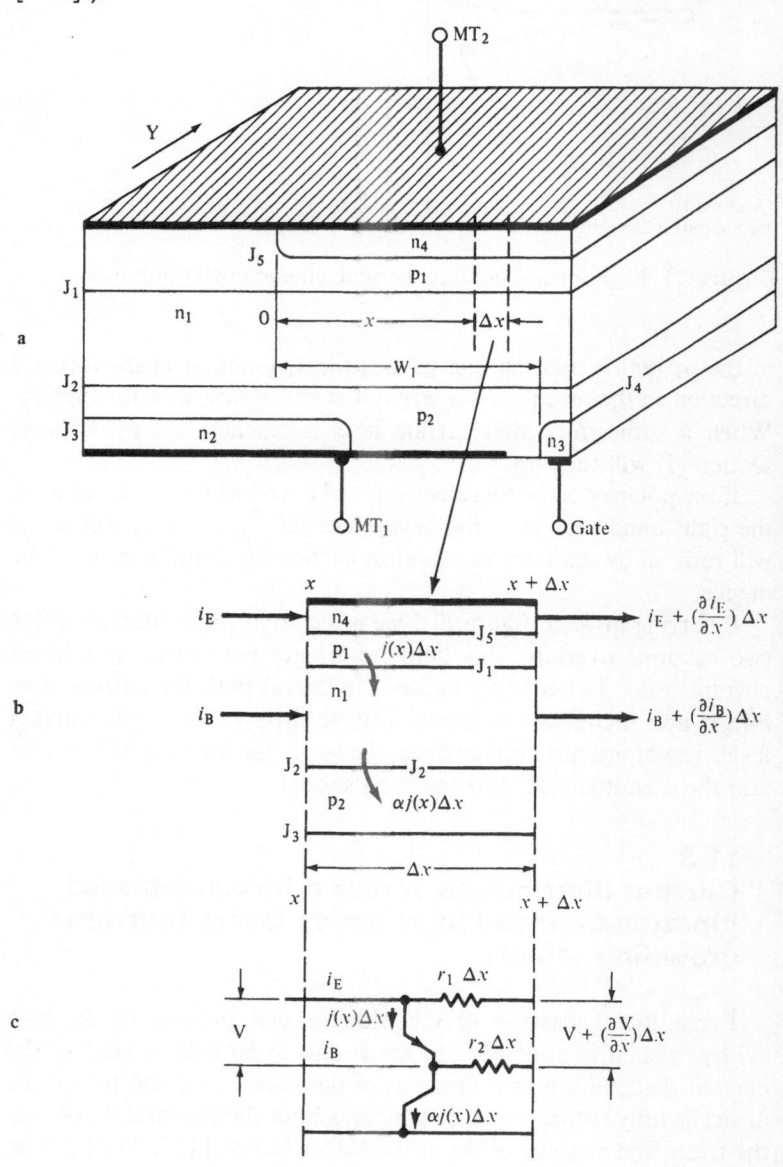

device represented by the figure is a bidirectional diode with an added junction gate (n_3-p_2 junction J_4). The function of the gate is to trigger the device when a potential is applied to it. The combination of the bidirectional diode and the junction gate constitutes a triac. The function of the gate is described fully later in this chapter.

The device thickness in the y-direction is assumed to be equal to unity for the sake of simplicity. An increment of length along the x axis, Δx, is blown up to the size of Figure 11.5b, where i_E is the current entering laterally the p_1 region and i_B is the current entering laterally the n_1 region. If the current density through the J_1 junction is $j(x)$ at the point of the abscissa x, counted to the right of the point of origin O (Figure 11.5a), then the emitter current entering the area $(\Delta x) \cdot$ (unity) is $j(x)\Delta x$. The current at $x + \Delta x$ (the right-hand side of Figure 11.5b) is $i_E(x)$ plus the negative current increment

$$\left(\frac{\partial i_E}{\partial x}\right)(\Delta x) = -j(x)\Delta x.$$

(11.1)

The positive increment of the base current is

$$\left(\frac{\partial i_B}{\partial x}\right)\Delta x = (1 - \alpha)j(x)\Delta x$$

(11.2)

with α the current gain of the p_1-n_1-p_2 transistor. The resistances per unit length of the p_1 and n_1 regions are designated r_1 and r_2, respectively. Consequently, the potential between the p_1 and n_1 regions, $V(x)$, at the point x along the x axis, can be represented by an equivalent circuit of Figure 11.5c. At x, the potential is $V(x)$ and at $x + \Delta x$ it is equal to $V(x)$ plus an increment $(\partial V/\partial x)\Delta x$. It can be shown on the basis of this equivalent circuit that

$$\frac{\partial^2 V}{\partial x^2} = -\left(\frac{\partial i_E}{\partial x}\right)r_1 + \left(\frac{\partial i_B}{\partial x}\right)r_2$$

(11.3)

and also that

$$\frac{\partial^2 V}{\partial x^2} = J_s r_0 \exp\left(\frac{qV}{kT} - 1\right)$$

(11.4)

J_s is the saturation current density and $r_0 = r_1 + (1 - \alpha)r_2$.

Equation (11.4) permits the determination of potential V as a function of position x and, vice-versa, of position x as a function of the potential V. The solution of the position–voltage relationship leads to the conclusion that *there exists a point of an abscissa x_0 at*

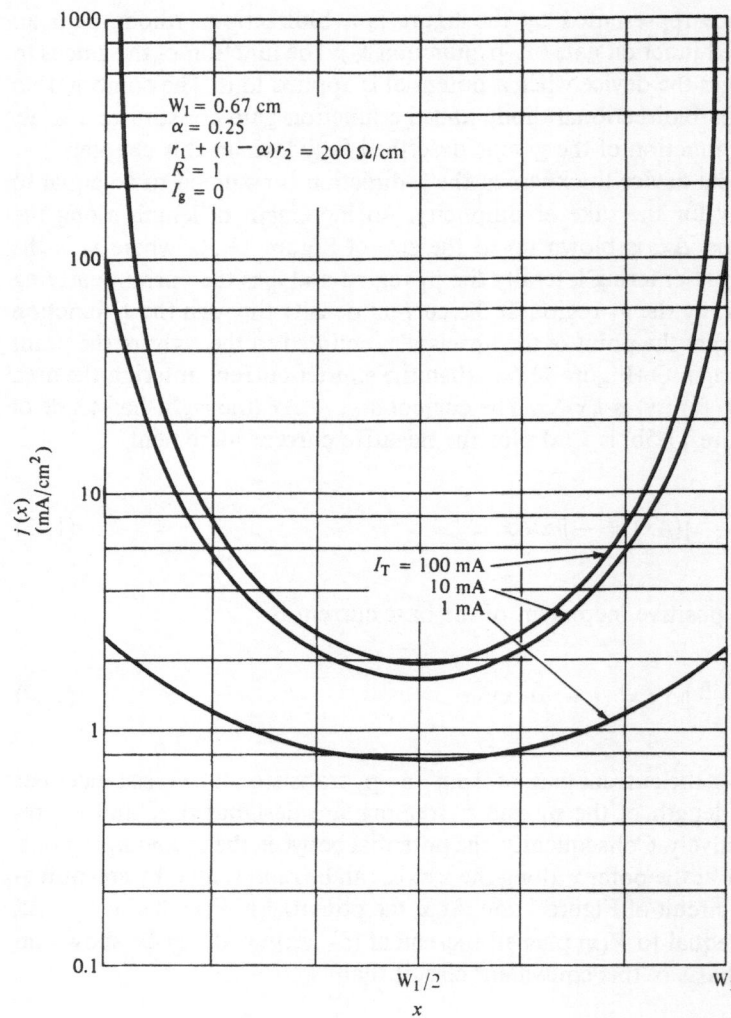

Figure 11.6 Injected current distribution for $R = 1$. (After Gentry *et al.* [11.2].)

which the potential V, *i.e., the potential across the junction* p_1-n_1 (J_1) *has a minimum value*

The total current I_T collected by the J_2 junction over the length W_1 (Figure 11.5a) is given by the expression

$$I_T = I_c = \alpha J_S \int_0^{W_1} \left[\exp\left(\frac{qV}{kT}\right) - 1 \right] dx \tag{11.5}$$

Relationships (11.4) and (11.5) permit the determination of the current levels to the left (I_{cL}) and to the right (I_{cR}) of the point x_0 *of*

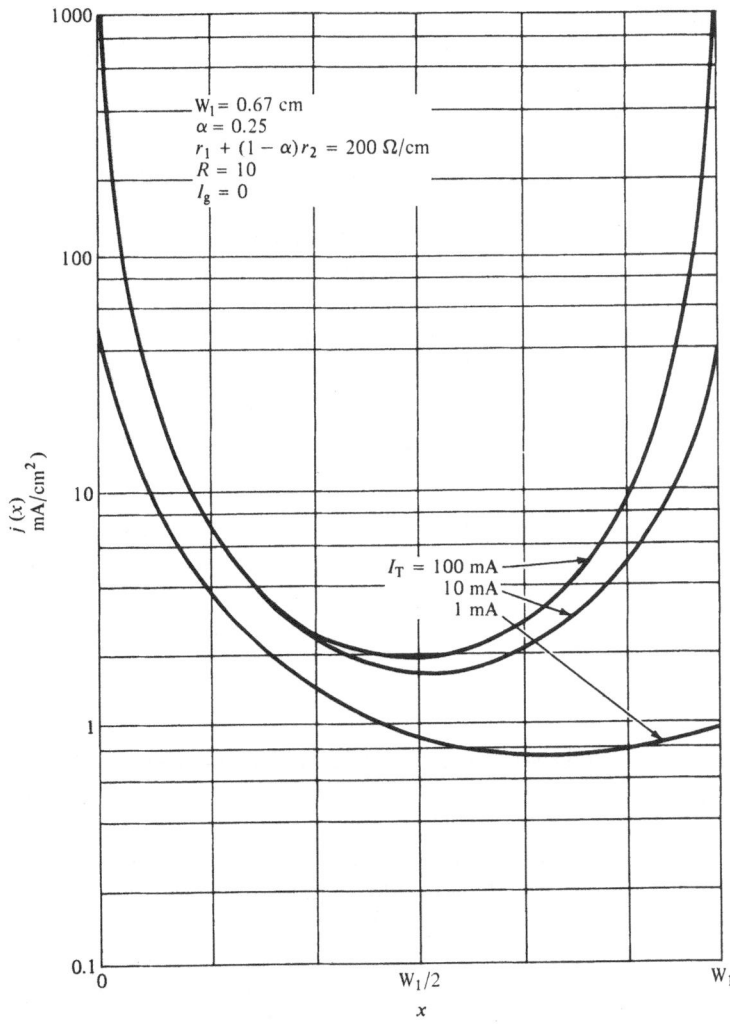

Figure 11.7 Injected current distribution for $R = 10$. (After Gentry *et al.* [11.2].)

the minimum potential. The ratio [11.2] of these two currents is

$$\frac{I_{cL}}{I_{cR}} = \frac{r_1}{(1 - \alpha)r_2} = R \qquad (11.6)$$

If $R = 1$ then $I_{cL} = I_{cR}$, and both are equal to $\frac{1}{2}\alpha I_T$ since $I_{cL} + I_{cR} = I_T$. If R is made equal to unity, i.e., $r_1 = (1 - \alpha)r_2$, then it can be shown that the minimum value of the potential V occurs in the center of W_1. Therefore, for $R = 1$, $x_0 = \frac{1}{2}W_1$. The potential V is minimum at x_0 so that the injected current is at its minimum at x_0.

Figure 11.6 illustrates the current density j distribution in function of position x for the case of $R = 1$ [11.2]. The device considered had $W_1 = -0.67$ cm, $\alpha = 0.25$, $r_0 = 200$ Ω/cm^1, gate current $I_g = 0$. The figure shows that a current crowding effect exists at $x = 0$ and $x = W_1$.

The effect of changing the value of R from unity to 10 is shown in Figure 11.7. The current density at $x = 0$ for a given total current I_T is much larger for the case of $R = 10$ than for $R = 1$ since R is the ratio of the current to the left of the x_0 to the current to the right of x_0. Since R depends on r_1 and r_2, which are resistances per unit length of the regions p_1 and n_1, respectively, it is possible to control the location of minority-carrier injection by proper choice of sheet resistances of p_1 and n_1 regions. Injection takes place in regions of a relatively small area. This aids the switching process by providing high local current densities even when the total current is comparatively small.

After the device is fully turned on, the current becomes uniformly distributed. The lateral current flow is of importance only during the transient before the switching occurs.

11.4
Junction gate

A conventional thyristor gate consists of a metallic electrode contacting usually the p-base of the device. On the other hand, the junction gate [11.2, 11.3] consists of a metallized silicon n layer forming an n-p junction with the p base. The advantage of the junction gate as compared to a conventional gate is that it makes it possible to turn on the triac by using either positive or negative gate current, if the metallization of the junction is extended beyond the n region to form a short with the p region in the same manner as in a conventionally shorted (shunted) emitter device.

11.4.1
Negatively biased junction gate

Figure 11.8 helps to illustrate the principle of the junction gate operation for the case when the junction is made negative with respect to terminal 1.

We assume that the device is in the forward-blocking state, i.e., terminal 2 is positively biased in respect to terminal 1. The junctions J_1 and J_3 are, therefore, forward-biased, while the junction J_2 is reverse-biased and supports the externally applied potential.

The sequence of events occurring before the SCR is switched from the blocking state into the conducting state is as follows: (1) A nega-

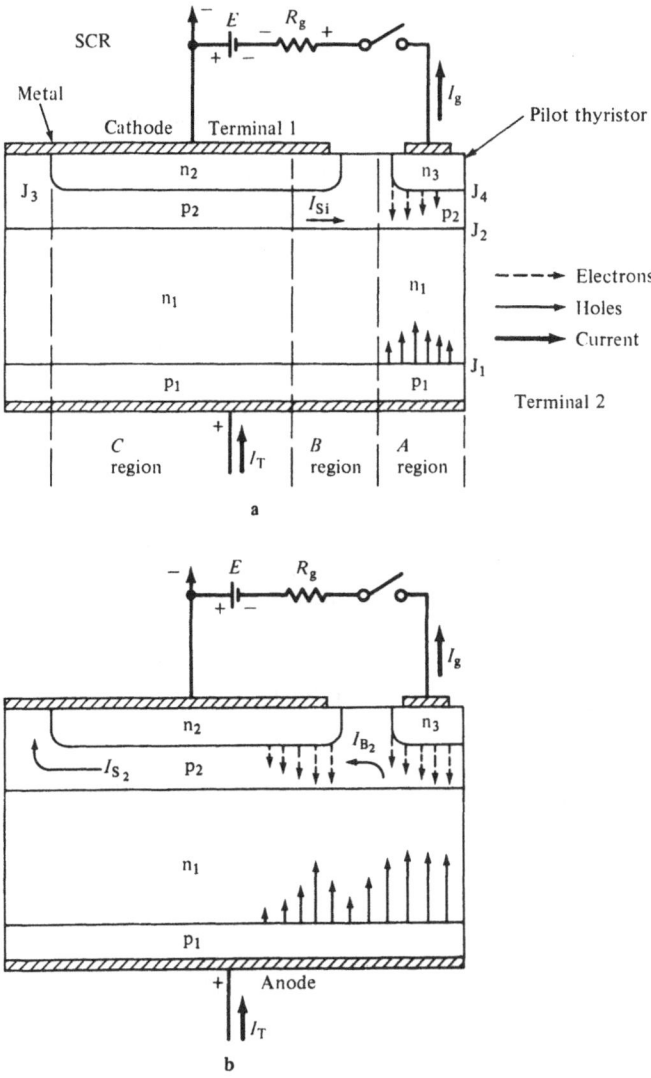

Figure 11.8 Junction gate thyristor. (After Gentry *et al.* [11.2].)

tive potential is applied to the gate (Figure 11.8a) and the J_4 junction of the gate becomes forward-biased. This is because the n_2 emitter of the SCR is shorted to the p_2 base, and the p_2 base is positively biased in respect to the n_3 emitter of the gate. The J_3 junction of the SCR becomes reverse biased and remains inactive temporarily. (2) A current I_{s_1} starts flowing laterally in the p_2 base from the emitter short toward the gate. I_{s_1} is the base (driving) current of the n_3-p_2-n_1 transistor, which is a part of the n_3-p_2-n_1-p_1 pilot (driving) thyristor

141

shown at the right-hand side of Figure 11.8. I_{S_1} increases as the gate voltage is raised, the current I_g through the pilot thyristor reaches a sufficiently high value to make the sum of alphas of the pilot thyristor equal to unity, and the pilot device turns on. The current I_{S_1} acts as a gate current for the pilot thyristor. (3) The principal current of the junction gate I_g will continue to grow during the current rise time of the n_3-p_2-n_1-p_1 pilot thyristor. The maximum value of I_g is limited, however, by the resistance R_g for a given biasing potential E. (4) Since the pilot device is turned on, the p_1 emitter of region A injects holes which are collected by the junction J_2 and flow into the p_2 base (Figure 11.8b). Part of this hole current flows into the cathode of the pilot thyristor. Since the external current is limited by the resistance R_g, not all hole current will be able to follow this path. Therefore, the excess holes will flow laterally along the p_2 base and supply current I_{B_2} to the p-gate of the main SCR, i.e., the n_2-p_2-n_1-p_1 device (region B). (5) The electrons that are injected by the junction J_4 of the pilot device are collected by the junction J_2 and reach the n_1 base. Some of them recombine with injected holes in the n_1 base, but some with flow laterally into region B as a base current drive I_{B_1} for the p_1-n_1-p_2 portion of the SCR (left-hand side). The magnitude of I_{B_1} depends on the sheet resistance of the n_1 region. In order to simplify the discussion, it is assumed that the sheet resistance of the

Figure 11.9 Junction gate thyristor equivalent circuit. (After Gentry *et al.* [11.2].)

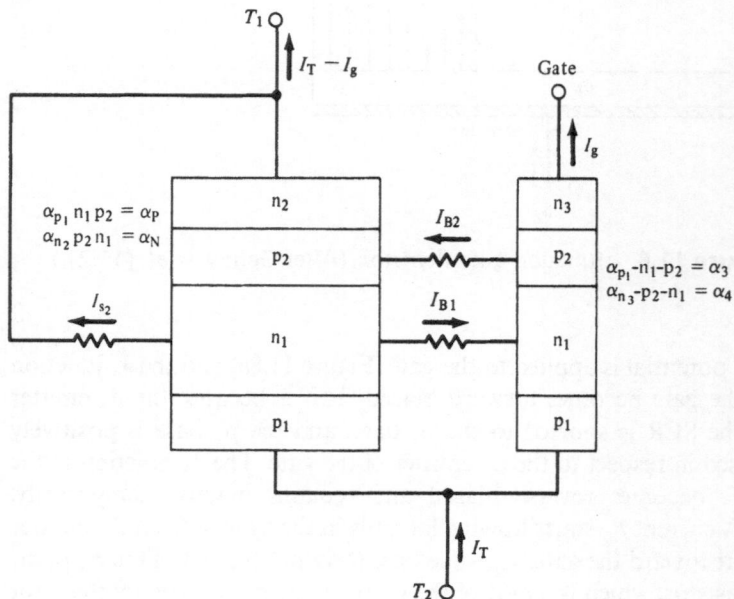

n_1 base is much larger than the sheet resistance of the p_2 base and, therefore, I_{B_1} can be neglected (see the equivalent circuit of Figure 11.9). (6) When the sum of the pilot alphas becomes greater than unity, the I_{S_1} *current reverses its sign.* The cathode current I_g of the pilot thyristor is shown in the Appendix [11.2] to be

$$I_g = \frac{I_{S_1}(1 - \alpha_3)}{1 - (\alpha_3 + \alpha_4)} \tag{11.7}$$

where α_3 is the common base current gain of the pilot thyristor p_1-n_1-p_2 transistor section and α_4 is the gain of the n_3-p_2-n_1 transistor section. Since I_g remains constant in magnitude and sign when $\alpha_3 + \alpha_4$ becomes greater than unity, the sign of I_{S_1} has to reverse itself (Figures 11.8b and 11.9). (7) The reversed I_{S_1} now becomes I_{S_2} and is the p_2-base driving current (base current) for region B of the main thyristor. Junction J_3 becomes forward biased and the main thyristor is activated. The total base drive which is provided for the p_1-n_1-p_2-n_2 device portion in region B depends on the magnitude of I_g. Both bases (p_2 and n_1) of the main SCR in region B are driven by pilot section A and will turn on when the sum of α_1 and α_2 becomes equal to or greater than unity. (8) Region C of the device is turned on by spreading of the on region, as in a conventional SCR.

11.4.2
Positively biased junction gate

If the metallization contacting the n emitter of the junction gate is extended beyond the n emitter, there will be a short (shunt) formed with the p_2 base. When a positive potential is applied between this metallization and the terminal T_1 (Figure 11.8a does not show the extended metallization), the J_4 junction of the junction gate is reverse biased and the J_4 becomes inactive. However, the metallization of the gate extended over the p_2 region converts the junction gate into a conventional gate.

11.5
Remote gate

With a remote gate a thyristor can be triggered without making a *direct* electric contact to either of its internal bases [11.2, 11.3].

The principle of the remote-gate operation can be explained by using Figures 11.10a, b. An additional n_4 layer has been added to the basic SCR structure at the lower right-hand side so that a new junction J_5 has been formed. Figure 11.10b represents an equivalent circuit of the remote-gate thyristor. The n_1 region is at the same time the *base* region of the p_1-n_1-p_2-n_2 SCR and the *collector* region of

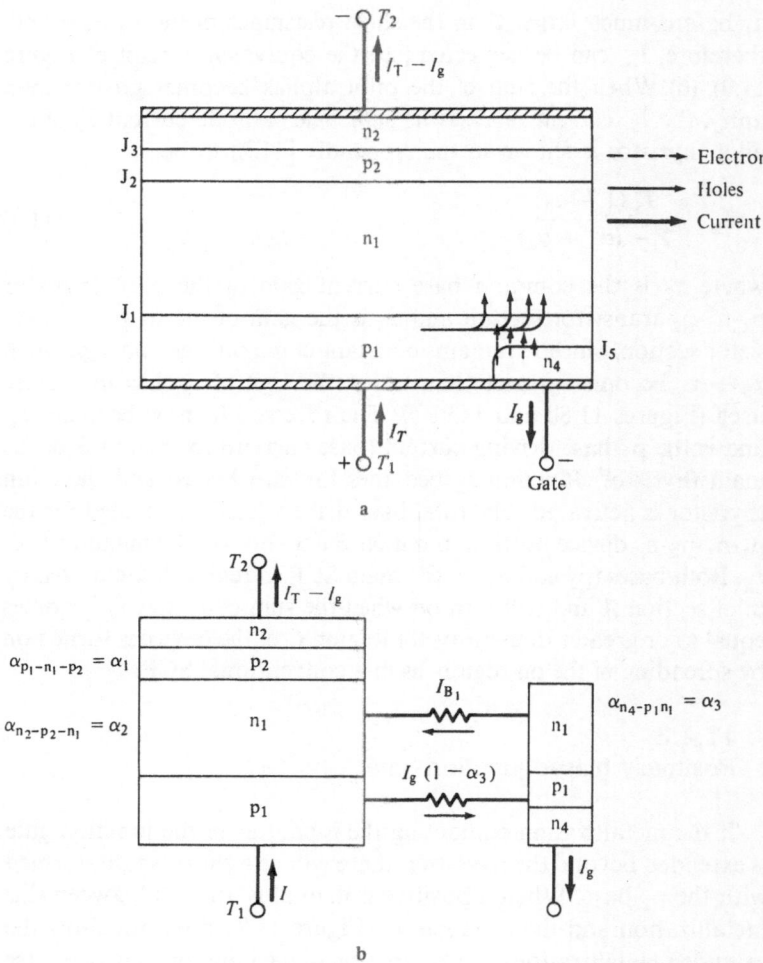

Figure 11.10 Remote gate thyristor. (After Gentry *et al.* [11.2].)

the n_4-p_1-n_1 transistor. When the J_5 junction is forward biased, the n_4 emitter injects electrons into the p_1 base. These electrons are collected by the J_1 junction and moved into n_1 region. This excess majority-carrier charge biases more negatively the n_1 region and causes an increase in the forward current flow through the J_1 junction. When a sufficient base drive is delivered to the n_1 and p_2 bases, the p_1-n_1-p_2-n_2 thyristor triggers.

The J_1 junction collects the electrons emitted by the J_5 junction despite the fact that it is forward biased whereas a conventional collector is reverse-biased. The collecting ability is due to the presence of the built-in potential and the electric field, which is in the same direction as the electric field of the reverse-biased junction. Also, the

144

magnitude of this field in the forward-biased junction remains comparatively high because, although the net potential is small, the depletion-layer width is very narrow.

The J_1 collector junction of the n_4-p_1-n_1 transistor collects a current of magnitude I_{B_1} which provides the n_1 base of the p_1-n_1-p_2-n_2 SCR with a driving (gate current):

$$I_{B_1} = \alpha_3 I_g \qquad (11.8)$$

Since the emitter current of the n_4-p_1-n_1 transistor is I_g, and its current gain is α_3. I_{B_1} is the gate current for the SCR; therefore,

$$I_T \text{ (anode current)} = \frac{\alpha_P I_{B_1}}{1 - (\alpha_P + \alpha_N)} \qquad (11.9)$$

Expressions (11.8) and (11.9) give

$$I_T = \frac{\alpha_P \alpha_3 I_g}{1 - (\alpha_P + \alpha_N)} \qquad (11.10)$$

where $\alpha_P = \alpha_{p_1-n_1-p_2}$, $\alpha_N = \alpha_{n_2-p_2-n_1}$, $\alpha_3 = \alpha_{n_4-p_1-n_1}$. The saturation current is neglected. Since α_1 and α_3 are smaller than unity, then it becomes obvious that the sensitivity of the remote gate is always smaller than the sensitivity of the conventional gate. The remote gate will require, therefore, considerably higher current for switching the thyristor than the conventional gate.

11.6
Triac

11.6.1
General

The triac (symbol shown in Figure 11.1) is a bidirectional triode with blocking and conducting characteristics as per Figure 11.4. The I–V characteristic is essentially symmetrical in respect to the point of origin. The device can operate in the first and third quadrant; it can block or conduct current of either polarity and can be triggered in either direction by the application of a current pulse of either polarity to the gate [11.2, 11.3, 11.6, 11.7]. This greatly simplifies the circuits required for the control of the full-wave AC power by reducing the number of power-handling components and by reducing the size and complexity of the gate-control circuit.

The triac geometry and structure ase shown in Figure 11.11 isometrically and in Figure 11.12 in an orthogonal projection. The two n-type emitters n_2 and n_4 at the top and the bottom of the device are metallized in such a way that the metallization extends onto the corresponding p regions. These metal contacts are the device main

145

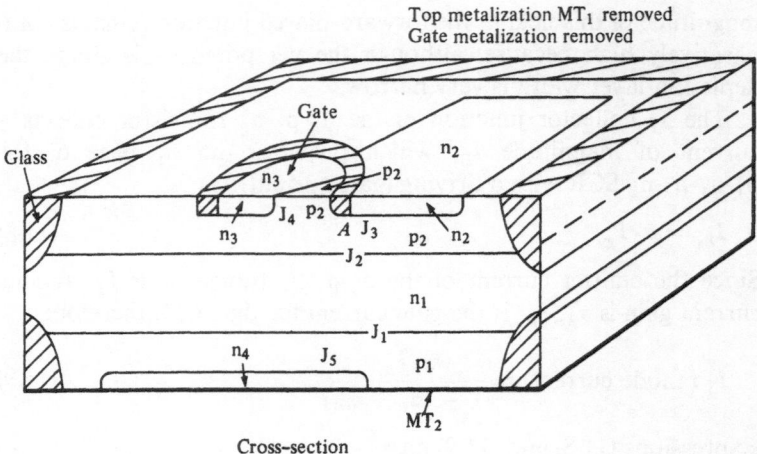

Cross-section

Figure 11.11   Glass-passivated triac, isometric view (not to scale).

terminals MT_1 and MT_2. The MT_1 metallization shorts (shunts) the n_2 layer to the p_2 layer and the MT_2 metallization shorts the n_4 to the p_1 layer.

The gate junction is formed by n_3 and p_2 layers. Its metallization extends beyond the junction outline shorting, therefore, the n_3 region to the p_2 region at the right-hand edge (Figures 11.11 and 11.12).

11.6.2
Four basic gate-triggering modes

The behavior of the triac in the four turning-on modes is discussed in some detail in this section, and the results are summarized in Table 11.1.

11.6.2.1   Triggering mode I(+): first quadrant: MT_2 positive, gate positive. Since the terminal MT_2 is positively biased, the J_5 junction at the bottom of Figure 11.12 is reverse biased and, therefore, is inactive, i.e., is not playing any significant role. The J_4 junction of the gate is also reverse-biased since the applied gate potential is positive in respect to the MT_1 terminal. The gate metallization extending beyond n_3 over the p_2 layer constitutes a conventional type of gate contact. The forward (positive) biasing of the conventional gate causes the n_2 emitter to emit electrons from the region close to point A which is near the gate contact. The turn-on of the

146

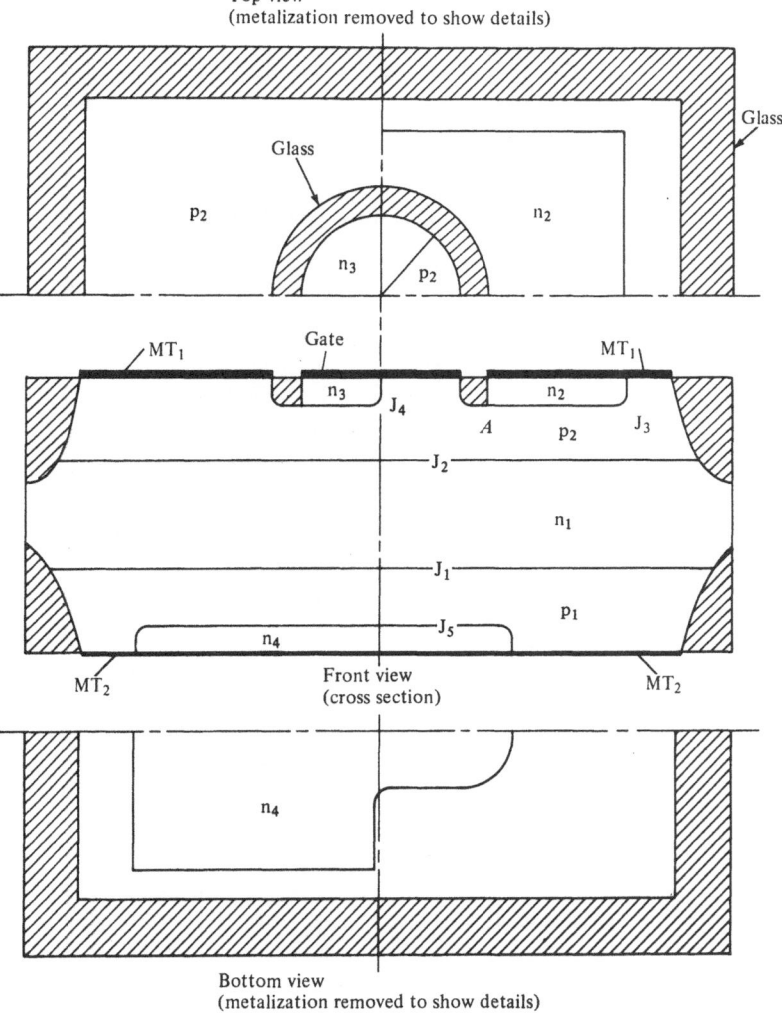

Figure 11.12 Glass-passivated triac, orthographic projection (not to scale).

p_1-n_1-p_2-n_2 thyristor proceeds like in an SCR with a shorted emitter and conventional p gate.

Table 11.1 and Figure 11.13 [11.11] summarize the state of various terminals, junctions, and current paths during the turn on for all four modes.

11.6.2.2 Triggering mode I($-$): first quadrant: gate negative. The gate potential is made negative with respect to the terminal

147

Table 11.1
The four triggering modes

Triggering mode	SCR triggered	MT_2 polarity	Gate polarity	Junction J_5 (under MT_2)	Junction J_4 (gate)	Gate action	Junction J_3 (under MT_1)	Gate sensitivity relative to mode I(+)
First quadrant I(+)	p_1-n_1-p_2-n_2	+	+	Reverse biased	Reverse biased	Conventional	Forward biased	1
I(−)	p_1-n_1-p_2-n_2	+	−	Reverse biased	Forward biased	Junction gate	Forward biased	~1/3
Third quadrant III(+)	p_2-n_1-p_1-n_4	−	+	Forward biased	Reverse biased	Remote gate (trans. n_2-p_2-n_1)	Forward biased	~1/4
III(−)	p_2-n_1-p_1-n_4	−	−	Forward biased	Forward biased	Remote gate (trans. n_3-p_2-n_1)	Reverse biased	~1/2

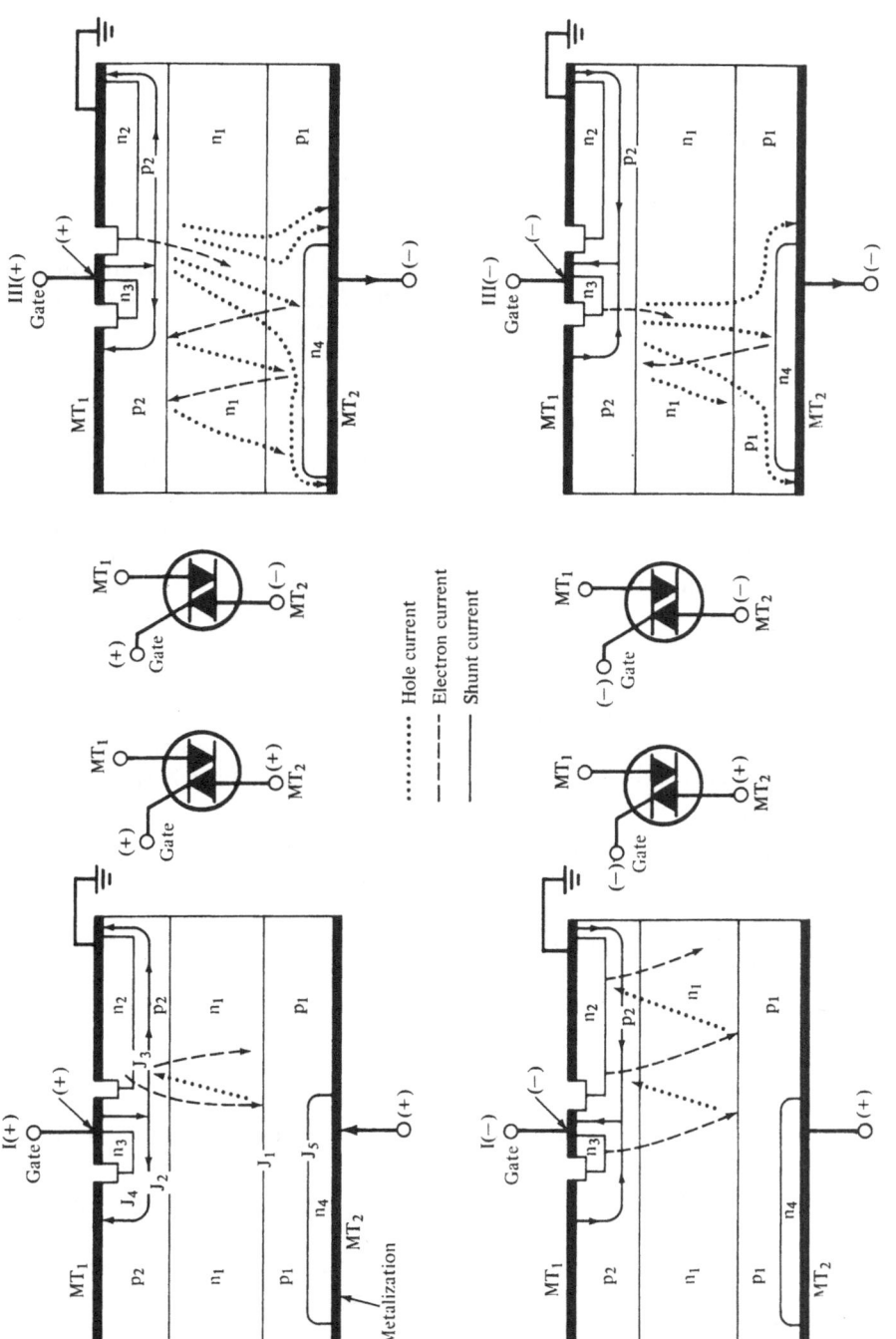

Figure 11.13 Current paths for various gating conditions. (After Neilson and Burke [11.8].)

149

MT_1, therefore, the gate junction J_4 is forward biased (Figure 11.12) and the n_3 emitter of the junction gate injects electrons into the p_2 region. The gate works this time like a *junction gate*. The injected electrons are collected by the J_2 junction, go into the n_1 region as majority carriers, and forward bias, therefore, the J_1 junction. The J_5 junction, between n_4 and p_1, is reverse biased since MT_2 is positive in respect to the terminal MT_1. Junction J_5 is essentially inactive. A voltage drop develops in the p_1 region above the n_4 layer so that holes are injected from p_1 into n_1, and the right-hand portion of the device (p_1-n_1-p_2-n_2) turns on.

11.6.2.3 Triggering mode III(+): third quadrant: MT_2 negative, gate positive. The gate is positive in respect to the reference terminal MT_1 and in respect to the n_2 emitter. Therefore, the J_3 junction of the n_2 emitter is forward biased and the n_2 emitter injects electrons into the p_2 base. The J_4 gate junction is reverse biased and thus rendered inactive.

The n_2-emitter–injected electrons are collected by the lightly forward-biased junction J_2, as discussed in the section on the remote gate, and are swept into the n_1 base as majority carriers, lowering n_1 potential with respect to the p_2 layer. Junction J_2 becomes, therefore, more forward biased and injects holes under the point *A*. Junction J_1 collects the holes which proceed into the p_1 base between the n_4 and n_1 layers, causing a lateral voltage drop in the p_1 base. When this voltage drop reaches about 0.5 V, the junction J_5 starts emitting and the p_2-n_1-p_1-n_4 portion of the triac is turned on (Figures 11.11 and 11.12).

In the triggering mode described, the device operates with a *remote gate*. The remote gate consists here of a n_2-p_2-n_1 transistor whose collector is common with the n_1 base of the p_2-n_1-p_1-n_4 thyristor.

11.6.2.4 Triggering mode III(−): third quadrant: MT_2 negative, gate negative. In the triggering mode III(−) the gate acts again as a *remote gate*. The terminal MT_2 is negative in respect to the terminal MT_1 and the gate junction J_4 is *forward* biased by the applied negative potential. The gate emitter n_3 injects electrons into the p_2 base. The electrons are collected by the lightly forward-biased junction J_2. The collected electrons proceed into the n_1 base and make it more negative in respect to the p_2 region. Therefore, junction J_2 becomes more forward biased and starts injecting *holes* more heavily into the n_1 and p bases. Thus the n_3-p_2-n_1 transistor (*the remote gate*) helps in turning on the p_2-n_1-p_1-n_4 thyristor.

11.6.3
Structure of a triac with a centrally located gate [11.9]

Figures 11.11 and 11.12 show a triac with a centrally located gate structure which is separated from the device active region by a glass-filled etched moat. The entire device is glass-passivated. It is of interest to note that the emitter n_4 has a shape as per Figure 11.11, bottom view. The purpose is to bring the n_4 emitter closer into

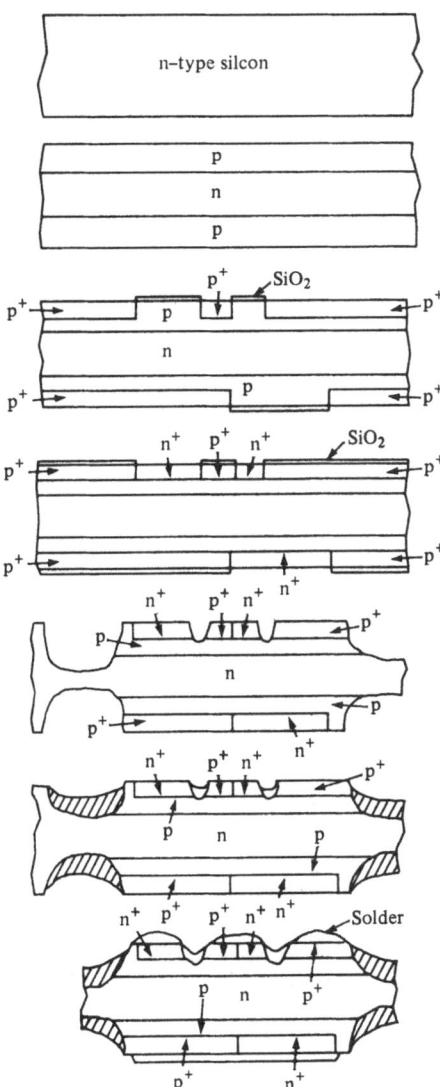

Figure 11.14 The seven basic steps required in the fabrication of a glass-passivated triac. (After McNulty [11.10].)

alignment with the n_2 emitter in order to increase the sensitivity of the triggering mode III($+$), since an increased lateral voltage drop will be developed in the p_1 base above the n_4 emitter. This mode operates with the n_2-p_2-n_1 transistor as a remote gate. It is, therefore, advantageous to have this transistor close to the device it helps to trigger on, i.e., the p_2-n_1-p_1-n_4 SCR.

11.6.3.1 Triac fabrication. Figure 11.14 illustrates the seven basic steps required to fabricate a glass-passivated triac with a centrally located gate [11.10].

1. A high-resistivity silicon wafer is cut of a single crystal silicon, lapped and polished to the required thickness.
2. p layers are diffused deeply into both sides using a boron source such as B_3N_4.
3. Silicon dioxide (SiO_2) masks are grown and p^+ regions are defined and diffused into the wafer.
4. A second diffusion mask is grown and n^+ regions are defined and diffused into the wafer.
5. An SiO_2 etch mask is grown and defined. Grids and gate moats are etched into the wafer.
6. A hard glass-passivating layer is applied in the grids and gate moats.
7. Contact areas are opened on the wafer, and nickel-lead-tin solder metallization is applied. The wafer is then laser scribed and separated into pellets (chips).

APPENDIX

The saturation and I_{B_1} currents are neglected. The cathode current of the pilot thyristor is, therefore,

$$I_g = \alpha_3 \cdot I_{AP} + \alpha_4 I_3 \qquad (11.A)$$

where I_{AP} is the pilot's anode current.
Also

$$I_g = I_{AP} + I_{S_1} \qquad (11.B)$$

where I_{S_1} is the base (gate) current. Combining equations (11.A) and (11.B) we obtain

$$I_g = \frac{I_{S_1}(1 - \alpha_3)}{1 - \alpha_3 - \alpha_4} \qquad (11.C)$$

References

11.1 T. C. McNulty. A review of thyristor characteristics and applications. RCA Solid State Division, Application Note AN-4242, 1973.

11.2 F. E. Gentry, R. I. Scace, and J. K. Flowers. Bidirectional triode P-N-P-N switches. *Proc. IEEE, 53*: 355–369, 1965.

11.3 F. E. Gentry, F. W. Gutzwiller, N. Holonyak, Jr., and E. E. Von Zastrow. *Semiconductor Controlled Rectifiers.* Englewood Cliffs, N.J.: Prentice-Hall, 1964.

11.4 E. K. Howell. The triac-gate controlled silicon AC power switch. *1964 IEEE International Convention Record,* part 9, pp. 86–91.

11.5 J. R. Hauser. The effects of distributed base potential on emitter-current injection density and effective base resistance for stripe, transistor geometries. *IEEE Trans. Electron Devices, ED-11*: 238–242, 1964.

11.6 Von Gunter Köhl. Steuermechanismus und Aufbau bilateral shaltender Thyristor. *Scientia Electrica, XII,* 1966 Facs. 4.

11.7 R. W. Aldrich and N. Holonyak, Jr. Two-terminal asymmetrical and symmetrical silicon negative resistance switches. *J. Appl. Phys., 30* (11): 1819–1824, 1959.

11.8 J. Neilson and D. Burke. RCA, Somerville, N.J., private communication, 1972.

11.9 *RCA Designer's Handbook.* Solid State Power Circuits Technical Series, Somerville, N.J., 1971.

11.10 T. C. McNulty. Power switching using solid-state relay. Application Note AN-6141, RCA Solid State Division, Somerville, N.J.

12

Commutation of triacs

Summary

In AC power-control applications, a triac must switch from the conducting state to the blocking state twice each cycle, at each zero-current point. This action is called commutation. Because of the charges stored in the triac during conduction, at high voltage-rising rates (dv/dt), the triac may fail to block the circuit voltage following the zero-current point. In such a case the control of the load power is lost, although the triac itself is not damaged in any way.

12.1
Introduction

The undesirable, false triggering of an SCR may occur when the displacement current flowing through the blocking junction capacitance becomes large enough to make the sum of the device current gains $\alpha_1 + \alpha_2$ equal to or greater than unity. This is the so called static dv/dt effect. The rate of application of the initial anode voltage should not exceed some critical value in order to avoid an undesirable triggering.

When the forward current of an SCR is reduced to zero at the end of the conduction period, the application of a forward, off-state voltage to the anode must be sufficiently delayed to avoid premature firing. A premature application of an excessive voltage or a voltage rising too rapidly will make the SCR revert to the on-state. This puts a limit on the maximum reapplied dv/dt.

In triacs there is a third type of a dv/dt effect, known as the commutating dv/dt effect. In AC power-control applications a triac must switch from the conducting state to the blocking state at each zero-current point, i.e., twice at each current cycle. This action is called commutation. If the triac fails to block the circuit voltage following the zero-current point, the control of load power is lost, although there is no damage done to the device.

Commutation for *resistive* loads presents no special problem because the voltage and current are essentially in phase. The time available for the triac to turn off extends from the time the device current drops below the holding current until the reapplied voltage exceeds the value of the line voltage required to allow latching current. However, *inductive* loads can cause a turn-off problem. If the triac is in series with an inductive load and an AC supply, the triac will attempt to turn off at the instant the current passes through zero. The supply voltage will lead the current by about 90 degrees and will thus have an appreciable value at that particular instant. The maximum rate of rise of this voltage that can be blocked without the triac reverting to the on state is termed the critical rate of rise of commutation voltage, or the commutating dv/dt capability of the triac. The commutating dv/dt capability is generally substantially below the triac's static dv/dt capability.

12.2
Current and voltage waveforms during commutation

Figure 12.1 shows a triac in a typical control connection with an AC power source [12.1–12.5] and an inductive load.

Figure 12.2 shows the triac principal voltage and current waveforms when the *load is resistive*. If the gate drive is removed at the

Figure 12.1 Series connection of a triac, an inductive load, and an AC power source.

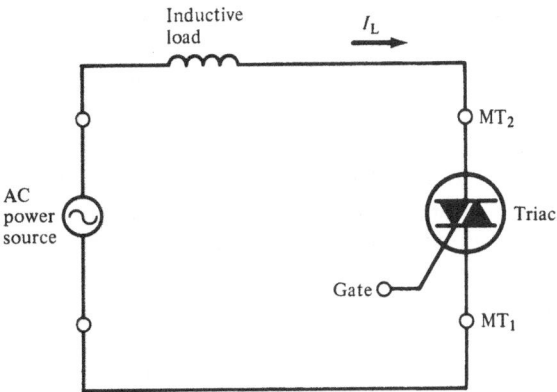

155

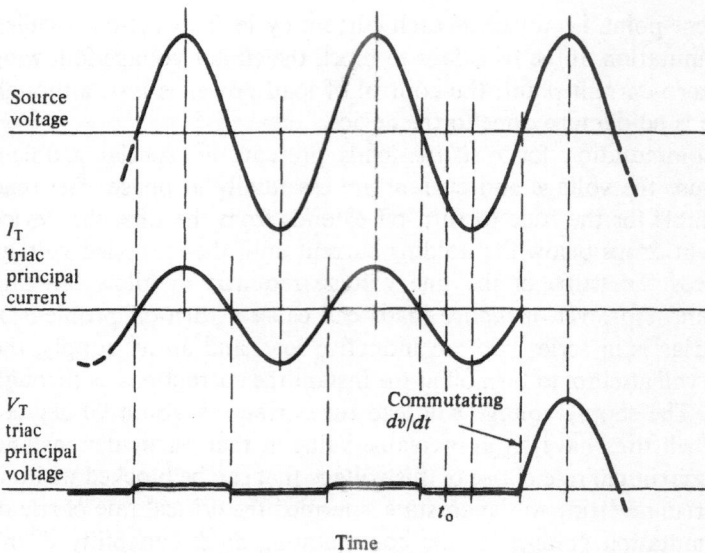

Figure 12.2 Principal voltage and current for a triac in operation with a resistive load. (After Wojslawowicz [12.1].)

time t_0, the device continues to conduct until the current attempts to reverse its polarity. The device then undergoes a reverse-recovery period and thereafter must support a main terminal voltage of the reverse polarity that is equal to the source voltage. The rate or re-application of this off-state voltage for a purely resistive load and 120-V 60-Hz source is typically 0.063 V/μsec. (This value is obtained by finding the dv/dt of a sine wave voltage at the time $t = 0$). This rate of reapplication does not cause turn-on of the device.

If a gate signal that allows continuous conduction is applied to a triac controlling *an inductive load*, the load current I_T lags the line voltage by about 90 degrees (Figure 12.3). During the triac's on-state conduction, the voltage across the triac, V_T, is in phase with the load current I_T and has typically an amplitude of 1.5–2.0 V.

When the gate signal is removed, the triac begins to commutate off near the end of the half cycle, i.e., at the point when the load current drops to a value below the holding current I_h of the triac. At the instant the triac commutates off, the voltage V_T across the triac reverses its direction and climbs to the peak of the line voltage —its rate of rise, i.e., the commutating dv/dt and overshoot, being a characteristic of the circuit components. After the triac successfully commutates off, the V_T voltage is in phase with the line voltage. If the commutating dv/dt of the circuit is greater than the commutating dv/dt capability of the triac, the triac does not turn off, but reverts to the on state (Figure 12.4) With no gate signal applied, the triac

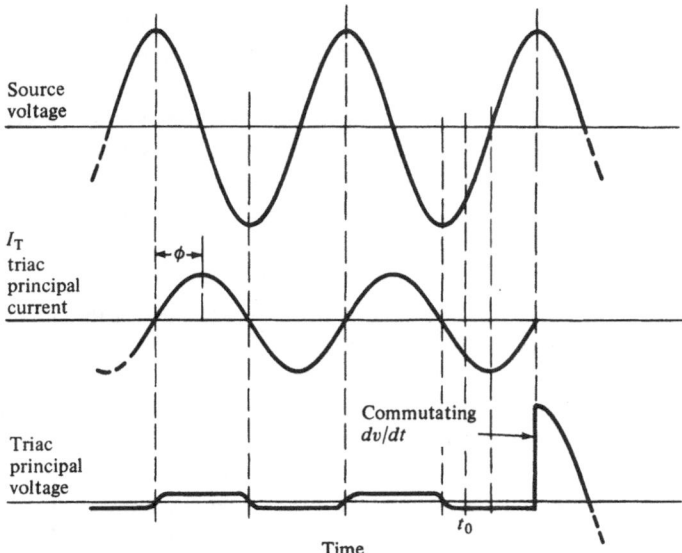

Figure 12.3 Principal voltage and current for a triac in operation with an inductive load. (After Wojslawowicz [12.1].)

Figure 12.4 Principal voltage and current curves showing triac malfunction that results from commutating dv/dt produced by inductive load. (After Wojslawowicz [12.1].)

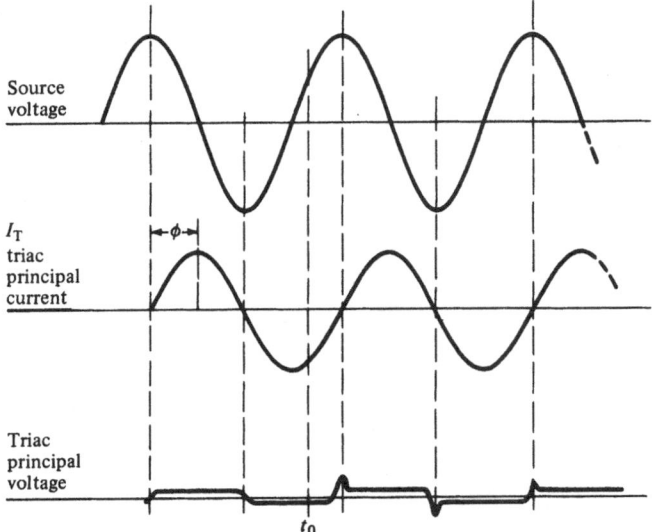

157

again attempts to turn off in the next half-cycle of opposite polarity. If it succeeds, it remains off; if its capability is again exceeded, it remains on until the circuit power is interrupted.

Figure 12.5 shows in more detail the current and voltage waveforms close to the turn-off point, i.e., close to the zero-current point. The triac current falls down with some rate equal to $-di/dt$. At the same time t_1 *the current direction of the triac is reversed,* and that part of the triac that was in the on state starts recovering analogically to a p-n junction diode recovering from the forward conduction into the blocking state. At some point in time the *voltage V_T across the triac changes its polarity* and grows rapidly in magnitude with a rate dv/dt.

The recovery current is due to the charge stored in the device. Superimposed on this current is the displacement current of the junction capacitance C_j, equal to $d(C_j V)/dt$. The capacitive current component appears at the moment $t = t_2$ when the triac begins to block the reverse polarity. The total current during the charge recovery period may be of sufficient magnitude to turn the device on. Before the turn-off point, for example, SCR-1 of the triac represented in Figure 12.6b, was on; if the recovery current is high enough, the second part of the device (SCR-2) will be activated. The recovery current plays the role of the triggering current and the device continues to conduct, although not intentionally, and the gate control is lost (Figure 12.4).

Figure 12.5   Triac current and voltage at commutation.

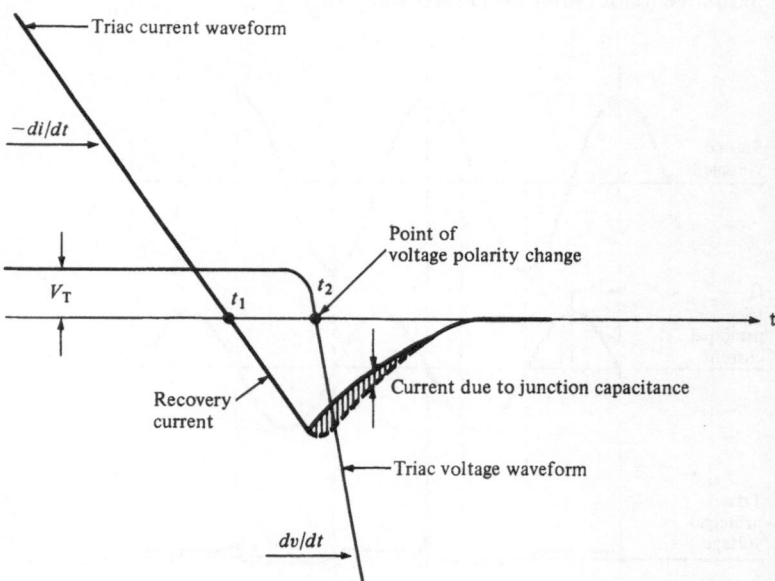

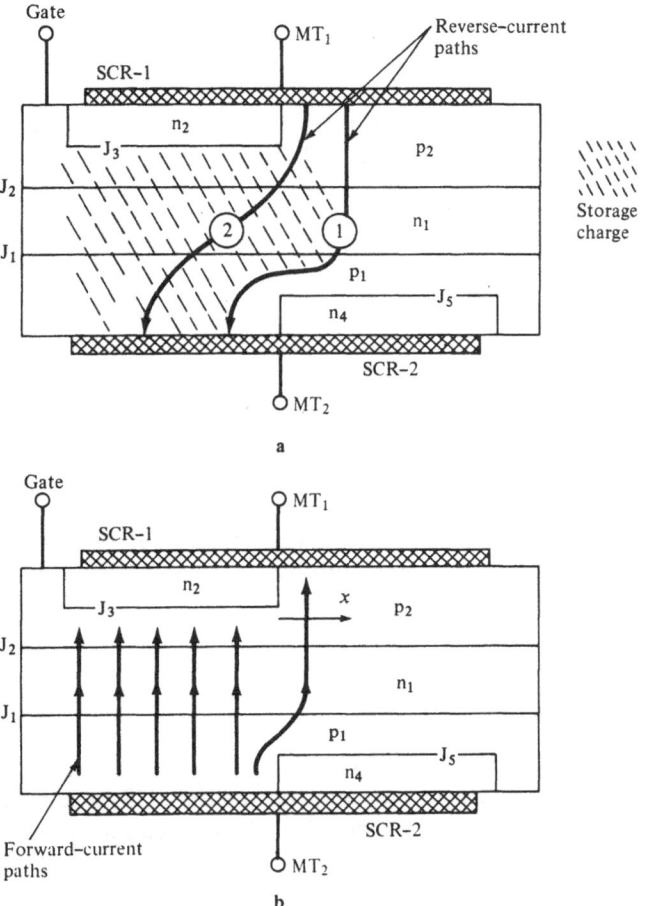

Figure 12.6 Two antiparallel SCRs representation of a triac. (After Bergman [12.6].)

12.3
Role of stored charges in triac commutation

When a triac is conducting in the first quadrant, the flow of current will be approximately as shown in Figure 12.6b. Most of the current will flow from the contact MT_2 toward the n_2 emitter. Some current will, however, be bypassing the n_2 emitter and flow sidewise to the MT_1 contact to the right of the n_2 emitter. The current density J_x decreases exponentially with distance [12.6] following the relationship

$$J_x = J_0 \exp\left(-\frac{x}{L_s}\right), \tag{12.1}$$

159

where J_0 is the current density at the emitter edge and L_s is a constant equal approximately to 0.7 W, where W is the total width of the five-layer structure.

The charge stored in the n and p bases of the device will depend upon the current density. Consequently, there will be a constant charge density over the n_2 emitter, falling off exponentially at the edge of the n_2 emitter (Figure 12.7). In reality, due to the concentration gradient, the electric charges in both bases will diffuse laterally and the charge distribution may be considerably modified. In the n base the charge will decrease exponentially in the x direction as

$$Q(x) = Q_0 \exp\left(-\frac{x}{L_p}\right) \tag{12.2}$$

where Q_0 is the charge at the emitter edge and L_p is the diffusion length for holes in the n base [12.6]. A similar expression applies to the charge distribution in the p base. Both of these effects will result in a complicated charge distribution in each base. In all probability, these charges will be decaying with some characteristic length $L_S = 0.7\ W$ or the diffusion length of the minority carriers, whichever is larger.

Immediately after the forward current in the first quadrant has been reduced to zero, the charge distribution in the device will be as shown shaded in Figures 12.6a and 12.7. When the potential applied to the SCR-1 (Figure 12.6a) reverses and the device is biased into the third quadrant, a reverse current (for the SCR-1) starts to flow with a rate of rise di/dt, dependent largely on circuit parameters.

Figure 12.7 Charge density versus distance. (After Bergman [12.6].)

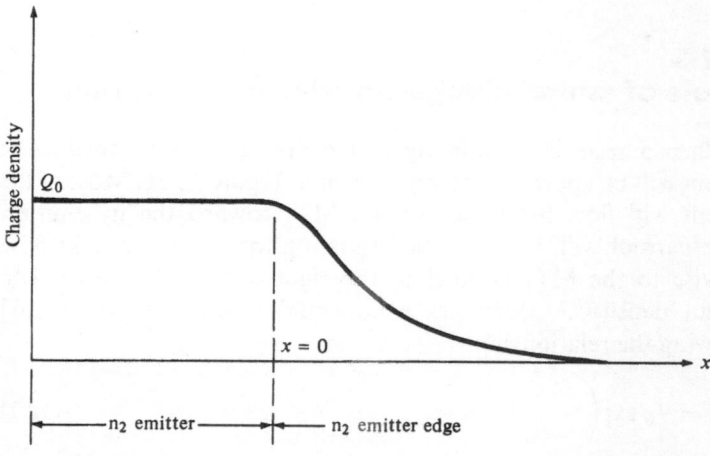

The current which is supported by the stored charge will flow following mainly path 1 offering the least resistance. As long as the excess charges exist in the n_1 and p_1 bases, the J_1 junction will be forward biased and will present practically no resistance to the reverse-current flow. As soon, however, as the junction J_1 recovers its reverse-voltage capability, the current flow will be impeded and the flow pattern will change, so that eventually it will follow path 2. The reverse current will behave to a large extent as a reverse current of a recovering p-n junction diode and its waveforms will be discussed a little later.

When the current follows path 1, the current reaches its maximum value just before the J_1 junction recovers its blocking capability. After that, the amount of stored charge declines and, in addition, the current follows a more resistive path. Therefore, from the commutation point of view, path 1 is the most significant since it forward-biases the n_4 emitter of the SCR-2 part of the triac. The failure to commutate will be due to this biasing effect.

We assume that the reverse current I_R, originating mainly from the n_2 emitter edge, spreads from this edge to a distance $L_S = 0.7 \, W$ or to the diffusion length L_p, whichever is greater. It is possible then to show [12.6] that the bias developed at the n_4 emitter is equal to

$$V_E = \frac{\rho_s I_R L_S}{4l} \tag{12.3}$$

where ρ_s is the sheet resistance over the n_4 emitter and l is the length of the n_2 emitter edge (SCR-1) from which spreading takes place. The SCR-2 will turn on, i.e., the device will fail to commutate if the reverse current I_R is sufficiently large so that the n_4 emitter will be forward-biased by a potential exceeding about 0.5 V for a time interval on the order of a few tenths of a microsecond. The situation will become worse when the displacement current $d(CV)/dt$ of the recovered junction will add up to the current due to the stored charges.

12.4
Recovery of a p⁺-n abrupt-junction diode

To understand, at least qualitatively, triac behavior at commutation, it is advantageous to simplify the problem by assuming that triac recovery is similar to that of a p⁺-n diode (Figure 12.8). Such a diode may be turned off by a current step or a current ramp. Both possibilities are reviewed below.

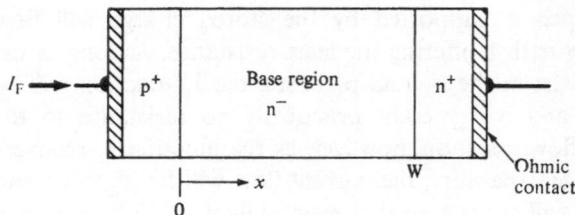

Figure 12.8 Structure of a p⁺-n-n⁺ diode.

12.4.1
Current-step recovery

In an ideal situation the recovery of a diode may be considered with a current being switched *instantaneously* from the forward flow to the reverse flow (step recovery). This situation is illustrated by Figure 12.9. Upon the application of the reverse external potential

Figure 12.9 Turn-off transient in a diode.

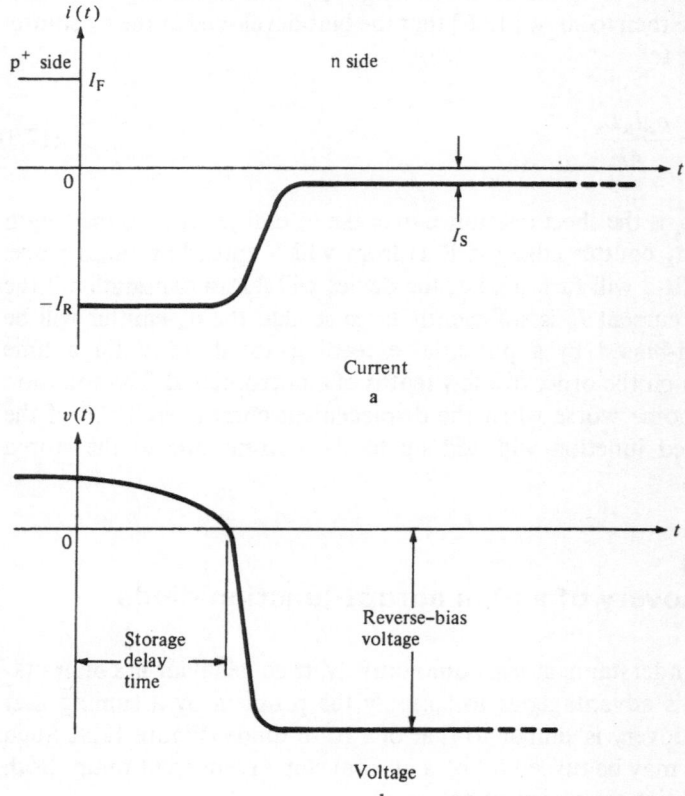

162

the positive current I_F is changed rapidly to the negative reverse current $-I_R$. The voltage drop across the diode, however, does not reverse instantaneously. It starts changing its polarity and magnitude only after the charge storage delay time. Figure 12.10 shows hole distribution in the n-type region following the application of a *constant* reverse current. The reverse current across the junction consists of holes being withdrawn from the n-type region into the p-type region. Initially, there is little change in the stored carrier density p_n at $x = 0$ and, therefore, little change in the voltage V across the diode. The diode remains with a forward-bias voltage, despite the fact that the current has been reversed. The hole concentration slope at $x = 0$ remains fixed since this slope determines the current I_R, which was assumed to be constant. The hole concentration declines everywhere, however, since holes are both being removed and recombining with electrons. At some point in time the hole concentration at $x = 0$ drops below the equilibrium value and the voltage V reverses (all excess charges above equilibrium value were removed at $x = 0$). After that, there are not enough holes stored to maintain the constant reverse current I_R; therefore, the reverse current starts falling off.

Figure 12.10 Hole distributions in the n-type region of a diode during the storage-delay transient.

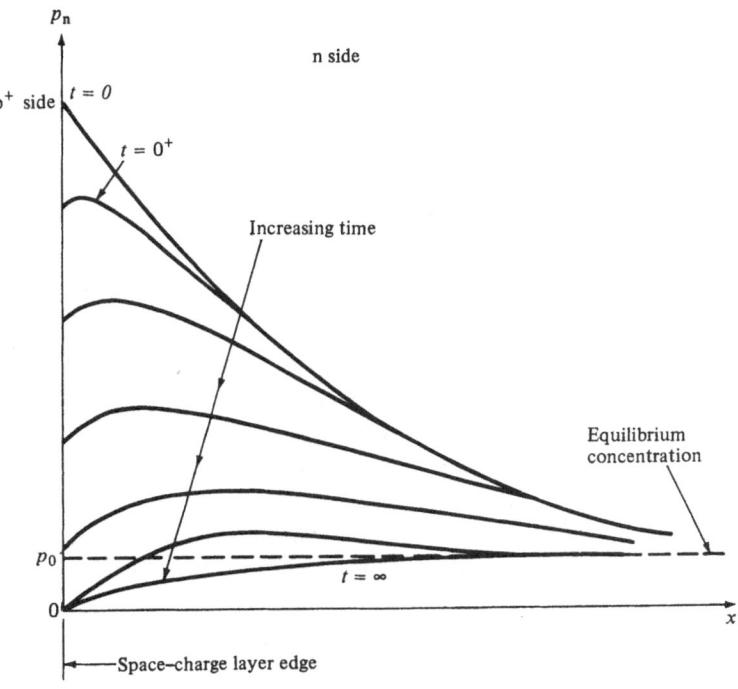

A true constant-current supply would make the reverse voltage swing to its maximum possible value, i.e., to the breakdown voltage. A supply with a moderate voltage and resistance would deliver less current as the reverse voltage rises in the absolute value. Figure 12.9 illustrates the form of the turn-off transient when I_R is supplied through a resistor from a voltage source whose open-circuit voltage is less than the diode-breakdown voltage. After the storage delay time, the voltage reverses and rises as the reverse current drops in magnitude. The reverse current drops to its steady-state (saturation-current) value I_S after all excess holes stored deeper in the n region are removed.

High-voltage devices usually contain diodes with wide bases, i.e., bases longer than one diffusion length of the minority carriers. This is necessary for the support of high breakdown voltages. For this kind of diode the total stored excess charge in a steady-state condition is the product of the total current I_F flowing in the device and the minority-carrier lifetime in the low-doped region.[1] The stored charge is, therefore,

$$Q = I_F \tau \tag{12.4}$$

When the *step recovery* takes place, all of the excess charges as per Equation (12.4) have to be removed. In the following, we discuss the case of a recovery by a current ramp, and we will see that by the time the current reverses, part of the stored charge has already disappeared due to carrier recombination.

12.4.2
Linear current-ramp recovery

In practice, the recovery takes place with a current ramp rather than a current step. The switching from the forward to the reverse conduction is not sudden, but gradual with some finite rate di/dt determined by the circuit parameters. Figure 12.11 shows a typical switching characteristic with a linear current ramp. The forward current I_F is initially constant. At $t = 0$ the current starts decreasing with a constant rate $-di/dt = R$. At the time T_1 it passes through zero and its direction reverses, but the value of R is maintained up to the time T_2 when the current is reduced sharply. At T_2, the p^+-n junction has recovered in its reverse potential blocking capability.

The recovery of a p^+-n diode by a current ramp was analyzed by Kao and Davis [12.7]. The diode is represented schematically by Figure 12.8. In the steady state the current in the diode is I_F. During

---

[1] In a *narrow*-base diode the charge stored is the product of the current and the transit time of the minority carriers in the low-doped region.

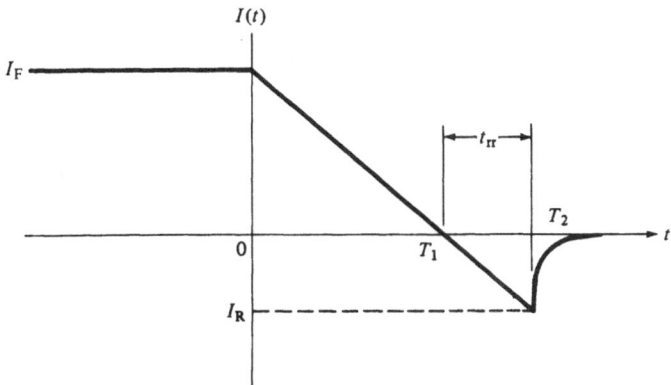

Figure 12.11 Switching characteristic with a linear current ramp when $T_2 < 2T_1$. (After Kao and Davis [12.7].)

any transient, however, the current will be time dependent and can be expressed as a charge equation

$$I(t) = \frac{dQ}{dt} + \frac{Q}{\tau} \tag{12.5}$$

where Q is the excess charge due to holes injected into the n base and τ is the minority carrier lifetime in the n base.

During switching, $I(t)$ is determined by the external circuit. The forward current, initially constant (Figures 12.11), starts decreasing at time $t = 0$ with a constant slope

$$R = -\frac{di}{dt} \tag{12.6}$$

At $t = T_1$, the current direction reverses and at $t = T_2$ the current drops sharply to a small value and the junction recovers its blocking capability. Therefore, prior to $t = 0$, $I(t) = I_F$, and the excess charge stored in the wide-base diode is $Q(0) = I_F\tau$. Between the $t = 0$ and $t = T_2$ the current follows the linear relationship in time, i.e.,

$$I(t) = I_F - Rt \tag{12.7}$$

At $t = T_1$ the current $I(t) = 0$, consequently

$$T_1 = \frac{I_F}{R} \tag{12.8}$$

165

Using the above relationships as initial conditions, one obtains an expression for the excess charge $Q(t)$ in function time, using the charge Equation (12.5):

$$Q(t) = I_F \tau \left\{ \frac{\tau}{T_1} \left[1 - \exp\left(\frac{-t}{t}\right) \right] - \frac{t}{T_1} + 1 \right\} \tag{12.9}$$

Since at $t = T_2$ the total excess charge in the base region is very small, we can assume that the charge at $t = T_2$ is $Q(T_2) = 0$, and from (12.9)

$$\frac{T_2 - T_1}{\tau} = 1 - \exp\left(\frac{-T_2}{\tau}\right) \tag{12.10}$$

For $T_1 \gg \tau$, i.e., for a shallow current slope R, the exponential term on the right-hand side is negligible and the recovery time t_{rr}, i.e., the time from the current reversal point at T_1 to the time $t = T_2$ when the current starts falling down rapidly, is

$$t_{rr} = T_2 - T_1 \approx \tau \tag{12.11}$$

If the current ramp rate R is much smaller than I_F/τ then the recovery time is about equal to the base lifetime and is independent of R. For $T_1 \gg \tau$ it is easy to show, utilizing (12.9) and neglecting again the exponential term, that

$$Q(T_1) = R\tau^2 \tag{12.12}$$

Referring again to Figure 12.11, we realize that the charge $Q(T_1)$ is recovered during the recovery time $t_{rr} \approx \tau$. From simple geometrical considerations we obtain, therefore, that

$$Q(T_1) \approx I_R \frac{\tau}{2} \tag{12.13}$$

where I_R is the peak recovery current. In reality, part of the charge stored will recombine during t_{rr}. Therefore, the computation of I_R below represents the worst possible case.

From (12.12) and (12.13) we obtain for the peak current.

$$I_R = -2\tau \frac{di}{dt} \tag{12.14}$$

This simple computation illustrates the fact that small di/dt values will result in small I_R values. It means that when the current ramp has a small slope, then by the time T_1, when the current changes its polarity, the charge left in the n region has already declined due to recombination and, therefore, little reverse current is required for its recovery.

If the ramp rate is much larger than I_F/τ, the diode recovery resembles more the step recovery for which $t_{rr} = \tau \ln\{1 + (I_F/I_R)\}$ [12.8, 12.9]. Therefore, at $t = T_1$ the charge in the base will be the steady-state charge $I_F\tau$ and the reverse current will have to be much larger to recover the charge in the same interval of time, let alone for shorter recovery times.

12.4.3
Sine wave recovery

Figure 12.12 shows the waveform for switching with a sine wave. By using the charge Equation (12.5) [12.7] and an expression for the current $I(t) = I_F \sin wt$, the charge at $t = T_1 \gg \tau$ is

$$Q(T_1) = \frac{(I_F\omega)\tau^2}{(1 + \omega^2\tau^2)} \tag{12.15}$$

Similarly to a current-ramp recovery, the recovery time is also found to be equal to the minority-carrier lifetime

$$t_{rr} \approx \tau \tag{12.16}$$

The ramp rate R for the sine wave can be found by finding a derivative of the current in respect to time at $t = T = \pi/\omega$. This rate is

$$R = \omega I_F \tag{12.17}$$

The charge $Q(T_1)$ is removed by current and recombination. The higher the ramp rate R the larger the stored charge $Q(T_1)$ will be, as follows from Equations (12.15) and (12.17), and the larger the reverse current I_R will be on the basis of the same considerations used in

Figure 12.12   Swtiching with a sine wave. (After Kao and Davis [12.7].)

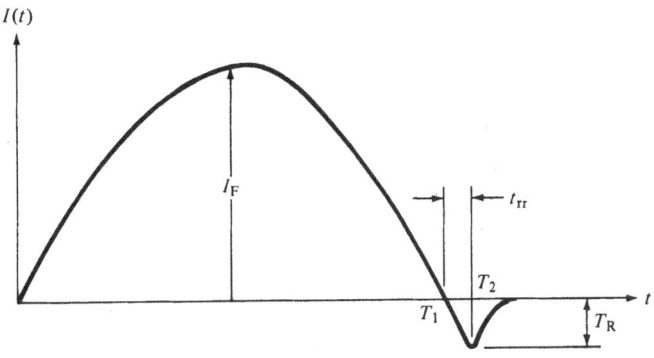

connection with the linear current ramp. Also, the higher the peak current and the higher the frequency, the larger will be the charge to be recovered.

12.4.4
Triac dv/dt capability

The discussion of the diode-recovery phenomena makes it possible to understand the experimental results obtained for the dv/dt capability of a triac as a function of the load current, rate of current removal, frequency, temperature, and lifetime. It was shown that the charge stored in the device partially recombines and is partially recovered due to the reverse-current flow I_R (see Section 12.4.2). The higher the forward current I_F in the device, and the larger the slope di/dt of the recovery current, the larger I_R will be. This current adds up to the displacement current $C(dv/dt)$ and forward biases the N_4 emitter turning on prematurely the SCR-2 (Figure 12.6), so that the gate loses its control. The allowed commutating dv/dt will, therefore, be smaller for large di/dt rates. Figure 12.13 illustrates the depen-

Figure 12.13 Commutating voltage as a function of commutating current [12.3].

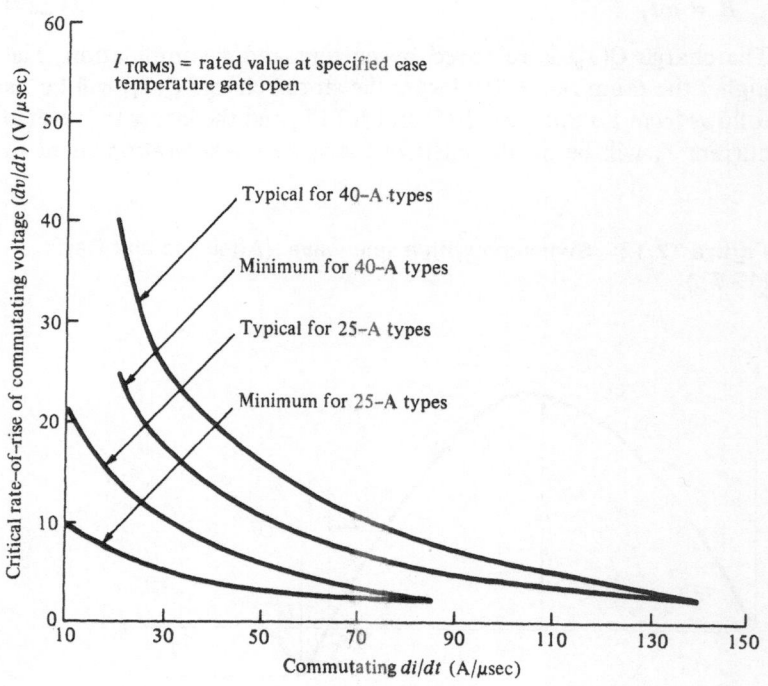

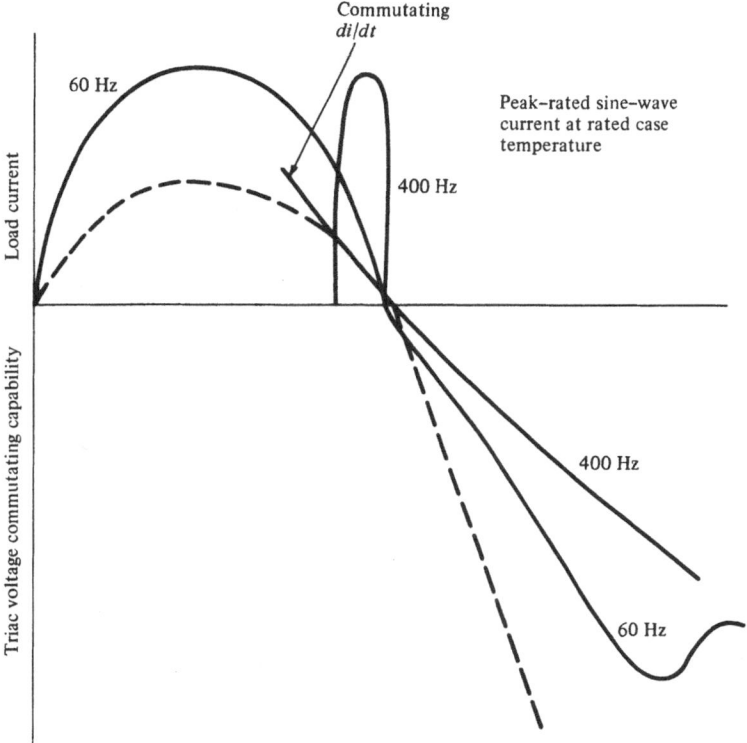

Figure 12.14 Dependence of triac commutating capability on current and frequency [12.3].

dence of the commutating dv/dt on the rate of removal of current di/dt at a case temperature of 75°C. Figure 12.14 [12.3] indicates how the commutating dv/dt capability of a triac depends on current and frequency. A triac has some specific commutating capability at the rated 60-Hz on-state current. If this 60-Hz current is reduced (dashed line), then its associated commutating dv/dt capability is increased for two reasons: lower current and smaller di/dt. For a 400-Hz on-state current of the same magnitude, it is evident that the current rate di/dt is much greater than at 60 Hz and, therefore, the commutating dv/dt capability is greatly reduced. A triac capable of 400-Hz operation or higher must have an extremely high commutating capability. Figure 12.15 is a plot of frequency capability of a triac in function of junction temperature [12.3]. The higher the temperature, the lower will be the allowed AC current frequency. This temperature effect is primarily due to the positive coefficient of the minority-carrier lifetime. Figure 12.16 is a plot of the frequency capability of a 400-Hz triac as a function of load current.

169

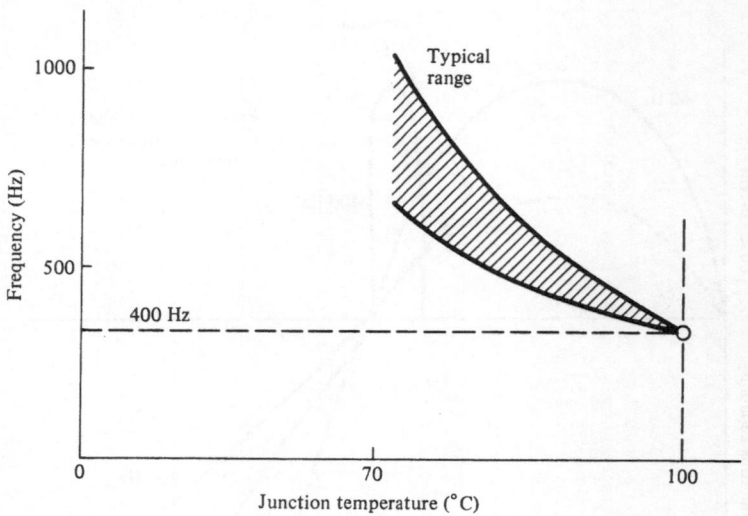

Figure 12.15   Frequency capability of a 400-Hz triac as a function of junction temperature [12.3].

Figure 12.16   Frequency capability of a 400-Hz triac as a function of load current [12.3].

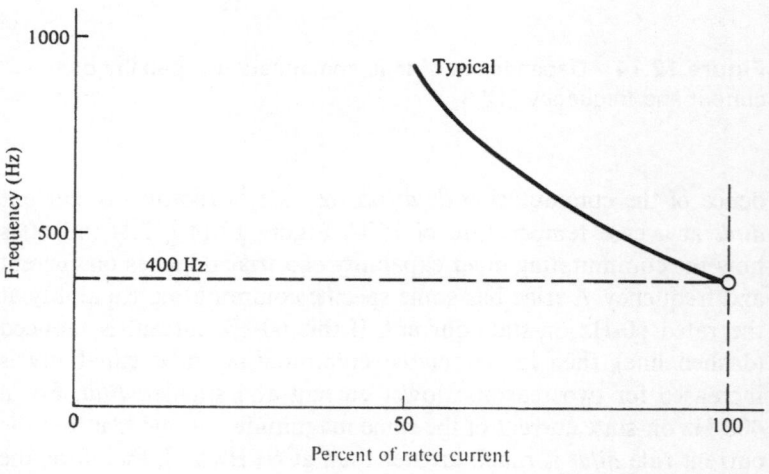

12.5
Snubber networks for *dv/dt* suppression

If the total recovery current, which is a sum of currents due to stored charges and to the *dv/dt* effect, is high enough, the device switches on. The use of shorted-emitter structures alleviates the

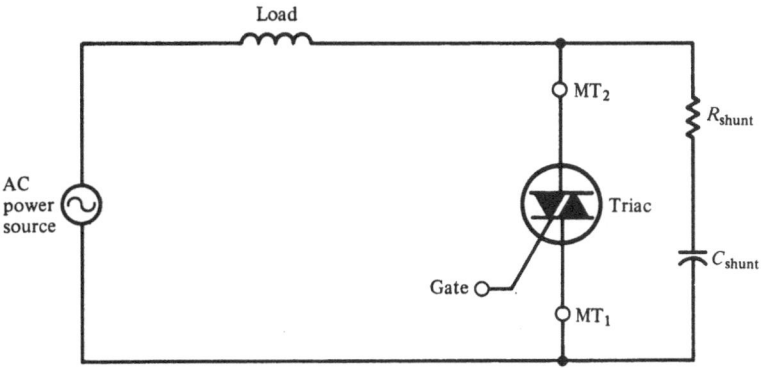

Figure 12.17 Circuit using a snubber network of R_{shunt} and C_{shunt} connected across the triac.

problem considerably, but does not solve it. The shunt behavior is similar to its action in an SCR. The dv/dt performance may be improved further by the use of the so-called snubber circuits [12.1] which control the displacement current flow in the device. An RC snubber network (Figure 12.17) limits the rate of reapplication of the off-state voltage. A snubber network in its simplest form consists of a resistance R_{shunt} and a capacitance C_{shunt} connected in series and placed across the main terminals of the device. For some R_{shunt} and C_{shunt} values combined with inductive loads, damped electrical oscillations (ringing) may occur. Since such a ringing may result in voltages exceeding device-blocking capability, the snubber circuits should be designed with special care.

References

12.1   J. E. Wojslawowicz. Analysis and design of snubber networks for dv/dt suppression in thyristor circuits. RCA Solid State Division, Thyristors, Application Note AN-4745, 1971.

12.2   *GE SCR Manual*, 5th ed., Syracuse, N.Y.: General Electric, 1972.

12.3   RCA Solid State Power Circuits Technical Series SP-52, RCA Solid State Division, Somerville, N.J. 08876, 1971.

12.4   J. F. Essom. Bidirectional triode thyristor applied voltage rate effect following conduction. *Proc. IEEE*, *55* (8): 1312–1317, 1967.

12.5   T. C. McNulty. A review of thyristor characteristics and applications. RCA Solid State Division, Somerville, N.J., Application Note AN-4242, 1962.

12.6   D. Bergman. Gate isolation and commutation in bi-directional thyristors. *Int. J. Electron.*, *21* (1): 17-35, 1966.

171

12.7 Y. C. Kao and J. R. Davis. Correlations between reverse recovery time and lifetime of p-n junction driven by a current ramp. *IEEE Trans. Electron Devices*, *ED-17* (9): 652–657, 1970.

12.8 J. L. Moll, S. Krakauer, and R. Shen. P-N junction charge stored diodes. *Proc. IRE*, *50*: 45–53, 1962.

12.9 H. J. Kuno. Analysis and characterization of p-n junction diode switching. *IEEE Trans.*, *ED-11* (1): 8–14, 1964.

Silicon surface

Summary

The silicon–silicon-dioxide system is of utmost importance to all silicon devices both unipolar and bipolar. A metal-oxide–silicon diode presents an excellent tool for the study and understanding of the behaviour of the silicon surface under various conditions. With the help of the capacitance–voltage characteristic of such a diode it was possible to shed light on the nature of the depletion, accumulation, and inversion layers at the silicon surface.

Extensive studies of the Si–SiO$_2$ system using C–V plot and, of course, other techniques revealed the nature of the various existing surface states and charges. The presence of the interface states and mobile and fixed charges modify profoundly the electrical properties of the silicon surface affecting seriously such bipolar device parameters as electric breakdown or leakage currents.

13.1
Introduction

The silicon–silicon-dioxide system is of utmost importance to all silicon devices and has been extensively studied in recent years in connection with the development of the MOS (metal-oxide–silicon) devices. A simple MOS diode proved to be of great help in the understanding and the investigation of the behavior of the silicon surface. Therefore, before we concentrate on various surface states and charges, we review the MOS diode characteristics.

13.2
Ideal MOS diode [13.1–13.7]

The MOS diode structure as shown in Figure 13.1 consists of an oxidized silicon crystal and a metal electrode (gate) deposited on the

173

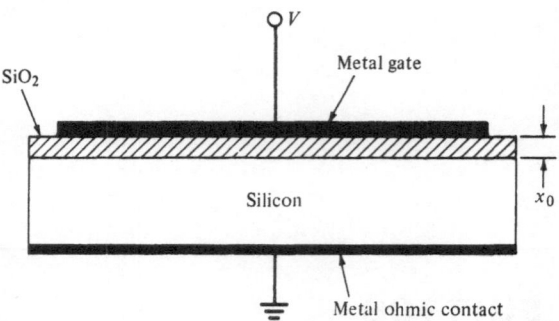

Figure 13.1   MOS diode.

top of the oxide. The band diagram of the ideal MOS diode is shown by Figure 13.2 for an applied potential $V = 0$. By an ideal diode we understand an MOS device in which the work function difference ϕ_{MS} between the metal and the semiconductor is equal to zero.

The work function is the energy required to remove the electron from the Fermi level in a given material to a vacuum. A certain amount of energy is required to move an electron from the metal Fermi level into the conduction band of SiO_2 in order to overcome the metal-oxide energy barrier. Also, in order to move an electron from the silicon-valence band to the oxide-conduction band, it is necessary to overcome the silicon-oxide energy barrier. The barrier energies are related to the work functions and can be measured by shining light of increasing frequency on the MOS structure, until the photon energy will be high enough to excite electrons into the conduction band of the oxide [13.8].

The ideal case assumes also that there are no electric charges on the silicon surface or in the oxide. The only charges are those induced by the applied gate potential. The effect of the presence of the work function difference and surface charges will be introduced later.

Under the above ideal conditions, the band diagram of Figure 13.2 for the p-type silicon is entirely flat in the absence of gate potential. This is the flat-band situation. ϕ_M and ϕ_{Si} represent the metal and the semiconductor work functions, respectively. ϕ_{BI} is the potential barrier between the metal and the insulator while ϕ_{BS} is the potential difference between the Fermi level of the silicon and the intrinsic level E_i. The difference ϕ_{MS} between the two work functions is zero. Also, there is no charge transfer possible in the ideal structure since the insulator has an infinite resistance. When ϕ_M and ϕ_{Si} are not equal ($\phi_{MS} \neq 0$) and we wish to maintain the flat-band situation, we have to apply a potential to the diode gate

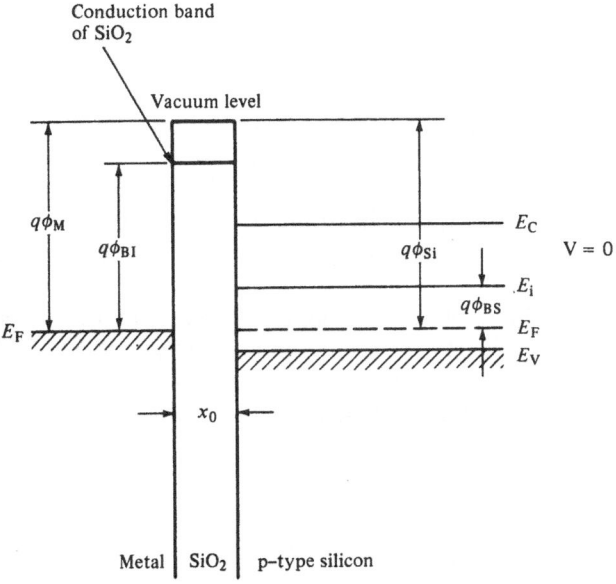

Figure 13.2 Energy-band diagram for an ideal MOS diode with no applied potential. E_C: conduction band energy; E_V: valence band energy; E_i: midgap energy; E_F: Fermi level energy.

equal and opposite in sign to ϕ_{MS}. This value V_{FB} (Figure 13.3) is called the flat-band potential:

$$V_{FB} = |\phi_{MS}| = \phi_M - \phi_{Si}$$

Figure 13.3 Energy-band diagram of an MOS diode with $\phi_{MS} \neq 0$.

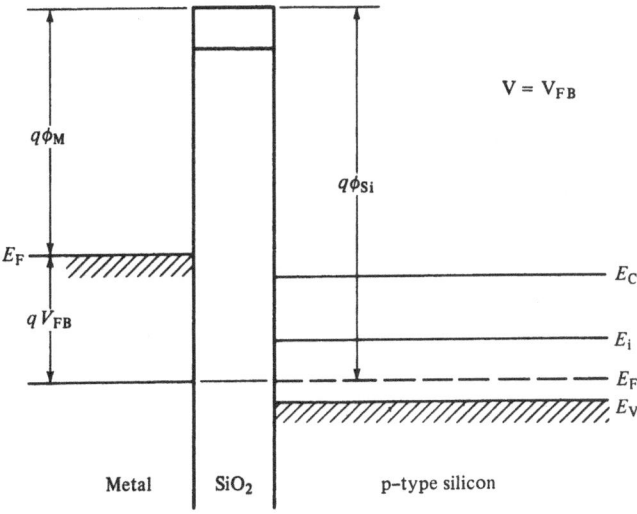

175

13.2.1
Ideal MOS diode with applied gate potential

When a potential is applied to the diode gate, the bands at the Si–SiO$_2$ interface undergo some bending depending on the applied voltage polarity and magnitude (Figure 13.4). We consider here the case of a p-type silicon only, but the results are applicable to n-type silicon as well, with appropriate changes in the direction of the band bending.

A *negative potential* applied to the gate will attract mobile holes in silicon toward the silicon surface and an accumulation layer will be formed at the Si–SiO$_2$ interface. The concentration of holes in this layer will be higher than the hole concentration away from the

Figure 13.4   Band bending for various applied potentials.

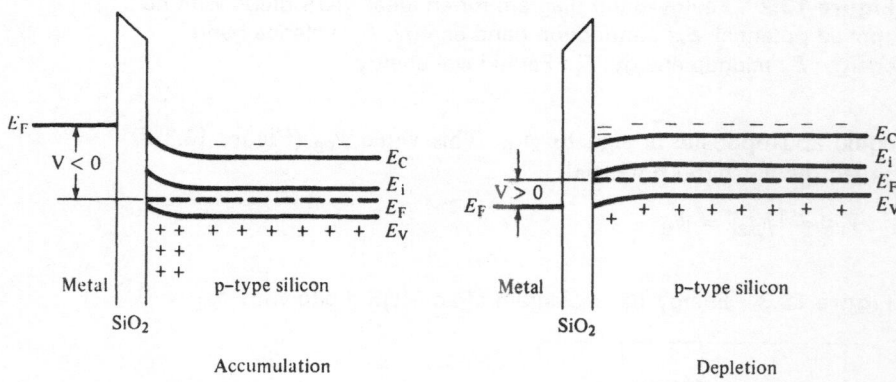

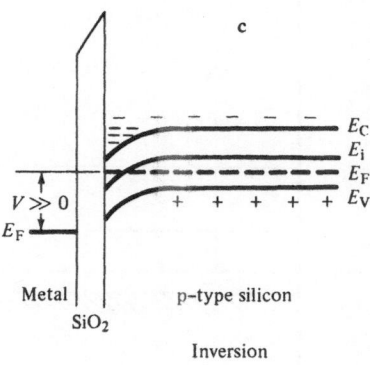

surface in the silicon bulk. A higher concentration of holes is equivalent to the formation of a P^+ region; therefore, the valence band in the vicinity of the surface will have to come closer to the Fermi level since the carrier density depends exponentially on $(E_F - E_V)$. This takes place by the upward band bending as per Figure 13.4a.

A small positive applied potential will cause the bands to bend downward since the holes at the silicon surface will be repelled (Figure 13.4b). The silicon at the surface will be *depleted* of holes, and as a result a net charge of negative acceptors will appear. The semiconductor will behave as if it were more lightly doped than it actually is, and the Fermi level will therefore be farther away from the valence band than would otherwise be the case. Thus, the bands have to bend downward. This is the depletion case.

If we apply still higher positive gate potential (Figure 13.4c), the holes at the surface will be repelled, but at the same time the electrons that are minority carriers in the p-type silicon will be attracted to the surface and, as a result, the silicon surface will behave like an n-type rather than p-type semiconductor. In this situation, which is called inversion, we assume that the bands bend appreciably downward so that the difference $E_F - E_i = q\phi_{BS}$ changes its sign Figures 13.4c and 13.5.

13.2.2
Surface potential

The potentials in the bulk and at the silicon surface are measured in respect to the *intrinsic* Fermi level E_i (Figure 13.5). In the bulk the crystal potential is designated by ϕ_{BS}; at the surface it is designated

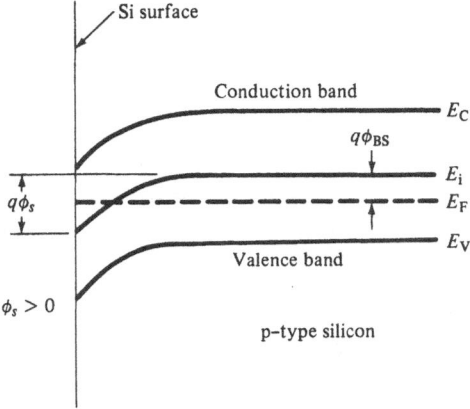

Figure 13.5 Band bending at the p-type semiconductor–SiO$_2$ interface at inversion.

177

by ϕ_s and it is called the *surface potential*. ϕ_s represents the extent of the band bending from the flat-band condition at which $\phi_s = 0$. On the basis of this definition, in combination with the relationships existing in the bulk of the silicon, the electron and hole concentrations can be expressed in function of the surface potential

$$n_{\text{surface}} = n_0 \exp\left(\frac{q\phi_s}{kT}\right) \tag{13.1}$$

$$p_{\text{surface}} = p_0 \exp\left(\frac{-q\phi_s}{kT}\right) \tag{13.2}$$

where n_0 and p_0 are the equilibrium values of electrons and holes in bulk silicon, respectively. It follows from (13.1) and (13.2) that for the p-type semiconductor, for instance, the concentration of carriers at the flat-band condition ($\phi_s = 0$), the carrier concentration at the surface will be the same as in the bulk. For $\phi_s < 0$, however, we will have more holes at the surface than in the bulk (accumulation), and for $\phi_s > 0$ the number of holes will be smaller at the surface than in the bulk of silicon (depletion). When $\phi_s > \phi_{\text{SB}}$, i.e., when the surface potential is greater than the bulk potential, the surface will change its polarity; it will undergo *inversion* so there will be more electrons than holes at the Si–SiO$_2$ interface. An inversion is considered strong when

$$\phi_s \approx 2\phi_{\text{BS}} \tag{13.3}$$

i.e., when the surface potential is twice as high as the bulk silicon potential. If we make the reference energy level E_i equal to zero, relationship (13.3) becomes

$$q\phi_s \approx 2E_{\text{F}} \tag{13.4}$$

13.2.3
Charge distribution for various gate voltages

Figure 13.6a,b,c, summarizes the charge distribution in an ideal MOS diode for the accumulation, depletion, and inversion conditittions. For the negative gate potential the surface is accumulated. The metal electrode charge $-Q_M$ and the positive induced silicon charge Q_{Si} are of equal magnitude and are separated by the oxide thickness x_o (Figure 13.6a).

When the device is positively biased the silicon surface is depleted of majority carriers (holes in our example) so that the silicon region

178

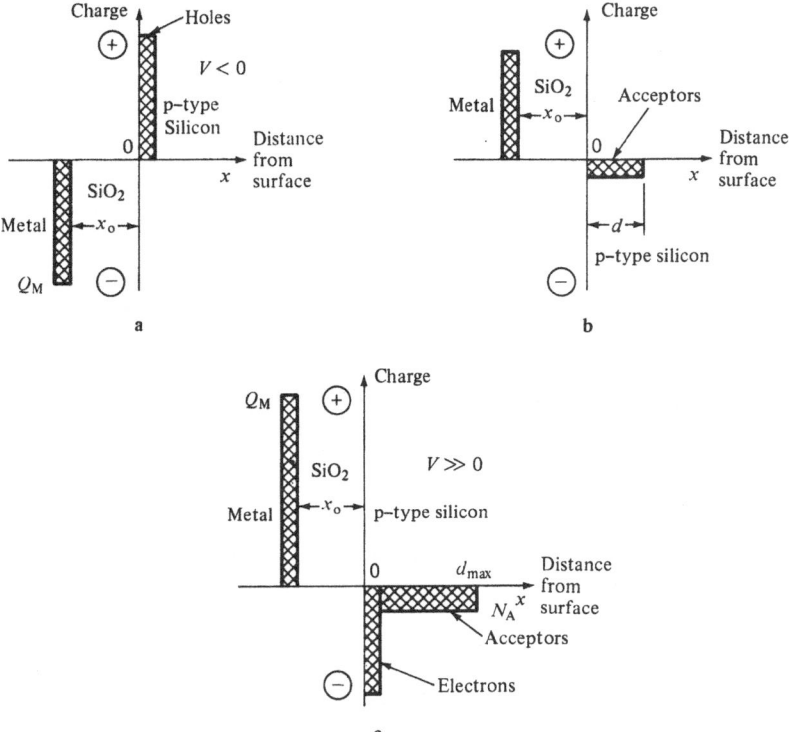

Figure 13.6 Charge distribution for an ideal MOS diode (p-type silicon). (a) $V < 0$ accumulation of holes; (b) $V > 0$ depletion of holes; (c) $V \gg 0$ inversion (accumulation of electrons at the surface).

at the surface contains a considerable number of uncompensated acceptors. The charge per unit area in the depletion region is

$$Q_{Si} = -qN_A d \qquad (13.5)$$

where d is the depletion-region width (Figure 13.6b).

For a much larger positive bias, the holes will be repelled as in the depletion case, but large numbers of electrons will be attracted also to the surface. Their concentration will increase rapidly with the positive potential. The negative charge of electrons in the inversion layer will be designated by Q_N. To the right of the inversion layer there will be a region depleted of holes (Figure 13.6c). The situation is similar to that in an n^+-p–type junction in which the n^+ region was, however, formed not by impurity doping but was instead field induced. The width of the depletion region does not vary significantly when the surface potential is increased beyond the $2\phi_F$ value since most of the induced negative charge necessary to maintain the overall charge for an increasing potential will be contained in the thin inversion layer. Therefore, when strong inversion takes place, the width of the depletion region reaches its maximum value d_{max}. The charge on the metal electrode Q_M is always equal to the total charge in the semiconductor Q_{Si}. Therefore,

$$Q_M = Q_{Si} = Q_N - qN_A d_{max} \tag{13.6}$$

where Q_M, Q_S, Q_N, and $qN_A d_{max}$ are charges per unit area.

13.2.4
Capacitance—voltage characteristics of an ideal MOS diode

When a potential V is applied to the device gate, one part of it, V_i, will appear across the insulator and the second part in the semiconductor. Therefore,

$$V = V_i + \phi_s \tag{13.7}$$

If the capacitance per unit area of the insulator is C_i, then

$$V_i = \frac{Q_{Si}}{C_i} = \frac{Q_M}{C_i} \tag{13.8}$$

Equation (13.7) suggests that the diode can be represented by an equivalent circuit consisting of the two capacitors connected in series. One capacitance, C_i, is due to the insulator and the other one, C_s, is due to the induced electric charges at the semiconductor surface. The total capacitance per unit area is, therefore,

$$C = \frac{C_i C_s}{C_i + C_s} \tag{13.9}$$

C_i and C_s are *incremental* (differential) or small-signal capacitances related to the charge by dQ/dV.

For a given insulator thickness x_o, the value of C_i is constant and represents the maximum possible capacitance of the diode. On the other hand, the semiconductor depletion-region capacitance is voltage-dependent like in a p-n junction. *In the depletion mode*, the approximate value for the differential diode capacitance [13.2] is given by

$$\frac{C}{C_i} = \left(1 + \frac{2K_o^2 \varepsilon_o V}{qN_A K_{Si} x_o^2}\right)^{-1/2} \tag{13.10}$$

where K_o is the oxide dielectric constant, $\varepsilon_o = 8.85 \times 10^{-14}\,\mathrm{F cm^{-1}}$, K_{Si} is the silicon dielectric constant, N_A is the p-type silicon doping concentration, V is the applied potential, and x_o is the oxide thickness.

While the surface is being depleted the capacitance C will fall with the square root of the applied potential. At zero potential there is no depletion region formed, and the expression (13.10) is not applicable, but it is obvious that the only capacitance left is that of the insulator, so that $C = C_i$. When a strong inversion is reached, the depletion-region thickness reaches its maximum value and the capacitance will become constant and independent of the applied potential, consequently the maximum value of the voltage that we can put into the equation is

$$V = V_i + 2\phi_{BS} \tag{13.11}$$

Figure 13.7 represents a plot of the normalized differential capacitance versus applied potential for the p-type silicon. At negative potentials there is hole accumulation and the capacitance $C = C_i$. When the magnitude of the negative potential is sufficiently reduced, a depletion region starts forming in the semiconductor. Since its capacitance is in series with C_i, the total differential capacitance starts decreasing. At some sufficiently high *positive* potential the capacitance reaches its minimum. Beyond this region there is an onset of an inversion layer consisting of a negative-charge (electron) layer, and the capacitance starts increasing again. However, the

Figure 13.7 Capacitance–voltage plot of an MOS diode. For explanation of *a* and *b* see text.

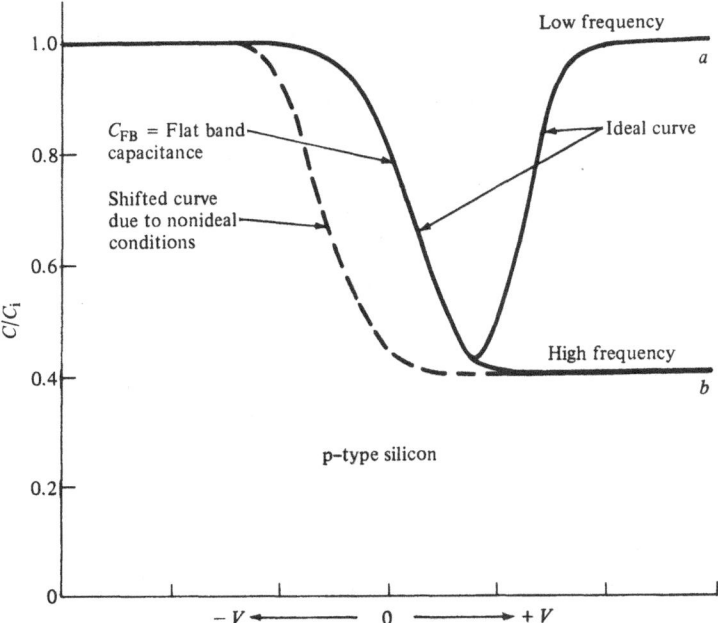

181

increase in capacitance depends on the ability of the electron concentration to follow the applied AC signal which is used to measure the differential capacitance. Only when the generation rate of minority carriers (electrons in our example) in the depletion region is fast enough compared to the AC signal frequency will the capacitance increase.[1] Usually, if frequencies higher than about 20–100 Hz are used for Si–SiO$_2$ system, the upper branch (marked a) of the $C–V$ plot (Figure 13.7) will not be obtained and only the branch marked b will be seen. The accurate theoretical $C–V$ plots for ideal MOS diodes were computer-calculated by Goetaberger of Bell Laboratories [13.9] and by Zaininger of RCA Laboratories [13.10] using *exact* formulations. The curves generated by computers include a wide range of silicon doping levels and oxide thicknesses. The reader is referred to the original memoirs since it is impossible to reproduce here, *in toto*, close to 200 plots.

13.2.5
Deviations from the ideal case

The ideal diode had no surface and/or oxide charges other than those induced in silicon by the applied potential and had equal metal–silicon and metal–SiO$_2$ work functions. In real-life devices we have to cope with less than ideal structures, and there will be almost always some work function difference ϕ_{MS} present. The ideal diode curve has a flat-band point at $V = 0$, since in the absence of an applied potential there is no band bending. For $\phi_{MS} \neq 0$ the applied potential for the flat-band situation has to be different from zero in order to compensate for the presence of ϕ_{MS}. The flat-band point will be, therefore, shifted as per Figure 13.7. The $C–V$ plot will be accordingly shifted, e.g., to the left side of the diagram.

As we will see in the next section, actual Si–SiO$_2$ systems contain many types of charges such as mobile ions, interface states, and fixed-surface charges. The effect of these additional charges is to modify and to shift the ideal $C–V$ plots. The degree of this shift determines the degree of deviation from ideal-case conditions. We will return to the $C–V$ plot displacement while discussing various Si–SiO$_2$ surface charges.

13.3
Silicon surface states and charges

The extensive studies of the Si–SiO$_2$ system showed that there exist in it several types of electric charges and states [13.1–13.19], including:

---

[1] When p–n junctions are present, there is no such limitation since the minority carriers may be injected rather than generated.

1. interface states, called also fast states;
2. fixed surface charges, also known as surface-state charges;
3. mobile impurity ions such as sodium and other alkali;
4. ionized traps due to irradiation by low- and high-energy electron beams, x-rays, and gamma rays.

The location of the various types of surface states and charges is summarized in Figure 13.8.

The electrical properties of the silicon–silicon-dioxide system are affected also by the redistribution of the silicon-doping impurities during the oxidation process and by the presence of metals and/or other insulators on the oxide surface.

13.3.1
Interface (fast) states

The periodicity of the silicon crystal is disrupted at its surface so that not all valence bonds are satisfied; as a result, a large number of states are introduced into the forbidden energy gap. Such states were studied first by Tamm and Shockley and were called initially Tamm states; later this was changed to "fast-surface states" since in many cases they can exchange charges rapidly with the substrate semi-conductor.

Figure 13.8 States and charges in nonideal Si–SiO$_2$ system.

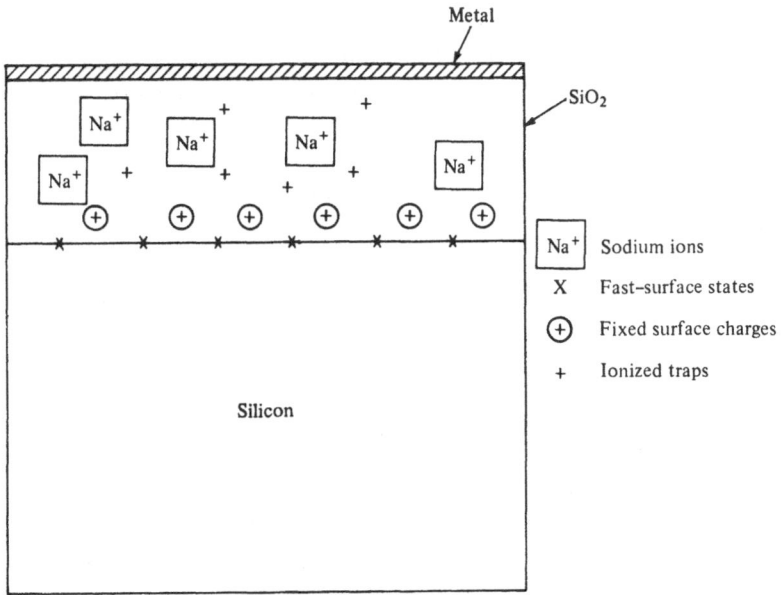

183

On the other hand, the states or charges existing in the SiO_2 or on the SiO_2 surface were referred to as slow states since their interaction with the silicon substrate was slower.

Theoretically, one can speculate that every atom at the silicon surface is going to contribute one fast state. In reality, however, the silicon surfaces are either covered by naturally existing oxide or are oxidized on purpose so that the number of states, rather than being as expected $\sim 10^{15}$ cm^{-2} is typically 10^{10}–10^{12} cm^{-2}.

Heat treatments in some gas atmospheres, e.g., H_2, lead to a very significant reduction of the fast-surface states. A state density of 10^{10} cm^{-2} may be presently realized with the existing annealing techniques.

The surface-recombination velocity, which is essentially the reciprocal of the surface lifetime, is directly proportional to the number of the existing fast-surface states and affects the magnitude of a p-n junction saturation current. Thanks to the low state density of thermally oxidized silicon surface, which reduces drastically the number of the fast states, it is possible to fabricate bipolar planar junctions with extremely low saturation (leakage) currents.

The presence of the fast states has a profound effect also on all MOS devices. Figure 13.9 illustrates the effect of the fast states on the C–V plot of an MOS diode with a p-type silicon substrate. The figure shows that the experimentally obtained C–V plot is shifted appreciably toward the negative potentials when a high density of fast-surface states exists.

There is sufficient evidence at this time that energetically the surface states lie in all the silicon energy gaps, but their distribution, before any heat treatment, has two high density regions: one close to the conduction band and one close to the valence band [13.12, 13.13]. The accurate nature of these two groupings is not yet quite clear.

Figure 13.9 Capacitance–voltage plot of an MOS capacitor (p-type silicon).

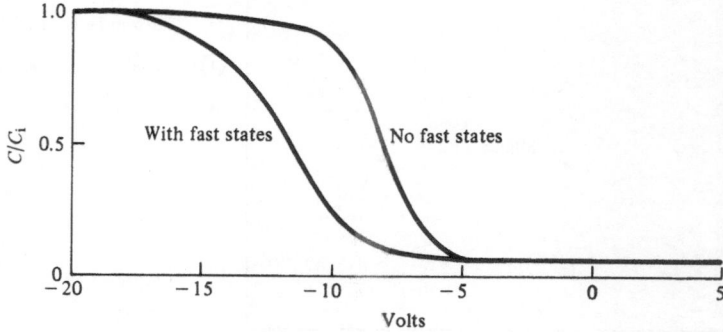

It has been found [13.14, 13.15] by using very sensitive surface-conductance techniques that a heat treatment of the Si–SiO$_2$ system in ambients such as hydrogen, forming gas, or wet nitrogen, in temperatures of 400–500°C, reduces the surface-state density to values down to about 10^{10} cm^{-2}. At the same time, the distribution of the states within the band gap becomes more uniform.[2]

It is almost certain that it is the hydrogen that reduces the surface-state density. It is theorized that the hydrogen diffuses to the Si–SiO$_2$ interface and fills the "dangling" missing bonds at the silicon surface. The effect of moisture is similar, since under typical heat treatment conditions, e.g., in the presence of aluminium metallization, the traces of water undergo decomposition with a release of hydrogen.

Oxides grown in dry oxygen exhibit higher surface-state densities than oxides grown either in wet oxygen or in steam.

13.3.2
Fixed-surface charge Q_{ss} (surface-state charge)

The surface-state charge has the following properties:

1. It is fixed and positive.
2. The energy levels of these states are located close to the edges of the silicon energy gap and possibly even outside it.
3. It depends on the crystal orientation and decreases in the order [111] > [110] > [100], approximately in the ratios of 3:2:1.
4. It is not much affected by the oxide thickness or by the type or doping of silicon crystal concentration.
5. By thinning the oxide step by step it was possible to determine that the charge is located within 200 Å from the Si–SiO$_2$ interface.
6. The charge is a strong function of the oxidation and postoxidation annealing conditions. Regardless of the previous history of the sample, the final heat treatment determines its magnitude [13.16].

The dependence of the surface-charge density on the ambient and temperature of the final heat treatment is illustrated in Figure 13.10. The charge Q_{ss} is shown after the heat treatment in dry oxygen, argon, and in nitrogen atmosphere. According to [13.16] these results indicate that the fixed-surface charges are related to the presence in the oxide of an excess of positively charged *silicon ions* left at the Si–SiO$_2$ interface in an unoxidized state. By very careful heat treatments, it is possible to reduce Q_{ss}/q to values on the order of 10^{10} cm^{-2}

---

[2] According to the Hall–Shockley–Read recombination-generation theory only the recombination centers with energy levels close to the energy gap center have a pronounced effect on the recombination process.

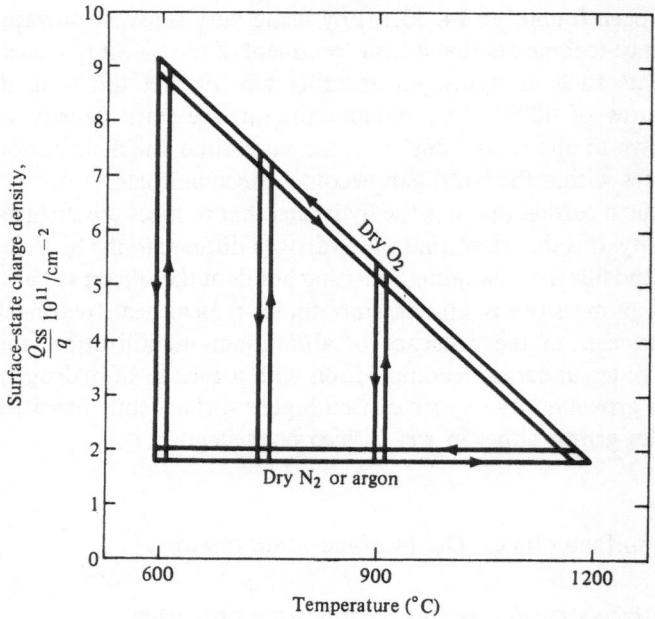

Figure 13.10 Reversibility of heat-treatment effects on the surface-state charge density Q_{ss}. (After Deal *et al.* [13.16].)

The presence of the surface charge leads to a parallel translation (Figure 13.11) of the capacitance–voltage plot of an MOS capacitor toward the negative potentials, confirming the positive nature of this charge. There is some disagreement about the effect of the oxide thickness on Q_{ss}. According to [13.16], the charge is unaffected by SiO_2 thickness. On the other hand, according to [13.17, 13.18], the thicker the oxide, the smaller the surface charge. For instance, an

Figure 13.11 Capacitance–voltage plot translation under the effect of the fixed-charge Q_{ss} or shifted ionic charge.

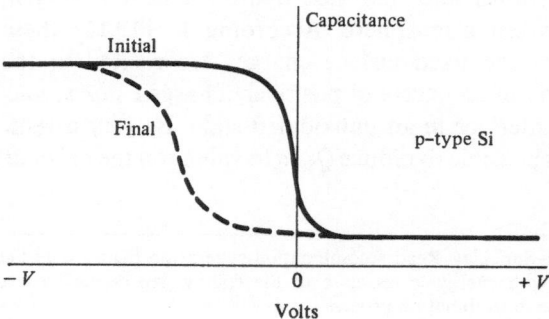

186

oxide 1000 Å thick may have a surface charge $Q_{ss}/q = 10 \times 10^{11}$ cm^{-2}, whereas a 4000-Å thick oxide grown the same way at 1100°C in O$_2$ and fast cooled will exhibit a charge density of 6×10^{11} cm^{-2}.

13.3.3
Mobile ions

By studying the early MOS devices it was observed that under the influence of elevated temperatures and applied gate bias the device characteristics, such as capacitance–voltage curve of an MOS diode, may drift appreciably from its initial value. This drift effect was shown to be due to the rearrangement of an ionic space-charge distribution within the oxide. The ions were shown to be usually *positive sodium*

Figure 13.12 Drift of positive ions in the SiO$_2$ of an MOS capacitor.

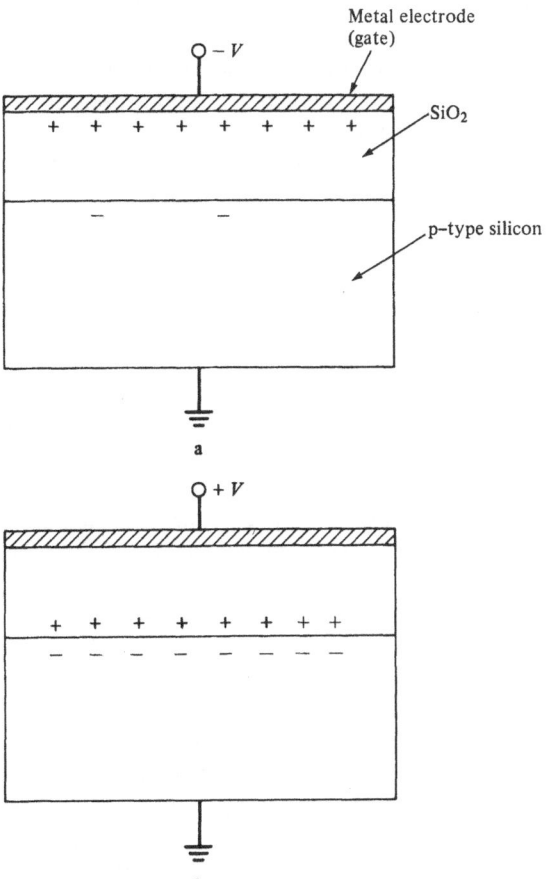

187

ions (Na^+), *other alkali ions*, or protons. The initial charge distribution at t = 0 may be confined entirely to the region at the metal–oxide interface (Figure 13.12a). When a positive potential is applied to the electrode at some elevated temperature, say 200°C, the ions acquire enough mobility and move toward the $Si-SiO_2$ interface under the influence of the field, so that after some time, dependent on the temperature and voltage, most of the existing charge is shifted to the $Si-SiO_2$ interface (Figure 13.12b). This rearrangement of the ionic charge distribution has a profound effect on device characteristics despite the fact that the total charge density per unit area has not changed. When all charge is located close to the metal–oxide interface, it is far removed from the surface by the oxide thickness. When the charge is drifted to the $Si-SiO_2$ interface, its distance to the Si surface is at its minimum and the order of few angstroms. In the first case, the image charge in the silicon is negligible; in the second, the image charge obviously becomes very large. In a case of an MOS capacitor, this will result in a drastic change in the capacitance at a given voltage (Figure 13.11), and a translation of the capacitance–voltage characteristic toward more negative potentials. If the polarity of the electrode is changed again to the negative potential, it is possible to revert, at these elevated temperatures, to the initial state. If, however, the alkali ions are frozen in the drifted position because of the lowering of temperature, the change in the characteristics becomes permanent.

13.3.3.1 Reduction of the mobile ion concentration. The mobile sodium ions are due to the contamination present on the silicon surface and are also introduced during the device processing.

If the device is oxidized under most stringent conditions of cleanliness it is possible to reduce the number of ions to a value of about 10^{10} per cm^2 and to make the device drift free [13.19]. In less than ideal conditions this number may be closer to 10^{11} and usually exceeds the number of all other positive space-charges which were reduced by appropriate heat treatments. In practical devices then, it is the mobile charge that will affect the silicon surfaces the most. A processing which will reduce the number of positive ions in the SiO_2 layer will have to include purification of all gases used, thorough cleaning of silicon surfaces, use of chemicals and containers sodium-free, proper handling of water in steam oxidation, etc.

It has been found [13.20] that if a phosphosilicate glass (SiO_2 rich in P_2O_5) is present on the outside of a silicon dioxide layer, the ionic charge in the oxide becomes significantly reduced. This is due to a much higher solubility of sodium in the phosphosilicate glass than in the silicon dioside, and as a consequence sodium will segregate into the glass; its concentration in the glass may become three orders of magnitude higher then in the underlying oxide. There is one drawback, however, of the phosphosilicate glass in that it exhibits dipole-type polarization which has an effect on the surface potential and, therefore, on the surface properties. The surface potential change may take place in either direction depending on the polarity of the gate voltage applied during the polarization process [13.21]. It was reported recently that the polarization effects may cause surface instability under aging conditions involving prolonged application of a gate potential.

Another glass of importance which has a pronounced effect on the S–SiO$_2$ system is the lead-silicate glass [13.22, 13.23]. When a potential is applied to the glass–SiO$_2$ system, positive lead ions originally uniformly distributed over the entire glass thickness will redistribute themselves due to a comparatively large lead ion mobility. The less mobile negative ions, on the other hand, will remain uniformly distributed. An uncompensated negative charge will, therefore, appear in the glass and will induce an image charge at the silicon surface, shifting the surface potential. The shift may be as in the case of the phosphate-glass in either direction depending on the polarity of the applied polarizing voltage.

The most valuable of other dielectrics used for surface protection against ionic contamination is the silicon nitride (Si$_3$N$_4$). It should be used preferably in a protective film on the top of SiO$_2$ [13.24]. Silicon nitride is practically impervious to the alkali atoms and, therefore, if deposited on the very clean, ion-free SiO$_2$, it will maintain the SiO$_2$ layer ion-free. Si$_3$N$_4$ has, however, much higher conductivity than SiO$_2$. Some hysteresis problems with Si$_3$N$_4$ may arise due to large trap density which fill and empty by communicating with the silicon substrate by a tunneling mechanism, if the nitride is deposited directly on the silicon substrate or on extremely thin SiO$_2$ films [13.25]. When hysteresis effects are present, the $C–V$ plot obtained by changing the potential from negative to positive values is different from that obtained by voltage changing in the opposite direction. Another dielectric in use presently is the aluminium oxide (Al$_2$O$_3$). Aluminium oxide films provide a barrier to sodium ions [13.26] and are also more radiation resistant than thermal oxide films [13.27]. Al$_2$O$_3$ films make the silicon surface more p type whereas SiO$_2$, as we saw, due to positive charges, make the silicon more n type. The negative charge in Al–Al$_2$O$_3$–SiO$_2$–Si double layers was found to be located at the Al$_2$O$_3$–SiO$_2$ interface. This charge is independent of the Al$_2$O$_3$ thickness, but is inversely dependent on the SiO$_2$ thickness. The traps are attributed to the presence of oxygen vacancies [13.28]. If the Al$_2$O$_3$ is used directly on the silicon surface, hysteresis effects in $C–V$ characteristics will

189

be observed as with Si_3N_4, although to a lesser degree. Therefore, like in the case of Si_3N_4, the most efficient way to use the aluminium is by depositing it as a second protective layer or on top of SiO_2.

13.3.4
Radiation-induced surface states

When an $Si–SiO_2$ system is exposed to an ionizing radiation, positive oxide charges are introduced into the dioxide and more interface states are created [13.29]. The number of surface states introduced by irradiation is dependent on the density of initially existing surface states; the number of surface states induced by radiation increases with the number of states originally present. Steam oxides will then result in much smaller number of states than so-called dry oxides.

The induced charge is a strong function of the MOS diode gate potential V_i with the negative bias present. The oxide charge is almost one magnitude lower than that generated with applied positive potentials. When the radiation dose is increased more and more a saturation point is eventually reached (Figure 13.13). Heat treatments at temperatures between 150°C and 400°C may remove most of the radiation effects.

Figure 13.13 Radiation-induced charge density versus radiation dose. (After Zaininger and Holmes-Siedle [13.29].)

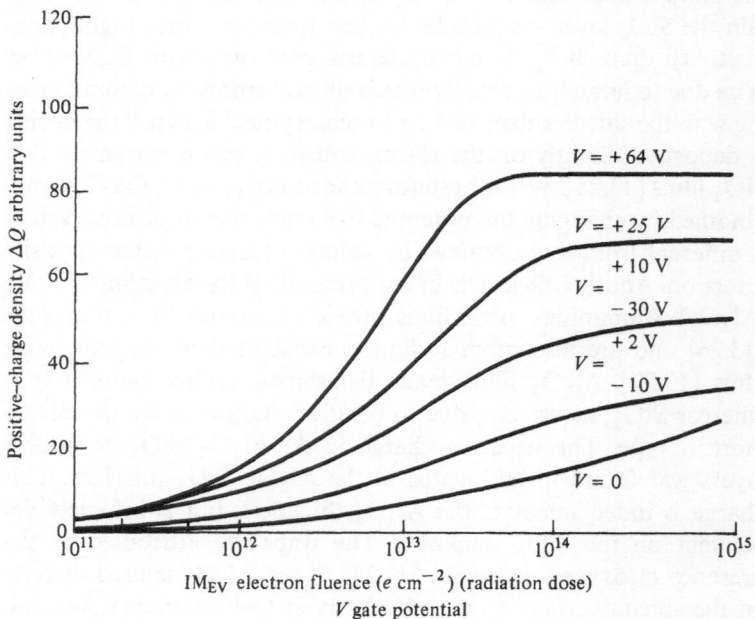

IM$_{EV}$ electron fluence (e cm^{-2}) (radiation dose)

V gate potential

A simplified model proposed by [13.30] explains the build-up of the charge in the SiO_2 by the interaction of electrons and/or holes with traps present in the oxide. When the Si–SiO_2 system is exposed to an ionizing radiation such as low- or high-energy electrons, x-rays, or gamma rays, electron–hole pairs will be created. In the absence of any electric field in the oxide, the electrons and holes will recombine and there will be no net electric charge left in the oxide. The situation will be entirely different, however, in the presence of an electric field in the SiO_2, since the field will separate the positive from the negative charges. Consequently, if in an MOS capacitor the metal electrode is *positively biased*, the electrons will move toward that electrode and will be swept out of SiO_2 if there is no injection from silicon, whereas the holes will move to the Si–SiO_2 interface and will be trapped there, building up a positive space charge in the SiO_2 at the Si–SiO_2 interface. This positive charge will induce a negative charge in the underlying silicon crystal. If the bias on the metal gate is *negative*, the holes will move toward the metal–SiO_2 interface and may be trapped there. The traps in the silicon oxide are thought to be due to the presence of Si-O groups [13.29]. If electrons from the silicon are injected into the oxide, the positive charge will be neutralized to some extent; therefore uv illumination of the Si–SiO_2 system and photoinjection into the oxide will remove a large part of the induced charge.

13.3.5
Impurity redistribution at the Si–SiO_2 interface

The oxidation of silicon proceeds by the motion of the oxidizing species (e.g., O_2) through the SiO_2 layer and not by the motion of silicon to the outer surface of silicon. As the silicon surface is converted into an oxide, the SiO_2–Si interface moves deeper and deeper into the silicon. If the silicon is doped with some impurity, this impurity in the region converted into SiO_2 will have to move somewhere. Where it will go will depend on such factors as the given impurity segregation coefficient for the Si–SiO_2 system, the relative magnitudes of impurity diffusivities in SiO_2 and Si, and on the oxidation rate of silicon. Experimentation showed [13.31, 13.32] that, in general, p-type impurities such as boron and aluminium are segregated into the SiO_2 during the thermal oxidation. As a result, the silicon surface becomes depleted of p-type impurities and has a lower impurity concentration than the silicon bulk. An opposite situation exists with n-type impurities so that they accumulate at the interface and the silicon surface becomes more heavily doped than the bulk. These effects are very much accentuated at lower oxidation temperatures. Also, oxidizing with wet oxygen results in greater segregation of impurities than oxidizing with dry oxygen at the same oxidizing temperature. This is so because wet oxidation is much faster than dry.

In the case of boron, it is not uncommon to observe an accumulation of boron atoms in the oxide with a concentration three times higher than the silicon surface concentration. In the absence of H_2 in the oxidizing atmosphere the depletion of the silicon surface, due to increased boron mobility, may be still considerably greater.

In many instances it is necessary to reduce the lifetime of silicon devices to obtain desired device characteristics, such as fast current switching. One of the most common methods is to dope silicon with elemental gold. Gold has two energy levels in the silicon energy gap; a donor level 0.35 eV above the valence band edge and an acceptor level 0.5 eV below the conduction band edge. Gold, therefore, always acts in silicon as a compensating impurity; in n type silicon it behaves as an acceptor and in p type silicon as a donor. It has been found [13.33] that diffusing gold into the back of silicon slice introduces a

Figure 13.14 MOS *C–V* plot for various metallic electrodes. (After Deal *et al.* [13.35].)

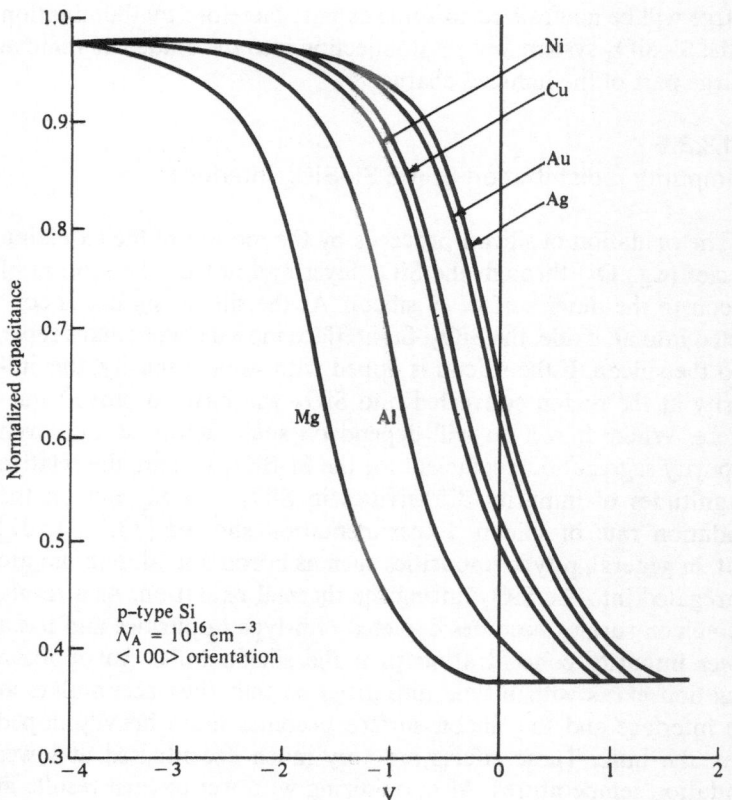

negative charge at the Si–SiO$_2$ interface which is of the same order of magnitude as the usually observed positive surface charge and causes the MOS C–V curve to shift toward the positive potentials. It has been suggested [13.34] that the gold piles up at the Si–SiO$_2$ interface, reaching concentrations far in excess of the solid solubility, and always acts as an acceptor at the silicon surface regardless of the type of the bulk silicon doping.

13.3.6
Metal work-function effects

The work-function difference, ϕ_{MS}, between the metal and silicon causes energy-band bending that depends on the position of the silicon Fermi level, i.e., on the impurity type and concentration. The work-function difference causes a displacement of the C–V plot

Figure 13.15 ϕ_{MS} versus silicon doping. (After Deal *et al.* [13.35].)

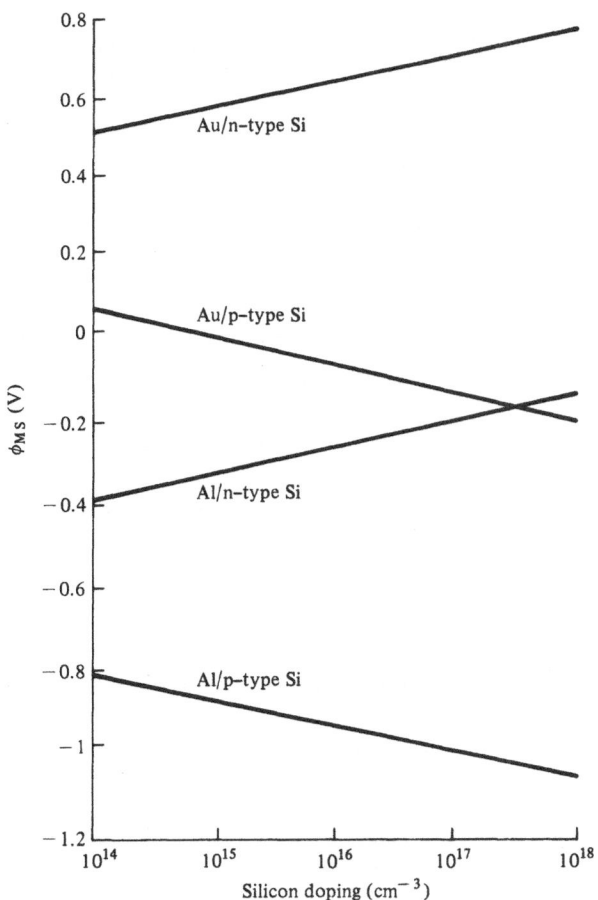

193

proportional to the magnitude of ϕ_{MS}. Figure 13.14 [13.35] demonstrates the effect of various metals on the capacitance–voltage characteristics of an MOS diode. Figure 13.15 gives the value of the work-function difference, ϕ_{MS}, in function of the silicon impurity concentration for aluminium and gold deposited on the SiO_2 surface [13.35].

In some cases, it may be of interest to use doped or undoped poly-crystalline silicon in place of a metal gate electrode. In the case of a silicon rather than metal gate, the ϕ_{MS} is designated by ϕ_{SS} and it is the difference between the work functions of the silicon on both sides of the SiO_2 layer. ϕ_{SS} is, therefore, equal to the difference between the Fermi levels of the silicon gate and the bulk silicon

$$\phi_{SS} = \phi_{FG} - \phi_{BS}$$

where ϕ_{FG} and ϕ_{BS} are the Fermi potentials of the silicon gate and bulk silicon, respectively [13.36]. Two extreme values of the ϕ_{SS} are obtained when the silicon gate is very heavily p-type doped and the substrate is heavily n-type doped so that $\phi_{SS} = 0.55 - (-0.55) = 1.1$ V. When the silicon gate is heavily n-type doped and the silicon substrate is heavily p-type doped, $\phi_{SS} = -0.55 - 0.55 = -1.1$ V.

References

13.1 J. T. Wallmark and H. Johnson, Eds. *Field Effect Transistors.* Englewood Cliffs, N.J.: Prentice-Hall, 1966.

13.2 A. S. Grove. *Physics and Technology of Semiconductor Devices.* New York: John Wiley, 1967.

13.3 S. M. Sze. *Physics of Semiconductor Devices.* New York: Wiley, 1969.

13.4 H. F. Wolf. Semiconductor surface depletion and inversion. In *Semiconductors.* New York: Wiley, 1971.

13.5 J. M. Feldman. *The Physics and Circuit Properties of Transistors.* New York: John Wiley, 1973.

13.6 A. Many, Y. Goldstein, and N. B. Grover. *Semiconductor Surfaces.* New York: John Wiley, 1965.

13.7 J. Olmstead, RCA, Somerville, N.J., private communication, 1972.

13.8 A. M. Goodman. Photoemission of electrons from silicon and gold into silicon dioxide. *Phys. Rev. 144*: 588, 1966.

13.9 A. Goetzberger. C–V plot ideal MOS curves in silicon. *Bell System Tech. J., 45*: 1097, 1966.

13.10 C. Zaininger, RCA Laboratories, private communication.

13.11 D. R. Lamb. Some electric properties of the silicon–silicon dioxide system. *Thin Solid Films, 5*: 209–246, 1970.

13.12 P. V. Gray and D. M. Brown. Density of Si-SiO_2 interface states. *Appl. Phys. Letts., 8*: 31, 1966.

13.13 A. Goetzberger, V. Herne, and H. E. Nicollian. Surface states from charges in the oxide coating. *Appl. Phys. Letts.*, *12*: 95–97, 1968.

13.14 A. G. Reversz, K. H. Zaininger, and R. J. Evans. Interface states and interface disorder in the Si-dioxide system. *Phys. Chem. Solids, 28*: 197–204, 1967.

13.15 E. H. Nicollian and A. Goetzberger. MOS conductance technique for measuring surface state parameters. *Appl. Phys. Letts.*, *7*: 216–219, 1965.

13.16 B. E. Deal, M. Sklar, A. S. Grove, and E. M. Snow. Characteristics of the surface-state charge (Q_{ss}) of thermally oxidized silicon. *J. Electro Chem. Soc., 114*: 266, 1967.

13.17 D. R. Lamb and F. R. Badcock. The effect of ambient temperature and cooling rate on the surface charge of the silicon/dioxide interface. *Int. J. Electron, 24*: 11–16, 1968.

13.18 A. G. Reversz and K. M. Zaininger. The influence of oxidation and heat treatment on the Si surface potential in the Si-SiO$_2$ system. *Trans. IEEE, EDB 13*: 246–255, 1966.

13.19 A. S. Grove, B. E. Deal, E. H. Snow, and C. T. Sah. Investigation of thermally oxidized silicon surfaces using MOS structures. *Solid State Electron., 8*: 145, 1965.

13.20 D. S. Kerr, J. S. Logan, P. J. Burkhardt, and W. A. Pliskin. Stabilization of SiO$_2$ passivation layers with P$_2$O$_5$. *IBM J., 8*: 376, 1964.

13.21 E. H. Snow and B. E. Deal. Polarization phenomena and other properties of phosphosilicate glass films on silicon. *J. Electro Chem. Soc., 113*: 263, 1966.

13.22 E. H. Snow and M. E. Dumesnile. Space charge polarization in glass films. *J. Appl. Phys., 37*: 123, 1966.

13.23 D. E. Carlson, K. W. Hang, and G. F. Stockdale. Ion depletion of glass at a blocking anode. RCA (Princeton), private communication.

13.24 R. R. Verderber, G. A. Gruber, J. W. Ostroski, and J. E. Johnson. SiO$_2$/Si$_3$N$_4$ passivation of high-power rectifiers. *IEEE Trans. Electron Devices, ED-17* (9): 797–799, 1970.

13.25 J. J. Wallmark and J. H. Scott. Switching and storage characteristics of MIS memory transistors. *RCA Rev. 30*: 335, 1969.

13.26 S. K. Tung and R. C. Caffrey. Pyrolific deposition and properties of aluminium oxide films. *J. Electro Chem. Soc., Solid State Science and Tech., 114*: 275C, 1962.

13.27 A. Waxman and K. H. Zaininger. Al$_2$O$_3$ silicon insulated gate field effect transistors. *Appl. Phys. Letts., 12*: 109–110, 1968.

13.28 J. A. Aboaf, D. R. Kerr, and E. Bassons. Charge in SiO$_2$–Al$_2$O$_3$ double layers on silicon. *J. Electro Chem. Soc., Solid State Sci. Tech., 120* (8): 1103–1106, 1973.

13.29 K. H. Zaininger and A. G. Holmes-Siedle. A survey of radiation effects in metal-insulator-semiconductor devices. *RCA Rev., 28*: 208, 1967.

13.30 A. S. Grove and E. H. Snow. A model for radiation damage in metal-oxide-semiconductor structures. *Proc. IEEE 54*: 89, 1966.

13.31 A. S. Grove, O. Leistico, Jr., and C. T. Sah. Redistribution of acceptor and donor impurities during thermal oxidation of silicon. *J. Appl. Phys.*, *35*: 2695–2701, 1964.

13.32 B. E. Deal, A. S. Grove, E. H. Snow, and C. T. Sah. Observation of impurity redistribution during thermal oxidation of silicon using the MOS structure. *J. Electro Chem. Soc.*, *122*: 308, 1965.

13.33 S. F. Cagnina and E. H. Snow. Properties of gold doped MOS structures. *J. Electro Chem. Soc.*, *114*: 1165–1173, 1967.

13.34 J. W. Adamic and J. E. McNamara. Paper presented at the New York meeting of the Electro-Chemical Society, Sept. 29–Oct. 31, 1963.

13.35 B. E. Deal, E. H. Snow, and G. A. Mead. Barrier energies in metal–silicon-dioxide–silicon structures. *J. Phys. Chem. Solids*, *27*: 1873, 1966.

13.36 F. Faggin and T. Klein. Silicon gate technology. *Solid State Electron.*, *13*: 1125–1144, 1970.

14

Avalanche breakdown enhancement by mesa contouring

Summary

In a thyristor there are two junctions that have to support high breakdown voltages. One supports the maximum forward-blocking potential and the other supports the reverse-blocking potential. For an idealized case in which the device shows perfect symmetry and has an almost perfect cathode-emitter short, the two blocking potentials are equal. In real-life devices this is rather an exceptional situation because the avalanche breakdown is limited by surface rather than by bulk breakdown and the two junctions have different edge contours. The reverse-blocking junction is usually beveled with a positive angle, whereas the forward-blocking junction is beveled, for practical reasons, with a negative angle. It is of interest to fully utilize the material's properties and to obtain full breakdown symmetry, with the bulk breakdown as the final limit.

The results reported recently by a few authors demonstrate by numerical calculations and by experimentation that a negatively beveled diffused junction, with a small bevel angle minimizing the surface field, develops a much higher field underneath and in the vicinity of the surface on the higher-doped side. Therefore, although the surface breakdown will not occur for small-angle negative bevels, silicon will still break down below the ultimately possible bulk breakdown due to the presence of this high field. The positively beveled junction does not exhibit this kind of limitation and presents a more desirable approach for mesa contouring.

It has been shown also that the smaller the negative angle, the smaller the surface field, and there is no optimum angle of 6 degrees as was thought until recently.

The normally existing positive surface charges increase the absolute maximum field underneath the surface and, therefore, lower the breakdown voltage. From the practical (mechanical and economical) point of view, the use of very small negative angles is highly undesirable. Some novel approaches avoiding the shallow-angle pitfalls were suggested recently and are reviewed in the last section of this chapter.

197

14.1
Introduction

There are two junctions in a thyristor (Figure 14.1) that support high breakdown voltages; junction J_1 supporting the maximum reverse-blocking potential and junction J_2 supporting the maximum forward-blocking potential. In Chapter 3 we discussed the maximum possible limits of the two blocking potentials, and it was shown that under idealized conditions the two blocking potentials are equal. When the device has a shorted cathode emitter, the thyristor behaves in the idealized case as a p-n-p symmetrical transistor, and the breakdown is that of an open-base transistor. If the device has, in addition, a shorted anode emitter, then the device maximum forward-blocking capability is equivalent to the breakdown of the center junction J_2. In real devices the equality of the two breakdowns is rather rare because the breakdown either occurs on the device surface or there is a lack of symmetry in the device geometry. In order to prevent the surface breakdown, it has become customary to bevel the device surface, as shown in Figure 14.1. Since the beveling angles are of opposite sign and magnitudes (Figure 14.2), the breakdowns normally obtained at J_2 are smaller than those obtained at J_1.

Davies and Gentry [14.1] have analyzed the field distribution along the beveled p-n junction surface and concluded that the maximum field tangential to the surface can be reduced by introducing either positive or very small beveling negative angles. According to [14.1] an angle of -6 degrees is an optimum angle for the maximum reduction of the surface field of a negatively beveled junction.

More recent work of Cornu [14.2, 14.3] and Bakowski and Lundström [14.4] shows that an optimum angle does not exist and the only limit to the angle value is determined by the mechanical stability and device cost. Further, it was found that in the case of a

Figure 14.1 Section through an SCR. Junction J_1 positively beveled; junction J_2 negatively beveled.

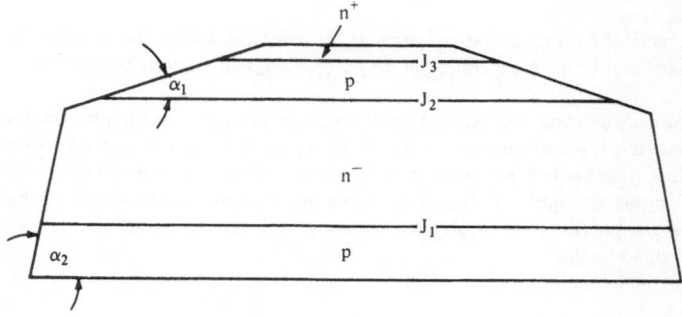

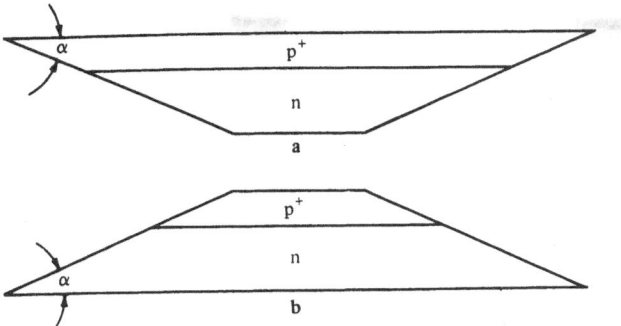

Figure 14.2 Bevel polarity convention. (a) Positive bevel: the highly doped region has a *larger* area than the low-doped region. (b) Negative bevel: the highly doped region has a *smaller* area than the low-doped region.

negatively beveled junction there exists an electric field just underneath the device surface whose magnitude is larger than the magnitude of the surface field and the bulk field. As a result, the device may not break at the surface for sufficiently shallow angles, but will break still underneath the surface where the field reaches its maximum value.

Of great importance to the device designer is the effect of the electric charges accumulated on the device surface and the influence of the dielectric properties of the coating used for the device passivation. The work of Cornu and of Bakowski and Lundstrom takes into account these phenomena.

14.2
Positive and negative bevel angles

The convention used for the polarity of the bevel angles is such that a positive bevel (Figure 14.2a) is obtained when the p-n junction has a decreasing area going from the more to the less heavily doped side. The negatively beveled junction, on the other hand, has an increasing area going in the same direction (Figure 14.2b).

14.3
Theoretical approach to the field computation
in the depletion region

The acceptor impurity concentration profile, which in most devices is either gaussian or complementary error function, was approximated [14.3] by a simple exponential expression

$$N_A(z) = N_D \ \exp\left(\frac{-z}{\lambda}\right) \qquad (14.1)$$

199

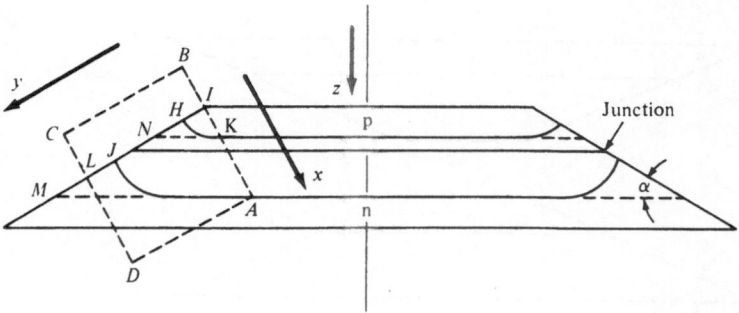

Figure 14.3 p-n junction model with negative bevel used in calculations of Cornu [14.3].

where z is the distance from the junction (Figure 14.3), $N_A(z)$ is the acceptor density at point z, λ is a space constant, and N_D is the donor concentration of the uniformly doped n base. The net impurity density at any point z can, therefore, be expressed by

$$N(z) = N_D\left[\exp\left(\frac{-z}{\lambda}\right) - 1\right] \tag{14.2}$$

In [14.4] a two-step acceptor diffusion with error function complement profile was used. In the first step the diffusion length of the impurities was L_1 and in the second step L_2. The diffusion length is given by the expression

$$L = 2(Dt)^{1/2} \tag{14.3}$$

where D is the given impurity-diffusion constant and t is the diffusion time. The purpose of the first diffusion, with low surface concentration, was to achieve very high bulk breakdown potentials; the second shallower diffusion, but with much higher concentration, was needed to avoid punching through.

When the p-n junction is reverse-biased, normally a very small saturation current flows and the potential drop outside of the depletion region can be neglected to all practical purposes. Since, in addition, outside the depletion region all charges are fully compensated, the net electric charge is zero and, therefore, the electric field intensity at the boundaries of and beyond the depletion region is zero.

In the depletion region, on the other hand, charges are separated and uncompensated by the mobile charges which were removed. The fixed-charge distribution in the depletion region is fully determined from the knowledge of the position of its boundaries and the doping profile used. Once we know the charge distribution, we can determine the field intensity at each desired point by solving the

Poisson equation which ties the charge to the electric field and potential.

Most actual devices have some degree of symmetry; therefore, a two-dimensional Poisson equation is quite satisfactory:

$$\frac{d^2V}{dx^2} + \frac{d^2V}{dy^2} = -\frac{\rho(x, y)}{K\varepsilon_0} \tag{14.4}$$

where V is positive dependent potential, x and y are the coordinates (Figure 14.3), K is the dielectric constant of silicon, and ρ is the electric charge density.

In the device dielectric coating (if there are no surface charges) $\rho = 0$ and Equation (14.4) reduces to the Laplace equation.

The numerical method used in [14.3 and 14.4] is iterative, and the field and potential are computed point by point starting by an arbitrary assumption of the position of the depletion-region boundaries. A correct guess will result in a field equal to zero at the boundary. An incorrect guess will require adding or subtracting positive or negative charges in order to bring this field to zero. After the correction is done, new depletion-region boundaries are obtained, the potential distribution is computed, and a new correction is introduced if the field is still not equal to zero at the new boundaries —and so on until the desired result, $E = 0$, is obtained at each computational point.

Figure 14.3 represents a negatively beveled junction. For the purpose of computation, a rectangle $ABCD$ was chosen encompassing the interior of the device, the silicon–dielectric interface, and the dielectric coating. Line AK is drawn in the bulk far away from the surface; therefore, along this line the Poisson equation can be solved in one dimension only. The applied potential V_A appears across the depletion region since we have assumed no voltage drop in the neutral regions. Line CB is chosen far enough from the silicon–dielectric interface; therefore, the effect of the assumptions made about the field distribution along lines IB, BC, and LC is negligible and, therefore, can be quite arbitrary. In [14.3], for example, the field component normal to the above lines was assumed to be zero.

Inside the depletion region $AKHJ$ the field at each selected point was calculated by putting the net charge density due to the impurities

$$\rho = q[N_D(x, y) - N_A(x, y)] \tag{14.5}$$

in Equation (14.4).

Inside the dielectric (region $LCBI$) the net charge, as mentioned before, is

$$\rho(x, y) = 0$$

The Poisson equation in the semiconductor and the Laplace equation in the dielectric were solved by finite difference approximation assuming the continuity of the tangential field at the silicon–dielectric interface and assuming the continuity of the flux normal to the interface. When there are no surface charges the continuity of the flux gives

$$K_1 E_1 = K_2 E_2 \tag{14.6}$$

where K_1 and K_2 are dielectric constants of the silicon and of the dielectric, respectively; and E_1 and E_2 are the electric fields in the silicon and in the dielectric, respectively, and normal to their interface. When a surface charge σ is present the continuity equation becomes

$$K_1 E_1 = K_2 E_2 + \sigma \tag{14.7}$$

14.4
Negative bevel angle

A typical form of a space charge (depletion) region in the vicinity of a negatively beveled surface is shown in Figure 14.4. The shape of the depletion-region boundaries is the result of the assumption that the electric field is zero at these boundaries and of the assumed net impurity profile [Equation (14.1)].

Figure 14.5 represents the field for the space-charge distribution of Figure 14.4. The field distribution is shown in function of the distance from the surface for three beveling angles [14.3].

The *point of the maximum field occurs always in the p-doped region*. The field deep in the bulk had a value of 2.08×10^5 V/cm. Figure 14.5 shows that for all three beveling angles the maximum field at the

Figure 14.4 Depletion region in the vicinity of the surface beveled with a negative angle $\alpha = -2.85°$ (not to scale). $V_A = 2000$ V; $N_D = 10^{14}$ cm^{-3}; $\lambda = 10$ μm. (After Cornu [14.3].)

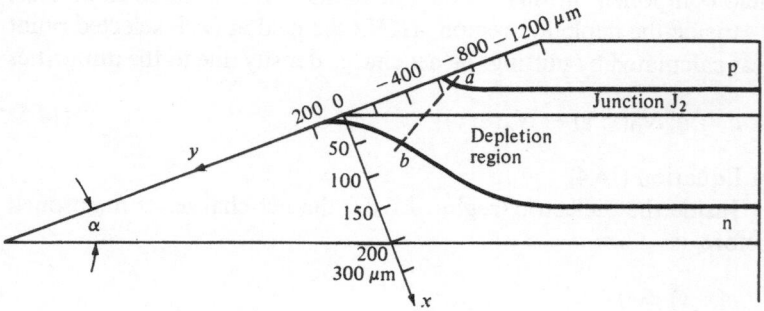

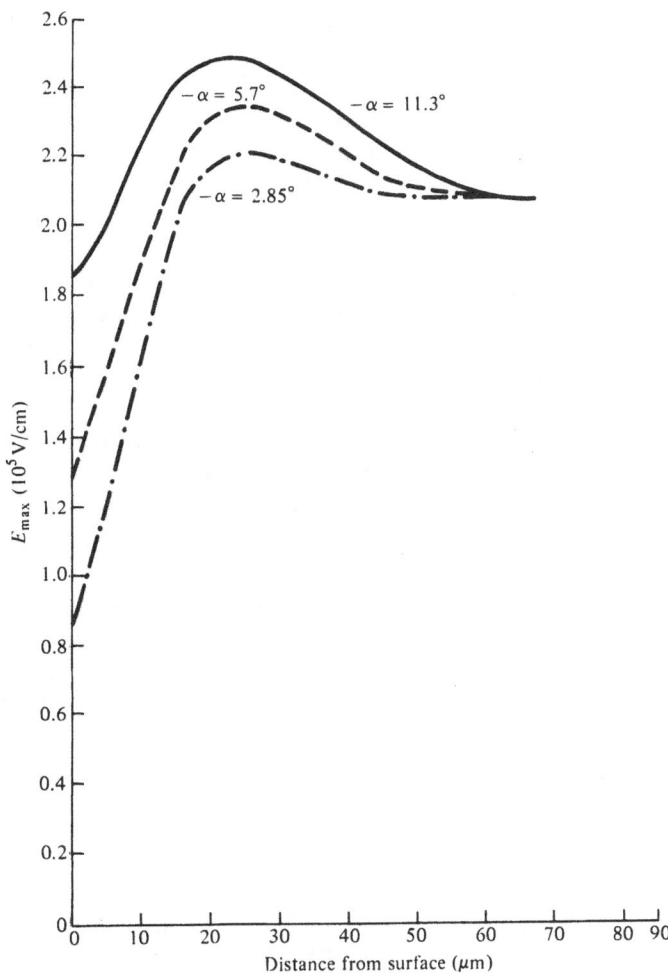

Figure 14.5 Maximum field as a function of the distance from the surface for a 2000-V structure with three different negative bevel angles. $\lambda = 10$ μm; $N_D = 10^{14}$ cm^{-2}. (After Cornu [14.3].)

surface ($x = 0$) was smaller than the field in the bulk. We can also observe that the fields decrease with decreasing magnitude of the angle and that there is no optimum at -6 degrees as predicted by an earlier theory [14.1]. Far away from the surface the field becomes constant and is independent of the distance. In this region the one-dimensional Poisson equation is applicable.

An absolute maximum of the field appears at about 25 μm from the surface for all angles studied. This absolute maximum is much higher than the tangential field at the surface and also appreciably higher than the field in the bulk.

203

Referring now again to Figure 14.4, we should note that the maximum field is occurring where the equipotential lines of the depletion-region boundaries come the closest together (line *ab*). Along this line the n region contracted already much more than the p region expanded. Therefore, the field in the n region increased appreciably since the applied potential V_A remained unchanged.

Figure 14.5 reveals a situation which has not been realized before —namely, that the breakdown may occur in an unexpected point beneath the surface. The bulk breakdown will not be achieved and breakdown will occur at about 25 µm beneath the beveled surface. Since, however, the breakdown occurs in the silicon rather than at the surface, the device will not be damaged.

For a change of the bevel angle from -2.85 to -5.7 degrees Figure 14.5, the maximum surface field changes by about 50 percent. The absolute maximum field below the surface, on the other hand, varies only by about 7 percent. An *experimental verification* shows that the change in the breakdown voltage is about 10 percent [14.3]. This confirms that the breakdown depends on the internal field. If the breakdown depended on the maximum surface field, the expected change in the breakdown voltage would be much greater since the breakdown voltage for a step junction is proportional to the square of the electric field at breakdown ($V_B \propto E_{mb}^2$) and for the graded junction we have $V_B \propto E_{mb}^{3/2}$ [14.5].

14.4.1
Effect of the background doping

For the case of Figure 14.5, for a 2000-V structure, the n-base concentration was 10^{14} cm^{-3}. The 4000-V structure studied by Cornu [14.3] had a concentration of 3.5×10^{13} cm^{-3}. In this last case, the maximum field beneath the surface lies at about 30 µm beneath the surface as compared to 25 µm for the lower resistivity case. As the base doping density N_D decreases, both the maximum bulk field and the maximum surface field also decrease with decreasing doping intensity. The lower the bulk concentration, the closer the magnitudes of the two fields will come together. Consequently, beveling will be less effective for higher resistivity materials, and vice versa. It was necessary to decrease the beveling angle by about 5 percent for the high resistivity case in order to obtain the same magnitudes of the surface fields as were obtained for lower resistivity silicon of Figure 14.5. There is a limit how small this angle can be made for mechanical breakage reasons. Therefore, due to this limitation, a 4000-V breakdown is the maximum one possible to obtain with negative-bevel angles.

14.4.2
Diffusion profile effect

Cornu [14.3] studied the surface field distribution for two different diffusion profiles: one with the space constant $\lambda = 10$ μm and the other with $\lambda = 2$ μm. In both cases $\alpha = -2.85$ degrees and $N_D = 3.5 \times 10^{13}$ cm^{-3}. When the space constant is only 2 μm long, the impurity concentration has a much steeper profile approaching that of a step junction, which exhibits a much higher maximum bulk field than does a graded junction since the depletion region on the highly doped side is very narrow. The same situation exists at the surface, and consequently steeper diffusions will require the use of smaller beveling angles.

14.4.3
Dielectric constant effect

Figure 14.6 shows the effect of the dielectric constant of the surface coating on the field distribution at the surface. The continuous curve refers to a dielectric constant of unity (air) and the dashed line to a

Figure 14.6 Influence of a change in dielectric constant on the field distribution on the surface. $V_A = 2000$ V, $\alpha = -11.3$ degrees, $N_D = 10^{14}$ cm^{-3}, $\lambda = 10$ μm. (After Cornu [14.3].)

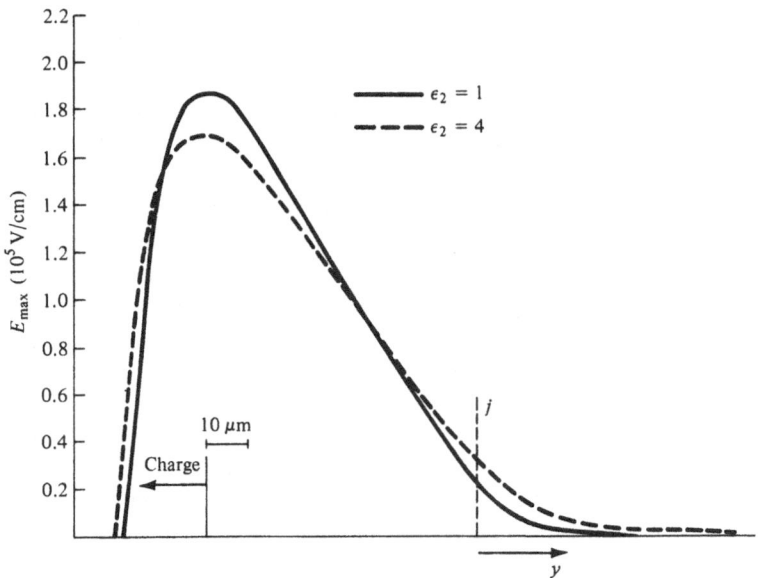

205

dielectric constant of four. The maximum field at the surface is reduced by approximately 10 percent for the higher dielectric constant coating. The effect on the absolute field maximum underneath the surface is much weaker.

To understand the effect of the dielectric constant on the surface field, we have to recall that under the influence of the applied reverse potential the elementary induced dipoles of the dielectric orient themselves along the electric field and the overall effect of this orientation is equivalent to the presence of one large dipole with positive charges at one end and negative charges at the other. The positive dipole charge induces electrons in the p-type silicon, while the negative charge induces holes in the n-type material. Consequently, the effective doping level of silicon on both sides of the junction is lowered and, as a result, the depletion region at the surface is increased and the field is decreased. The surface breakdown is improved, but the breakdown underneath the surface, however, is not significantly affected.

14.4.4
Effect of the surface charges

We consider first the positive surface charges which are most commonly encountered, particularly in SiO_2-coated devices. The prevalent positive charge is usually attributed, in such a case, to the sodium ions or to the SiO_2 charge (surface-state charge). This last charge is positive and fixed. It is located in the oxide within 200 Å of the interface and is a strong function of crystal orientation, oxidation, and annealing conditions. The final temperature the device sees determines largely the magnitude of this charge. The origin of this charge appears to be due to the presence of the excess ionic silicon in the oxide. The typical range of the total positive charges may be between $N_{SS} = 10^{10}$ and $N_{SS} = 10^{12}$ per cm^2.

Figure 14.7 shows the calculated electric fields [14.4] as a function of the bevel angle for two values of the surface charge density $N_{SS} = 0 \ cm^{-2}$ and $N_{SS} = 10^{12} \ cm^{-2}$. The uniform doping of the n base was $N_D = 6 \times 10^{13} \ cm^{-3}$; the p regions were two-step diffused with surface concentrations $N_{A_1} = 3 \times 10^{19} \ cm^{-3}$ and $N_{A_2} = 1.5 \times 10^{16} \ cm^{-3}$, with respective diffusion lengths $L_1 = 27 \ \mu m$ and $L_2 = 48 \ \mu m$. Both diffusions had complement error-function profiles. Figure 14.7 shows both the maximum tangential surface field and the absolute maximum field underneath the surface in function of the negative bevel angle for the two charge densities.

The positive surface charge induces a negative charge (electrons) in the p- and n-type silicon. In the p-type silicon this leads to an increased resistivity; in the n-type silicon the resistivity is decreased,

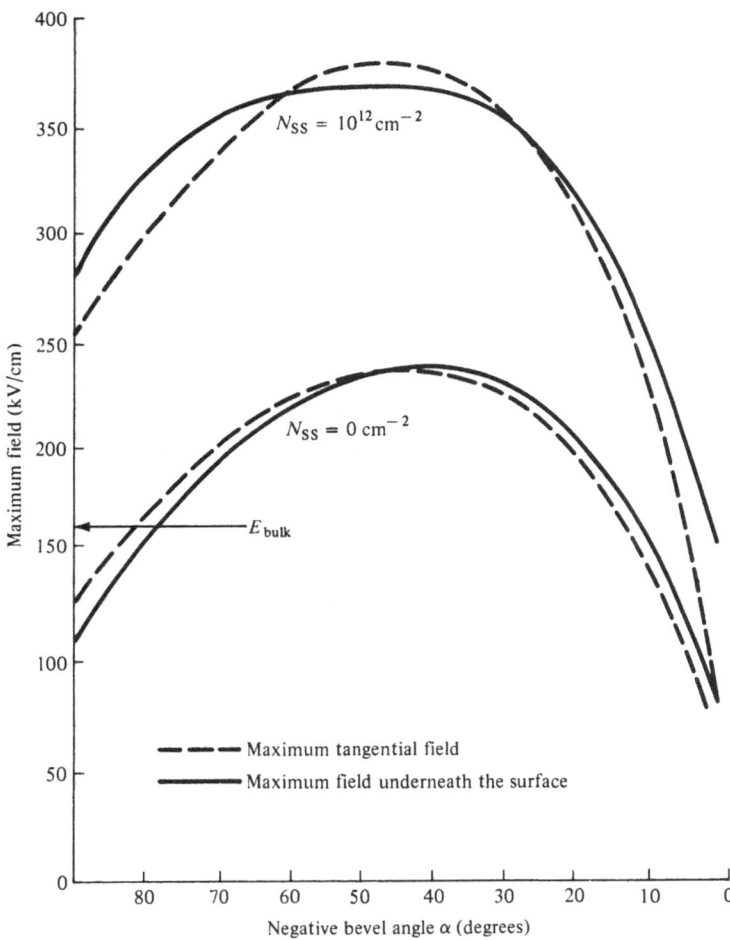

Figure 14.7 Calculated maximum field tangential to the surface and underneath the surface for two surface charges. (After Bakowski and Lundström [14.4].)

however. Therefore, the depletion region contracts in the vicinity of the surface on the n side and expands on the p side. This change in the depletion-region width at the surface is not very large, although it is large enough to change the field (Figure 14.7) by a factor somewhat less than 2 (the maximum field for $N_{ss} = 10^{12}$ cm^{-2} is twice as high as that for $N_{ss} = 0$ cm^{-2}). This leads to a lower breakdown voltage of the negatively beveled junction. The figure shows that the surface charge effect is much smaller for very small negative angles. It is so because when the bevel angles are made small, the depletion region on the p side reaches the highly doped p$^+$ region even without surface charges, and this will decrease the influence of surface

207

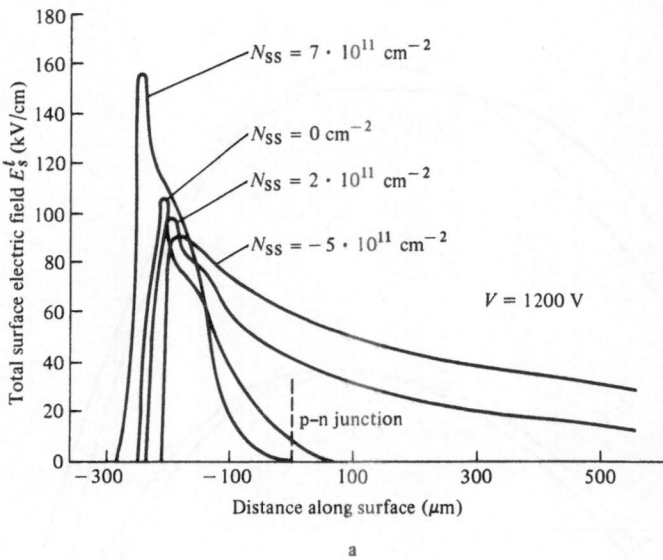

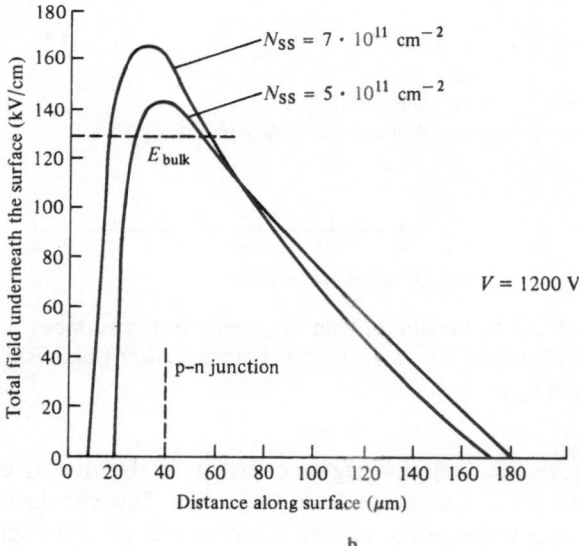

Figure 14.8 Calculated total surface field (a) and the total field under the surface (b) showing the influence of a negative surface charge for a structure with a 6-degree negative bevel angle. (After Bakowski and Lundström [14.4].)

charge on the depletion-region boundary on the p side [14.4]. In addition, at small bevel angles the depletion-region boundary on the n side comes extremely close to the metallurgical p-n junction even without any surface charge (Figure 14.4). A surface charge density of 10^{12} cm^{-3} is sufficient to pin the depletion-region edge on the n side for all bevel angles less than 15 degrees for $N_D = 6 \times 10^{13}$ cm^{-3} with the error function complement doping profile used by [14.4].

Since the positive surface charge induces negative charges in silicon, an inversion layer may be formed on the p side close to junction where the p-type inpurity concentration is small. This affects significantly the surface field, but only insignificantly the field in the region just underneath the surface. For some sufficiently large value of N_{ss} the surface field may become greater than the field underneath the surface.

In [14.3, 14.4, and 14.6] computations were also made with *negative* surface charges. The results of [14.4] are shown in Figure 14.8. Figure 14.8a represents the calculated total surface field (the tangential plus the normal component) versus the distance along the surface. Figure 14.8b shows the total maximum field in the bulk just in the surface vicinity versus distance along the surface. N_{ss} is a parameter for both plots. It can be seen that the maximum fields are reduced for the negative-surface charges as compared to the positive- or zero-surface charges. However, an inversion layer (p$^+$ channel) can be easily formed on the surface of the n region. If the channel reaches the p region on the other side of the n base of the thyristor, a very leaky conducting path will be formed across the forward-blocking junction J_2. It is of interest to observe, however, that the negative surface charge may be desired to keep the depletion region of the forward-blocking junction away from the highly doped p region that is normally entered as a result of bending up the depletion region at the surface.

14.5
Positively beveled junctions

Figure 14.9 shows a depletion region of a p-n$^-$ junction with a positive bevel angle of 45 degrees. For positive angles the space-charge region expands on the high-resistivity side close to the surface and becomes narrower on the low resistivity side. It was shown by Davies and Gentry [14.1] that for positive angles the surface field is reduced appreciably in respect to the field in the bulk for all values of the positive angles. The maximum field at the surface decreases monotonically with the beveling angle.

209

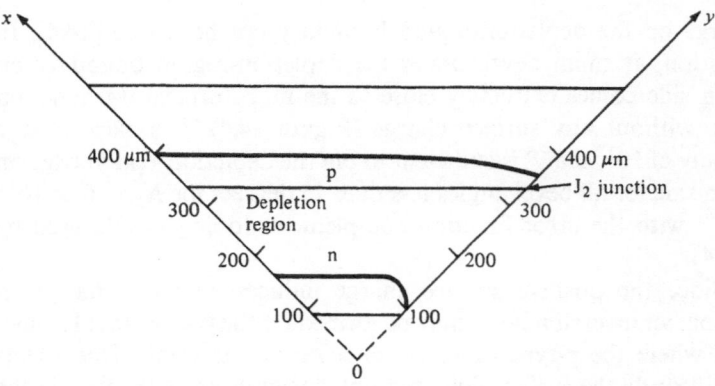

Figure 14.9 Depletion region in the vicinity of the surface beveled with the positive angle $\alpha = 45$ degrees. $V_A = 2000$ V, $N_D = 10^{14}$ cm^{-3}, $\lambda = 10$ μm. (After Cornu [14.3].)

The surface field does not have its peak value at the p-n junction [14.3], but some distance away from it on the lightly doped n-type side.

For a positive-beveled junction there is no peaking of the field going from the surface of the junction into the interior of the device until the bulk is reached; the avalanche breakdown will occur, therefore, in the bulk. The magnitude of the electric field along the surface for positive bevel angles was calculated also by Davies and Gentry [14.1]. Their results agree essentially with those of Bakowski and Lundstrom [14.4].

Experimental data of [14.1, 14.3, 14.4] show that for the positively beveled junctions it is possible to obtain the true bulk breakdown.

14.5.1
Effect of the surface charges [14.3, 14.4, 14.6]

The reverse-blocking junction of a thyristor usually is beveled with a positive angle. Very low n-base doping is normally used to obtain high breakdowns; concentrations between 0.5×10^{13} cm^{-3} and 2×10^{14} cm^{-3} are quite common. Consequently, at sufficiently high reverse bias the depletion region of the reverse-blocking junction can reach the forward-blocking junction at the device surface. If there is a positive charge present at the surface, this can be avoided, because although the highly doped p region will not be affected the low-doped n region will become more n type at the surface. Therefore, the depletion region at the surface will not widen as much under the influence of the applied potential and the punch-through will not take place.

14.6
Novel approaches to beveling

We have seen that the usually negatively beveled forward-blocking junction may achieve high breakdown voltage only with very small angles. There are, however, practical limits how far one can go in minimizing this angle from the mechanical integrity point of view and from the "real estate" point of view. With very small angles one is left with only a very small available cathode emitter area to carry the current unless one uses exceedingly large silicon pellets. Therefore, negative bevel angles are limited from the practical point of view to breakdown voltages below 4000 V. Attempts are made all the time to avoid the pitfalls of the negative-bevel approach.

Köhl [14.7] has proposed a mesa-like contouring (Figure 14.10) on the p-type base of the thyristor in such a fashion that the required small angle is only necessary near the periphery of the depletion region on the p side. With this contour, thyristors were produced having nearly symmetrical blocking characteristics of 4000 V. The desired edge profile was obtained on individual circular pellets by first mechanically grounding and etching the lower bevel and then, in the second step, jet-etching the mesa-like low-angle bevel.

Köhl's approach permits better area utilization. It does not remove, however, the need for very small, mechanically undesirable, beveling angles. Otsuka's approach [14.8] avoids entirely the use of negative bevel angles by mechanically cutting a groove into the silicon as per Figure 14.11 and thereby achieving a positive bevel angle. This solution requires individual handling of each device.

Cornu *et al.* [14.9] recently have proposed the use of a double-positive beveling for high-voltage device contouring as shown in Figure 14.12. Both the forward- and the reverse-thyristor junctions are beveled with a positive angle. The double-positive edge was made by sandblasting with subsequent etching to remove the mechanically

Figure 14.10 Section across the edge of a four-layer system of a thyristor with a mesa-like beveled p^+-n junction. (After Köhl [14.7].)

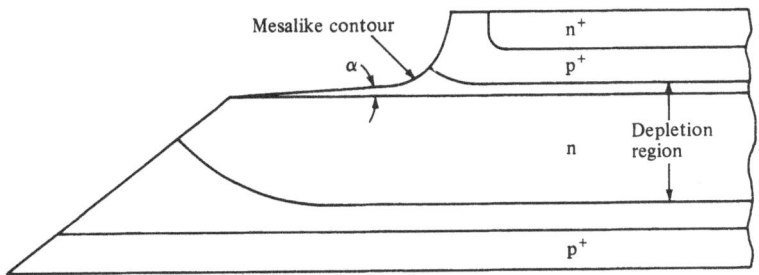

211

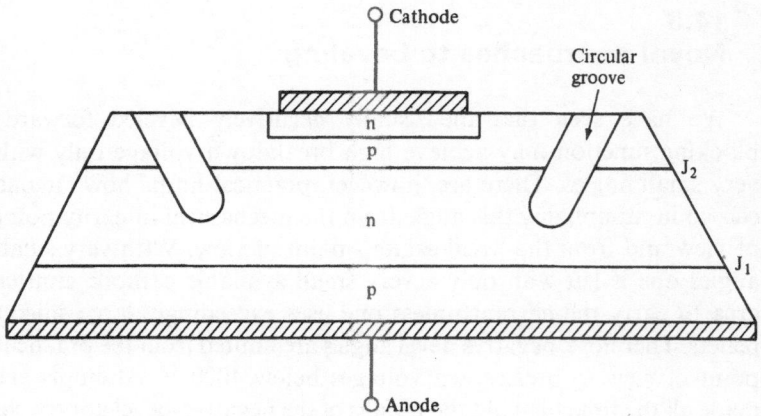

Figure 14.11 Junctions J_1 and J_2—positively beveled. (After Otsuka [14.8].)

Figure 14.12 Depletion region in the vicinity of surface with double-positive bevel. (After Cornu *et al.* [14.9].)

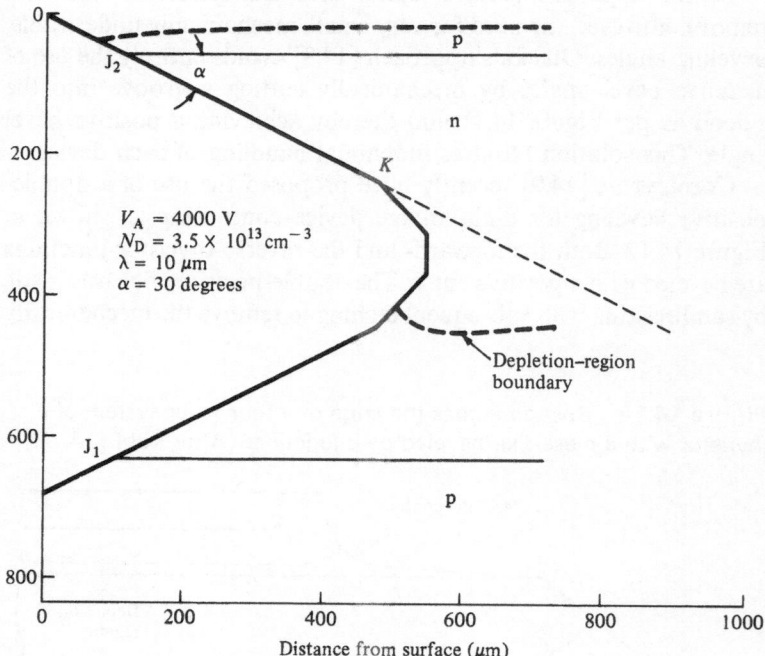

damaged region. As in Otsuka's approach, each device has to be handled individually.

To determine the field distribution on the surface of the double positively beveled structure, a very similar iterative computational approach has been used [14.3], as discussed above. Figure 14.12 represents a double-positive beveled structure with 4000-V break-down. The donor concentration of the n base was 3.5×10^{13} cm^{-3}, the space constant for diffusion profile was $\lambda = 10$ μm, and each angle had a value of $+30$ degrees. The dash-dot lines represent the boundaries of the space-charge region. The upper junction J_2 is forward blocking and the lower junction is reverse blocking. The field distribution on the surface is projected on the vertical axis and is shown in Figure 14.13. The dotted line in this figure shows the distribution of the field for the case of a single-positive-bevel angle

Figure 14.13 Field distribution at the surface for the double-positive-edge geometry of Figure 14.12 and for the corresponding single-positive-edge geometry. (The field distributions are projected on the vertical axis.) (After Cornu *et al.* [14.9].)

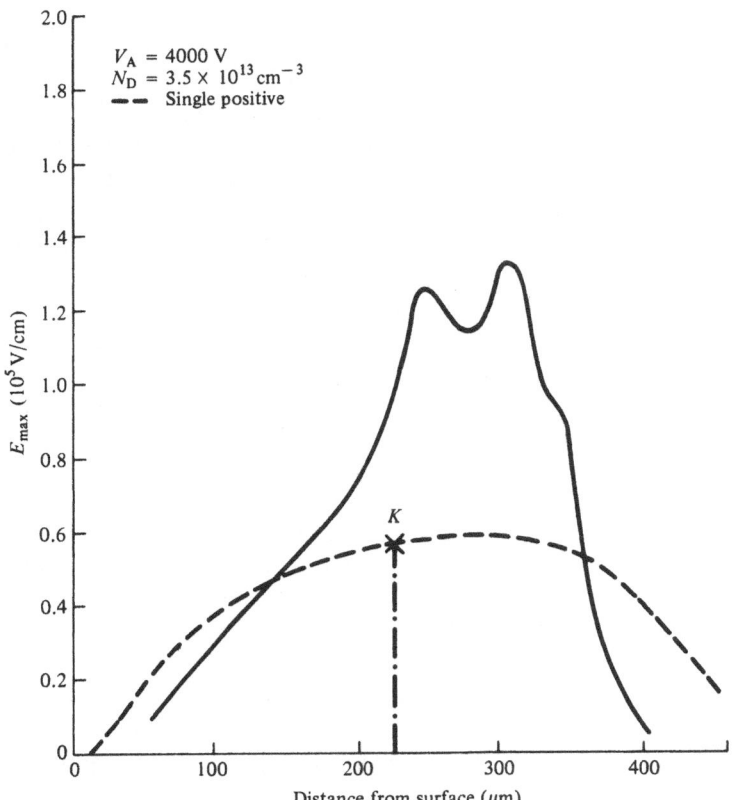

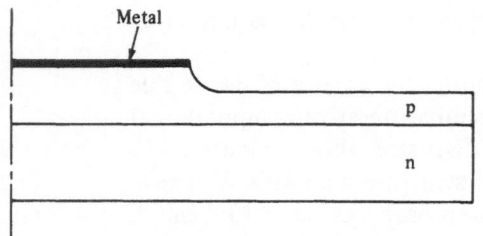

Figure 14.14   Etch contour for plane (mesa) junction. (After Temple and Adler [14.10].)

extending up to the end of the space-charge region. The maximum field in the case of the double-positive angle is higher than in the case of the single-positive angle. This is attributed to the fact that the depletion region of the J_2 junction (forward blocking) extends beyond the point K (Figure 14.12) where the geometries start to differ. From Figure 14.12 we can see that beyond the point K in the double-positive structure, the depletion region expands to a lesser distance than it would with a single-beveled surface. This smaller spreading increases the field intensity for the same applied potential. At some other angles, such as 45 degrees, the field distribution may be even less advantageous because of the sharp field peaking resulting from the coincidence of the high field-strength region (region at the surface beyond the point K) with the point of minimum radius of curvature of the geometry. However, further study has shown that generally there is no problem in reducing the maximum field on the surface to a value of about 1.2×10^5 V/cm^3. According to [14.3] the surface fields up to 1.5×10^5 V/cm do not affect adversely the reverse characteristics. Nevertheless, more careful passivation is definitely in order as compared to small-angle negative-bevel junctions. The big advantages of the double-positive bevel structure are that it can be used for breakdown voltages above 4000 V without running into the mechanical strength problems and leads only to rather insignificant reduction of the active device area.

Temple and Adler [14.10] were able to obtain high avalanche breakdowns for both type of junctions—mesa (plane) or planar—by extending the heavily doped side of the junction beyond the metallic contact and by partially removing it through chemical etching (Figure 14.14). This approach has a very important advantage over the negatively beveled junction in that it not only avoids mechanical beveling but, in addition, also uses only a small silicon area. By etching away part of the heavily doped junction side the depletion region is allowed to spread out on both of the junction sides so that the field component parallel to the junction plane is reduced. The avalanche breakdown voltage in this approach depends on the ability to etch the heavily doped region consistently to the optimum depth.

References

14.1 R. L. Davies and F. E. Gentry. Control of electric field at the surface of p-n junctions. *IEEE Trans. Electron Devices, ED-11*: 313–323, 1964.

14.2 J. Cornu. Electric fields at and near the surface of p-n junctions with negative bevel angles. *Electronic Letts., 8* (7): 169–170, 1972.

14.3 J. Cornu. Field distribution near the surface of beveled p-n junctions in high voltage devices. *IEEE Trans. Electron Devices, ED-20*, (4): 347–352, 1973.

14.4 M. Bakowski and K. I. Lundström. Depletion layer characteristics at the surface of beveled high-voltage p-n junctions. *IEEE Trans. Electron Devices, ED-20* (6): 550–563, 1973.

14.5 A. B. Phillips. *Transistor Engineering.* New York: McGraw-Hill, 1962.

14.6 M. Bakowski and B. Hanson. Influence of bevel angle and surface charge on the breakdown of negatively bevelled diffused p-n junctions. *Solid State Electron., 15*: 651–657, 1975.

14.7 S. G. Köhl. A mesa-like edge contour for Si high voltage thyristors. *Solid State Electron., 11*: 501–502, 1968.

14.8 M. Otsuka. A new edge contour for Si high voltage thyristors. *IEEE Conf Publ. 53*: 32–38, 1969.

14.9 J. Cornu, S. Schweitzer, and O. Kuhn. Double positive beveling: A better edge contour for high voltage devices. *IEEE Trans. Electron Devices, ED-21* (3): 189–194, 1974.

14.10 V. A. K. Temple and M. S. Adler. A simple etch contour for near-ideal breakdown voltage in plane *and* planar p-n junctions. International Electron Device Meeting, December 1975, Washington, D.C. *Technical Digest*, pp. 171–174.

Planar-junction avalanche breakdown improvement

Summary

Planar junctions are much easier to passivate and to protect from external contaminants than are etched mesa devices. They compare unfavorably, however, with the mesa-type structures from the breakdown voltage point of view if special designs which include guard rings, field-limiting rings, or field plates are not used. When they are used, though, planar-junction breakdowns as good or better than those of mesa devices can be achieved.

15.1
Introduction

A planar p-n junction is formed by diffusing an impurity through an oxide window so that the diffusion takes place not only in the direction normal to the silicon surface but also laterally in the x direction (Figure 15.1). To a first approximation it is adequate to assume that the diffusion proceeds at the same rate in both directions. Consequently, the junction geometry may be assumed to be circular-cylindrical near the window edges. If the junction is wide enough, most of it, with the exception of the periphery, may be considered plane. Under these conditions, the breakdown voltage will be determined by the curved cylindrical portion since the electric field there will be much higher than at the flat region. Figure 15.2 represents a plot of the calculated breakdown voltage in function of the junction depth x_j, equal to the radius of curvature at the edges [15.1, 15.2]. For reasonable junction depths the breakdown is considerably reduced as compared to the breakdown of a flat, plane junction with infinitely large radius of curvature. Figure 15.3 shows

216

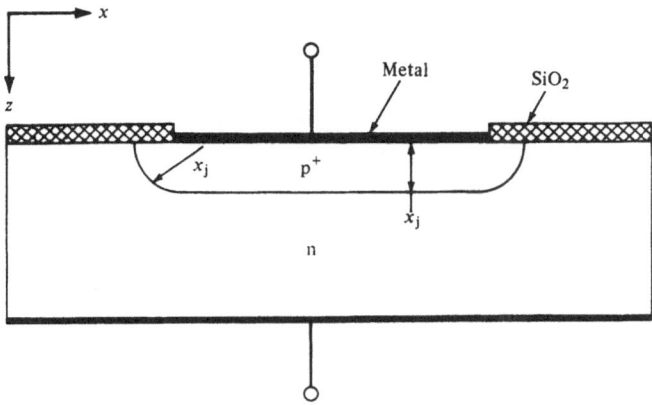

Figure 15.1 Planar junction.

the depletion layer width at breakdown versus x_j; the smaller the junction penetration, the smaller the depletion layer width and, consequently, the larger is the electric field. Temple and Adler [15.3]

Figure 15.2 Avalanche-breakdown voltage versus impurity concentration for one-sided abrupt doping profile with cylindrical and spherical junction geometries where x_j is the radius of curvature. (After Sze and Gibbons [15.1].)

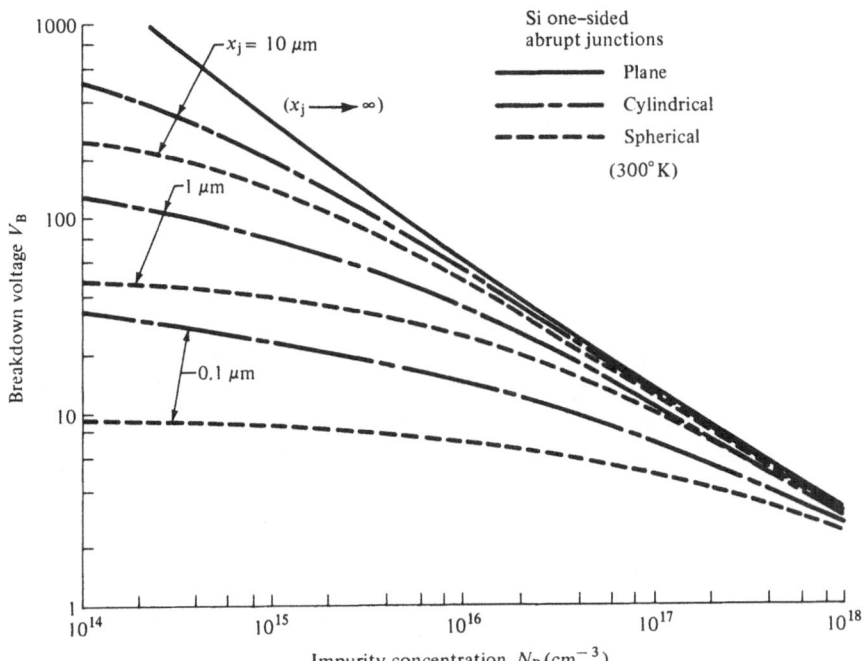

217

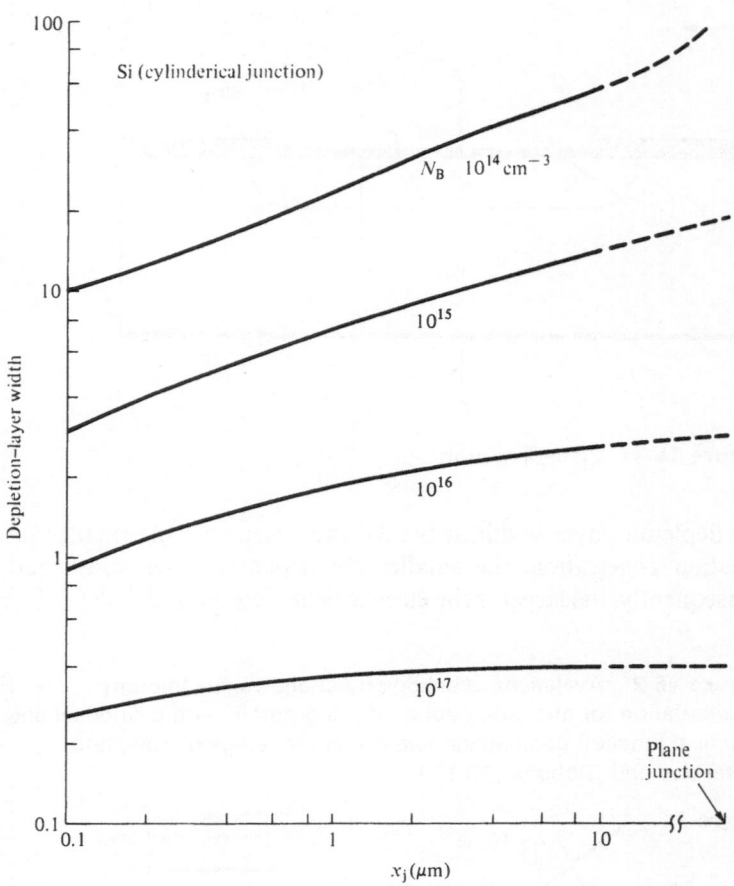

Figure 15.3 Depletion-layer width at breakdown for Si cylindrical junction. (After Sze and Gibbons [15.1].)

computed breakdown voltages of very deeply diffused p$^+$-n junctions with n-side concentrations down to 10^{-13} cm^{-3}.

Another problem arising with the use of planar junctions comes from the presence of extraneous surface charges which affect the shape of the junction depletion region close to the surface. A positive surface charge causes an accumulation of electrons in the comparatively lightly doped n-type material (Figure 15.4) and brings about the narrowing of the depletion region in the vicinity of the surface, with a resulting increase in the electric field magnitude. The positive surface charge, if large enough, may also invert a part of the p$^+$ region, forming an n-type channel (see Figure 15.8), and the breakdown, under such conditions, will occur between the p$^+$ region and the n-type channel.

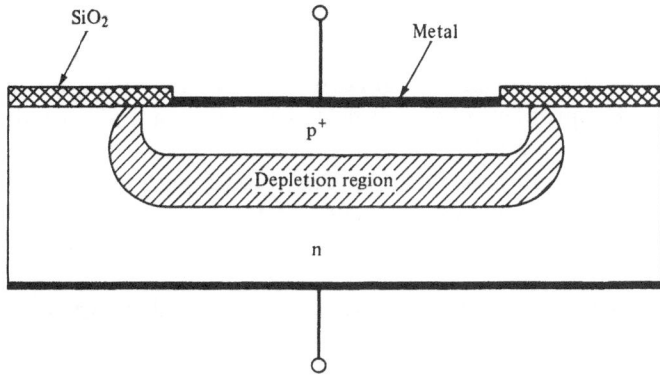

Figure 15.4 Narrowing of the depletion region at the surface due to surface accumulation.

The big advantages of planar junctions are their inherent good stability, low leakage currents (if properly fabricated), and easier passivation and surface protection techniques by various dielectrics than for the mesa-type junctions. It is of interest, therefore, to achieve planar junctions with high breakdown voltages. In the following, we will discuss a few approaches that are helpful in overcoming the junction curvature and surface problems.

15.2
Diffused guard ring

As it was mentioned above, the larger the planar junction depth and its radius x_j, the larger its breakdown voltage. It is useful then to surround a shallow planar junction with a deeply diffused ring as

Figure 15.5 Planar junction with a diffused guard ring.

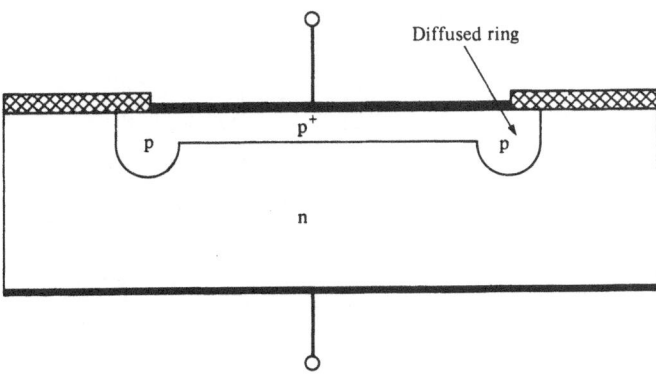

219

per Figure 15.5 to enhance the breakdown properties of the composite structure.

Although the problem of curvature may be overcome by this technique, the problem of the surface accumulation and inversion in the presence of positive electric charges in the $Si-SiO_2$ system remains unsolved unless junction field plates, counteracting the surface charge and junction-curvature effects, are used.

15.3
Junction field plate, annular ring, and channel stopper

The junction field plate (Figure 15.6) consists, in essence, of a metal overlay on SiO_2 overlapping the junction and is used usually in conjunction with a metal annular ring and a diffused channel-stopping ring. By using this approach it was possible to design low-leakage oxide-passivated diodes with breakdowns of about 1500 V [15.4]. Figure 15.6 shows the structure of a p^+-n^--n^+ diode provided with a field plate, annular ring, and channel stopper which was investigated by Zoroglu and Clark [15.4]. The field plate consists of a metallization on the oxide surface, extending beyond the junction by a length equal to $W_O - x_j$ and connected electrically to the p^+ region. Another electrode, in the form of a metallic annular ring, deposited also on the SiO_2 surface, surrounds the p^+-n junction and is electrically connected to the high resistivity n^- region or to an n^+-diffused channel stopper if it is also used. The annular ring

Figure 15.6 Field plate, annular ring, and channel stopper. (After Zoroglu and Clark [15.4].)

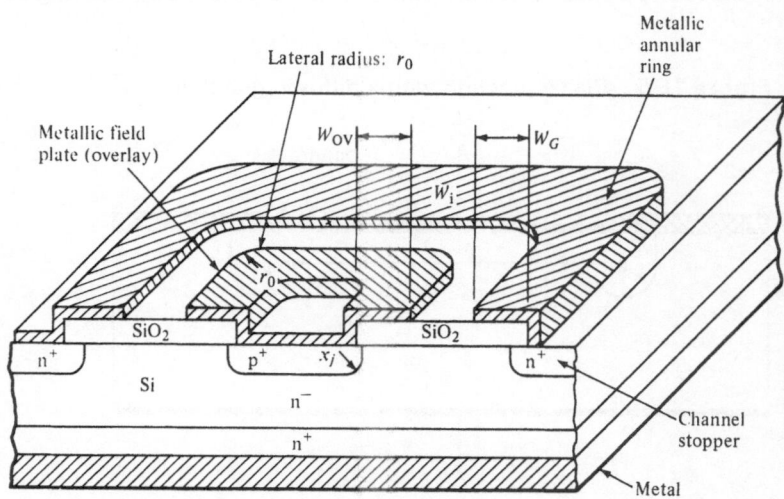

220

extends a distance W_G over the edge of the SiO_2. The spacing between the two metallizations is designated by W_i. Both circular and square geometries with large corner radii were studied by Zoroglu and Clark [15.4] in order to determine the effect of the *lateral* radius on breakdown. The presence of the passivation oxide over the junction protects it, in principle, from the external contamination. As discussed in Chapter 14, it is very easy, however, for the contaminants, such as sodium ions, to penetrate the oxide during the device fabrication. It may be possible to achieve partial removal of the mobile charge by extremely clean processing. But, even with the most careful preparation, some positive mobile ions may be left to cause a considerable device instability; when the positive ions are present and a voltage is applied to the device, the positively charged ions are attracted to the negative potential (Figure 15.7) and remain in the vicinity of the negatively biased electrode. The alkali ions located in the oxide over the p region will repel the positive holes of the p-type silicon just under the oxide and will attract the negative charges, the electrons. Part of the p region will then have more electrons than holes, and behave, therefore, like n-doped silicon; an inverted region (channel) is formed (Figure 15.8). If the inversion reaches the negatively biased electrode, the electron current may flow directly from the metal electrode through the inverted n-type region directly to the positively biased electrode, by passing the p-n junction and enormously increasing the leakage current. Since the electron concentration of the inverted region may be much larger than that of the n$^-$ substrate, the breakdown of the p-n junction formed by the p region and the n channel may be much lower than that expected from the resistivity of the n$^-$ material used.

Figure 15.7 Planar junction with contaminated SiO_2.

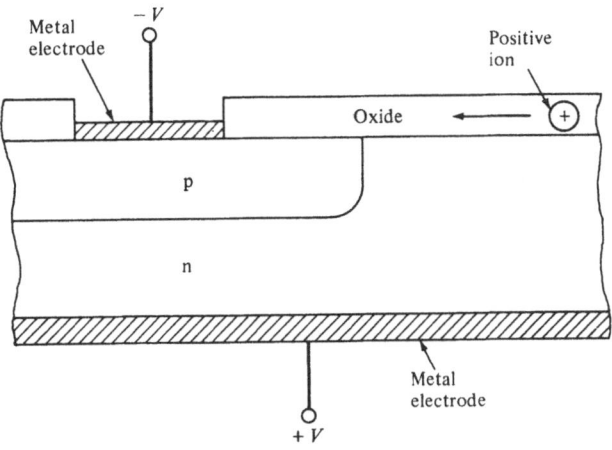

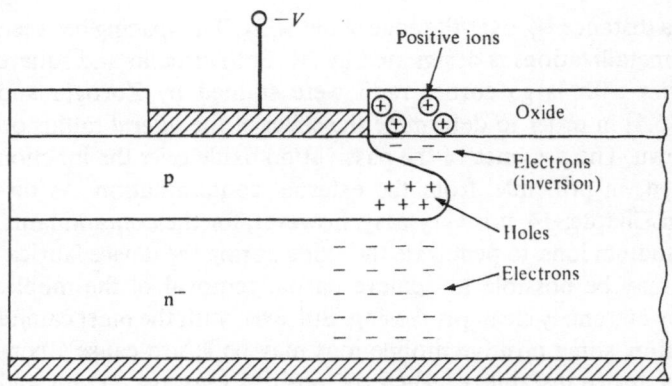

Figure 15.8 Surface inversion due to mobile positive ions.

A field plate consisting of a metal overlay may be used to prevent the formation of a channel (Figure 15.9). The metal contact to the p region extends over the oxide and overlaps the junction. As soon as a positive ion reaches the edge of the metallization, it is attracted to it and is kept away from the p region since the field between the plate and the n-type silicon is now normal to the Si–SiO$_2$ interface. Because the positive ions are drawn away from the silicon surface toward the negative plate, the inversion of the p-type region does not take place. When clean oxides with very few positive charges

Figure 15.9 Field plate.

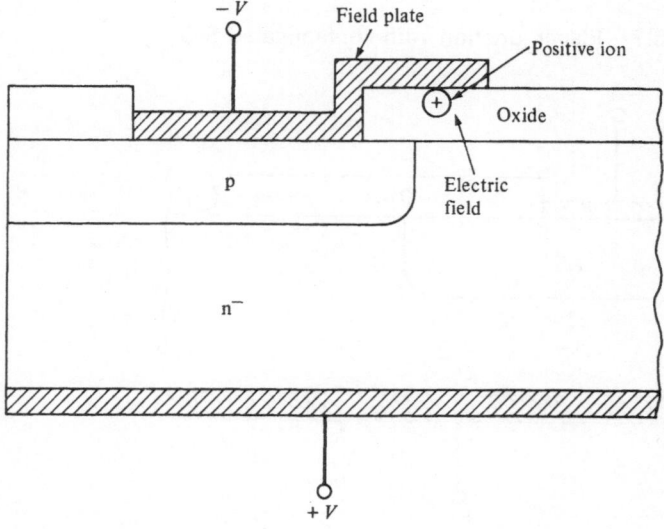

were used, it became obvious that the electric field created by the extended metallization was causing the appearance of an inversion layer in the n^- region in the area under the extension at the $Si–SiO_2$ interface, since the negatively biased electrode was repelling electrons and attracting holes. In order to avoid this effect, another electrode, the annular ring, was added (Figure 15.10), consisting of a metallization deposited on the oxide surface and connected electrically to the n region so that it became positively biased, while, on the other hand, the field plate was negatively biased. The positively biased annular ring attracts electrons to the interface and thus counters the tendency of the field plate to invert the n^- region.

A further improvement consists of the addition of an n^+-type diffused ring known as a channel stopper (Figure 15.6). The doping concentration of the ring is so high that the negatively biased field plate is not able to invert the n^+ region, so that the channel is stopped.

Under optimized conditions, the field plate depletes the surface in the vicinity of the junction edge, makes the n^- silicon less n-doped, and thereby reduces the electric field so that the breakdown is increased. At the same time the field plate contributes to the device surface stability. The major design parameters for the field plate and the annular ring (Figure 15.6) are: the insulator thickness, the width W_O of the metal overlay, the spacing W_i between the two electrodes, and the optimal use of the n^+ channel stopper.

Figure 15.10 Field plate and annular ring.

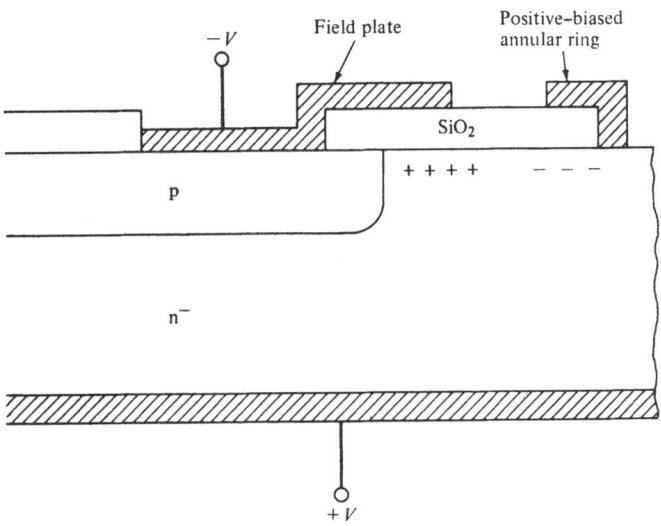

15.3.1
Experiments with the junction field plate

Zoroglu and Clark [15.4, 15.5] have performed a series of experiments to determine the effect of various geometrical parameters on the device breakdown. The n-type silicon used had a resistivity of 360–500 Ωcm. The wafer thickness was 152 ± 5 µm and the p$^+$ diffusion was 18 µm deep, with a sheet resistance of 4.2 Ω per square. The n$^+$ layer used for good ohmic contact was 5 µm deep and had a sheet resistance of 1 Ω per square.

15.3.1.1 Breakdown versus electrode overlap and shape. Figure 15.11 shows how the breakdown voltage changes with the varying overlap $W_{ov} - x_j$ while keeping the junction diameter and the

Figure 15.11 Breakdown voltage versus field-plate overlap $W_{ov} - x_j$. (After Zoroglu and Clark [15.4].)

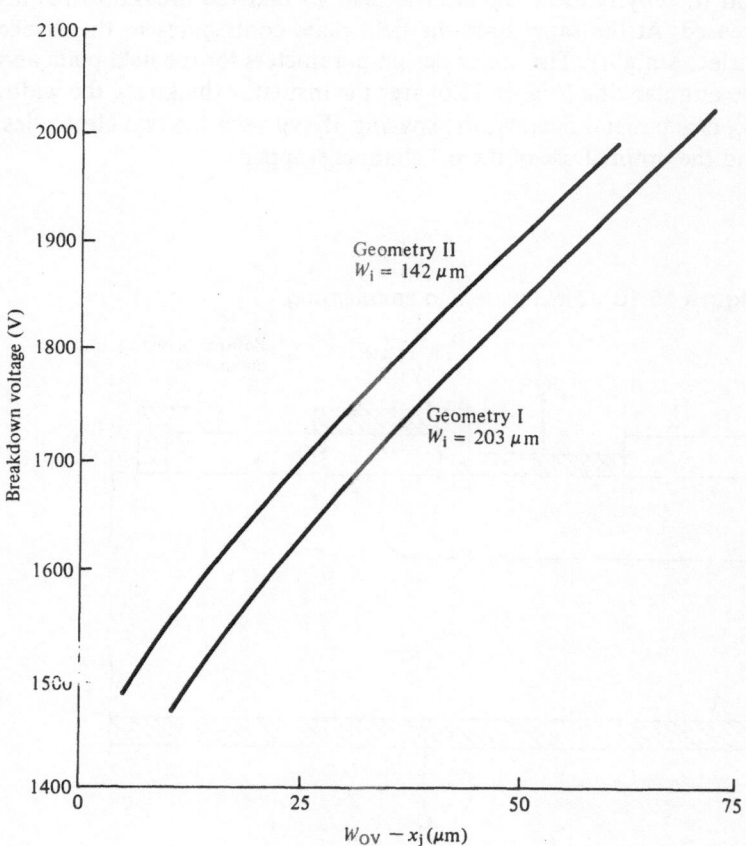

spacing W_i constant. In order to prevent the arcing between the two electrodes, the surface was coated with the Dow Corning 704 silicon oil. The spacing W_i was on the order of 100–200 μm, so the electric field magnitude between the two electrodes at the surface was on the order of 10^5 V/cm.

The electrodes were observed under a microscope; since the silicone oil is electrostrictive, it was possible to locate the high-field regions by observing the constricted areas. The highest field was found to be just at the edge of the overlay plate.

Geometry I of Figure 15.11 refers to the square-shaped field plate whereas geometry II refers to the circular field plate. The circular geometry had a breakdown voltage by about 80 V higher than the square geometry. The larger the extension of the overlay $W_O - x_j$, beyond the junction, the larger the breakdown voltage. Every micrometer added results in about a 9-V improvement on the breakdown. The increase in breakdown shown in the curves is mostly a result of an increase in the transverse radius of curvature of the overlay. Devices with no metal electrodes exhibited a breakdown voltage of only 900 V, whereas those with overlays had breakdown voltages of up to 2000 V.

In another series of experiments [15.4], breakdown voltage was plotted versus the field-plate overlap $W_O - x_j$ and versus the lateral radius r_O of the square-type plate (Figures 15.12 and 15.14a). For

Figure 15.12 Breakdown voltage versus ($W_{ov} - x_j$) and versus lateral radius of curvature r_O. (After Clark and Zoroglu [15.5].)

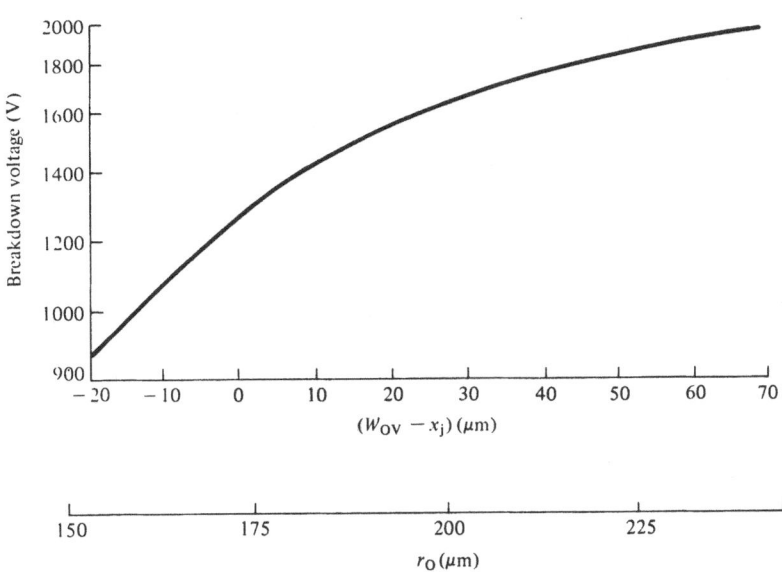

225

very small plates which do not overlap the junction, the breakdown increases at about 17 V/μm of the plate size. An extension of the plate just by a few microns beyond the junction yielded devices with breakdowns above 1500 V. For large overlaps, only about 9 V is gained for each micrometer of the plate size added, which indicates that the primary mechanism is an increase in the lateral corner radius.

15.3.1.2 Breakdown versus electrode spacing. The effect of increasing the field-plate–annular-ring spacing W_i is shown in Figure 15.13. The junction diameter as well as the field-plate and annular-ring diameters were kept constant while W_G of the annular ring was decreased from 102 to 51 μm. The breakdown was increased by about 4 V/μm of the electrode spacing.

15.3.1.3 Channel stopper. A group of diodes was fabricated with an optional n^+ annular-diffused channel stopper with and without the metal annular ring. Both types showed high breakdown voltages, but the diodes without the metallization had a very high leakage, which appeared a few seconds after the potential was applied and continued increasing with time. This was not observed, however, when both the diffused and the metal rings were present, probably due to the removal of the negative charges from the SiO_2 surface by the metallic electrode; the accumulation of the negative charges on the SiO_2 surface was thought to be responsible for the large leakage currents.

Figure 15.13 Breakdown voltage versus electrode spacing W_i. (After Zoroglu and Clark [15.4].)

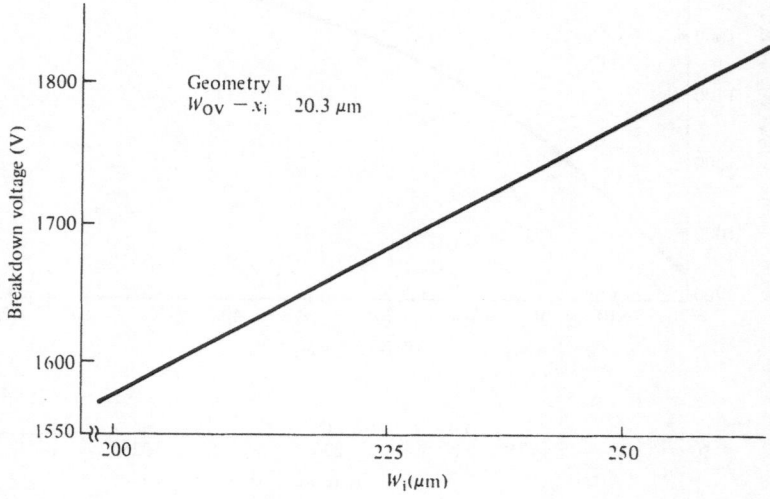

15.3.1.4 Effect of the resin type. When the diodes were passivated by other resins than the Dow Corning 709 silicon fluid, such as cured viscous silicone resins necessary for the practical devices that have to be handled during manufacturing and packaging, the breakdown was reduced by about 50 percent of the maximum 2000 V observed with the fluid silicone. This reduction was attributed by the authors to the possible presence of positive charges in the passivating substance and to an excessively high field near the edge of the field plate resulting from the inhibition of the negative-charge migration on the surface.

Very thick oxides (7 μ), which are normally impractical in production because of possible crazing, were used to overcome this problem partially, so that 1500-V breakdowns were achieved with the cured passivants.

15.4
Resistive field plate

In order to alleviate the problem of the highly nonuniform fields appearing at the field-plate edge, it is necessary to use very large electrode-to-electrode spacings which lead to large device areas and are therefore uneconomical. Moreover, even with the large spacing it is necessary to use passivating resins to avoid surface arcing. This often leads to further breakdown reduction if the resins are not completely devoid of ionic charges.

To overcome the problem of high localized fields, Clark and Zoroglu [15.5] proposed the use of a resistive polycrystalline film bridging the space between the field plate and the annular ring. The function of the resistive film is to homogenize, i.e., to make more uniform, the electric field at the surface, thereby increasing the breakdown voltage and suppressing the surface arcing.

The resistive film used by Clark and Zoroglu consisted of undoped polycrystalline silicon. Such film can be deposited on the silicon dioxide surface by using techniques similar to those used in depositing silicon epitaxial layers. The polycrystalline silicon films are as much refractory as silicon and are compatible with device processing such as photolithography. They are finding increasing applications, also, in dielectrically isolated integrated circuits, in silicon gate MOS technology, etc. By controlling the deposition parameters, polycrystalline films can be observed with properties ranging from those of the dielectrics to those of semiconductors. The polycrystalline silicon films have positive temperature coefficients of conductivity [15.5]. Therefore, an increased leakage current can be expected at elevated temperatures since the film is shunting the p-n junction. However, the junction saturation current increases at

227

about the same or even higher rate; therefore, the additional conductive path may not be objectionable at the elevated temperatures if the conductance at room temperature is acceptable.[1]

15.4.1
Experimental data

In the experiment performed by Clark and Zoroglu an SiO_2 layer 2 µm thick was covering the junction. The polycrystalline silicon film was 2000 Å thick with a sheet resistance of $5 \times 10^{10} \Omega$ per square ($\rho = 10^6 \Omega/cm$). The field plate had a square geometry similar to that shown in Figure 15.14a. The annular ring was also square, but the corners were not rounded in order to achieve greater spacing in the high field corner regions. Everywhere else $W_i = 305$ µm. Two types of diodes were fabricated with and without a polycrystalline silicon film added between the electrodes. The breakdown voltage of the polycrystalline silicon film was above 1700 V without resin coating. The same result has been achieved without the polycrystalline silicon film, but with the use of the Dow Corning 704 silicone oil. What is important is that the polycrystalline silicon coated devices did not degrade in the breakdown voltage when covered with a cured type resin.

Figure 15.14b shows the effect of the electrode spacing W_i on the breakdown voltage of the polysilicon passivated diode. The dimensions of the p-n junction and of the annular ring were kept constant so that when the field plate size was increased, the spacing W_i was decreased. As the spacing between the two electrodes was made smaller, the lateral radius r_0 increased. For small W_i spacings the slope of the voltage versus W_i curve is about 17 V/µ due to the reduction of the average field in the interelectrode space. This is comparable to a value of 20 V/µ expected in the high resistivity silicon breakdown. Consequently, *the required surface spacing for a given breakdown voltage is similar to that required for bulk breakdown.*

As W_i is increased, with the size of the external electrode remaining unchanged, the field-plate diameter and the lateral radius of curvature r_0 becomes smaller. Therefore, although the field at the straight sides of the electrode is decreased, the field at the plate corners is increased. When W_i is about 100 µm, the breakdown voltage starts falling off due to the increased plate corner field.

---

[1] More recently, T. Aoki *et al.* [15.6] passivated silicon surface by *direct* deposition on silicon of polycrystalline silicon films containing 10–25 weight percent oxygen obtained from SiH_4-N_2O-N_2 system at 650°C and found it to be superior to the directly deposited, undoped polycrystalline films from surface stability and leakage point of view (see Section 15.4.2).

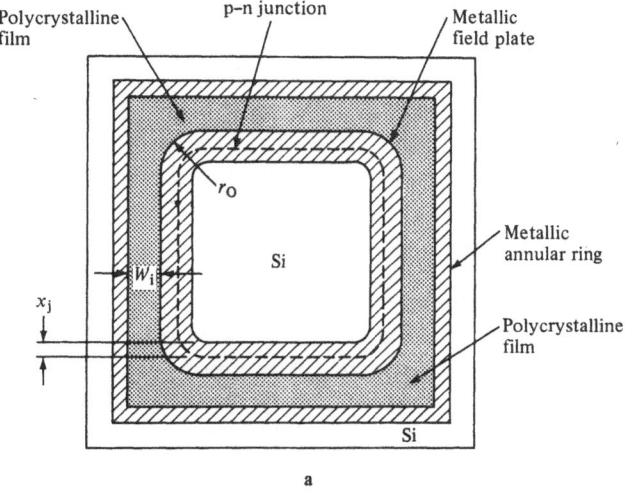

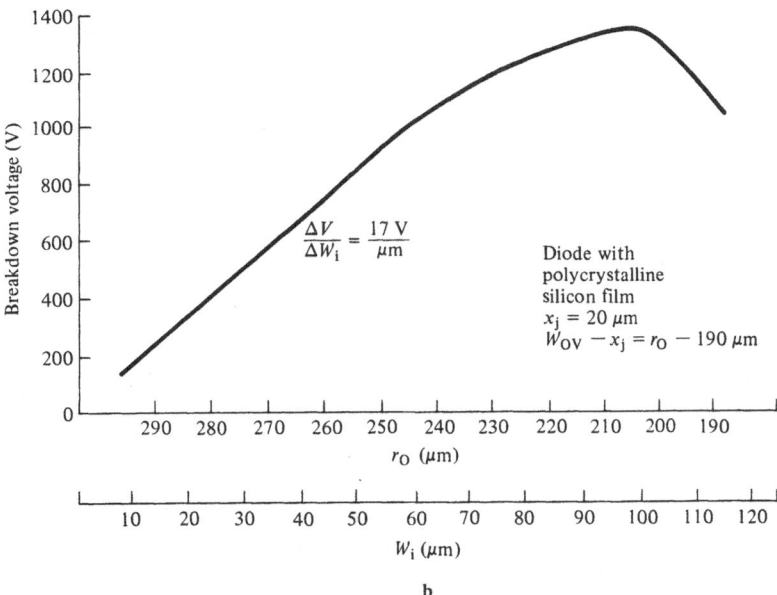

Figure 15.14 (a) Diode geometry with polycrystalline silicon film. (b) Breakdown voltage versus lateral radius curvature and versus electrode spacing. (After Clark and Zoroglu [15.5].)

Experimentation performed at RCA has shown that polycrystalline film is not impervious to such impurities as alkali ions. Additional protective layers such as silicon nitride may be necessary to make the structure truly impervious to the external contaminants.

In 1971, Spaden and Hower [15.7] reported the application of the resistive field plate technology to the fabrication of planar bipolar transistors with $V_{CEO(sus)} = 1200$ V. In their devices the leakage currents of the junctions were mainly due to the current flowing through the polysilicon film, and its value was dependent on the conditions of the growth of the polycrystalline layer. The leakages were on the order of several microamps at 1000 V.

The electronic properties of undoped polycrystalline silicon were recently studied and reported by Muñoz et al. [15.8]. The best results from the leakage and stability point of view were obtained with a polycrystalline layer 0.5 µm thick and deposited in a resistance-heated reactor at 680°C. The polysilicon film deposited at 680°C showed the highest resistivity of about 8×10^6 Ωcm as compared to the films deposited at temperatures above 680°C up to 1000°C. At 1000°C the resistivity was about 4.6×10^6 and the films were exhibiting surface arcing, while the 680°C films did not.

15.4.2
Planar junction passivation by oxygen-doped polycrystalline silicon film

Aoki et al. [15.6] and Matsushita et al. [15.9] have used oxygen-doped polycrystalline films to passivate silicon surfaces. This type of film is semi-insulating and almost neutral electrically. Matsushita et al. [15.9] have employed this film in place of the silicon dioxide layer in planar devices. In order to prevent sodium or other ions from reaching the silicon surface, a layer of nitrogen-doped polycrystalline film (silicon-rich nitride) has been deposited on top of the silicon crystalline film. In turn, the silicon nitride has been covered by an SiO_2 film to avoid dielectric breakdown in the high-field regions. The surface, therefore, is protected by a triple-layer film.

When a p-n junction is reverse biased, a small current flows due to the very small conductivity of the oxygen-doped polycrystalline layer. The structure therefore behaves like a resistive field plate which counteracts the effects of junction curvature and the effects of surface charge.

The reverse leakage current of junctions passivated by this method may be made (according to [15.9]) as small as the reverse current of planar junctions passivated by SiO_2 if the oxygen concentration in the silicon polycrystalline film is sufficiently high.

Matsushita *et al.* [15.9] have fabricated n-p-n and p-n-p transistors with limiting rings, with breakdown voltages of 800 V and 2500 V utilizing triple-film passivation. In the case of p-n-p transistors, they used 50-Ωcm silicon substrate for the 800-V transistor and 150-Ωcm silicon substrate for the 2500 V transistor. The triple layer consisted of a 5000-Å-thick oxygen-doped polycrystalline silicon film (20 at.% oxygen), 1500-Å-thick nitrogen-doped polycrystalline silicon film (52 at.% nitrogen), and a silicon-dioxide film 1 μm thick. The oxygen-doped film has been deposited at 650°C from SiH_4–N_2O–N_2 system, the nitrogen-doped film from the SiH_4–NH_3–N_2. The reverse leakage currents I_{CEO} were about 0.4 μA for the 800 device and 70 μA for the 2500 V device; they showed good stability when exposed to water vapor at 100°C for 1 hr and sodium contamination at 200°C for 2 hrs.

15.5
p-n junction with field-limiting rings

In order to reduce the field intensity at the p-n junction surface and thereby increase its breakdown voltage, Kao and Wolley [15.10] have proposed the use of an additional ring-type junction concentric with the main junction which can be diffused at the same time as the main junction (Figure 15.15). This field-limiting ring is floating electrically and is spaced from the main junction so that when one starts increasing the applied reverse main junction potential, the depletion region will eventually punch through to the ring junction much before the voltage reaches the breakdown of the flat (plane) portion of the junction. After the punch-through has occurred, further increase in the applied potential will bring about a further increase of the *ring-depletion region*.

Figure 15.15 p-n junction with field-limiting ring. (After Kao and Wolley [15.10].)

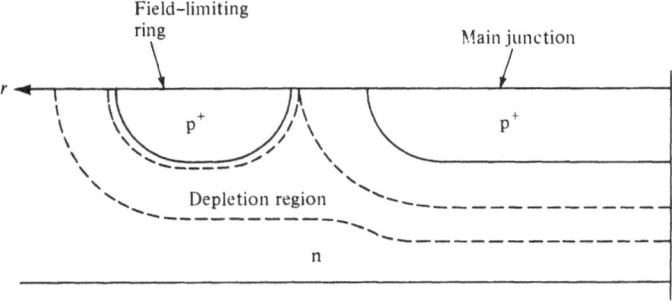

231

The maximum electric field across the depletion region of the main junction at the surface is determined by the punch-through voltage. The spacing between the ring and the main junction, as was mentioned above, limits the field much below the breakdown field. The ring junction acts like a *voltage divider*, and the voltage between the main junction and the first ring is essentially determined by their spacing. After the punch-through to the ring junction occurs, the field intensity at the main junction edge is appreciably reduced since the radius of curvature becomes now much larger.

15.5.1
Experimental results

The device investigated by Kao and Wolley [15.10] consisted of a circular junction surrounded with *several* concentric field-limiting ring junctions as per Figure 15.16, which shows the geometry and some dimensions of the experimental device. Variable spacing between rings was selected in order to study the variation of the punch-through voltage on a single structure. Both p^+-n and n^+-p type junctions were fabricated with the same geometry with n-type material of 80–220 and 450–650 Ωcm for p-type material. Typical junction penetration was about 50 μm. The final surface concentration was 5×10^{18} to 1×10^{19} cm^{-3} for both the p- and n-type diffusions. The authors were able to obtain voltages up to 2000 V for p^+-n structures with n-type silicon of 80 Ωcm and 3200 V with 220 Ωcm. The breakdown voltages of the highest voltage devices were considerably greater than that of a 60-μm cylindrical junction which was the near equivalent of these junctions. One of the methods used by Kao and Wolley [15.10] for the experimental confirmation of the proposed voltage-dividing model consisted of the point-by-point measurement of voltage–current characteristics of all test device junctions and fitting the experimentally obtained reverse

Figure 15.16 Geometry of the experimental device. (After Kao and Wolley [15.10].)

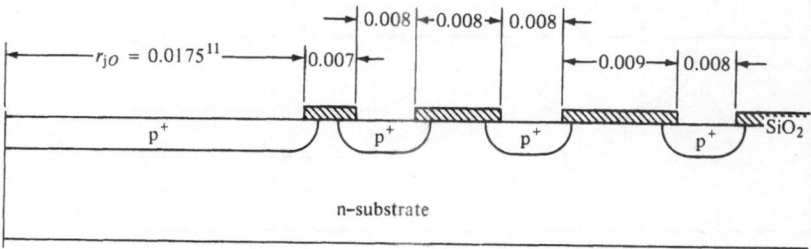

currents to a simple model of generation as a function of the voltage and the device geometry. It was assumed that the reverse currents originated entirely from the junction-depletion regions according to Sah–Noyce–Shockley theory [15.11]. The generation-recombination current was computed using the relationship

$$I_{rec} = \frac{qn_i}{\tau_{eff}} U \tag{15.1}$$

where U is the total volume of the depletion region, q is the electron charge, τ_{eff} is the effective lifetime of carriers, and $n_i = 1.4 \times 10^{10}$ cm^{-3}, the intrinsic carrier concentration at 300°K. The effective lifetime was determined by curve fitting and was found to be $\tau_{eff} = 8$ μsec. The measurements were performed in a very carefully controlled ambient (dry nitrogen). The results of measurements on one such p$^+$-n device fabricated on 220-Ωcm silicon are plotted in Figure 15.17. The results show that the largest part of the applied potential is supported by the main junction and that each succeeding ring supports a smaller part of the total potential.

At voltages below about 50 V the reverse current is proportional to the square root of the applied voltage. This is expected from the

Figure 15.17 *V–I* characteristics of the main p$^+$-n junction and the rings. (After Kao and Wolley [15.10].)

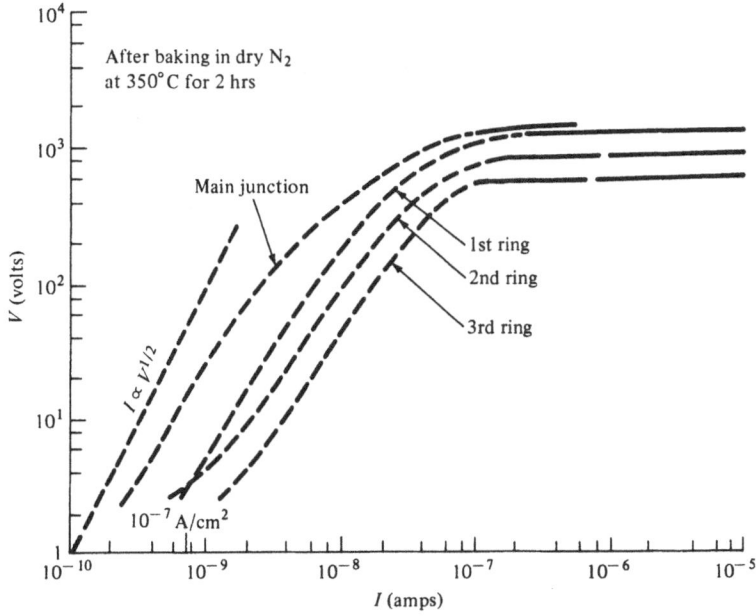

Sah–Noyce–Shockley recombination-generation theory since the depletion-region width where the generation takes place is proportional to $V^{1/2}$ for an abrupt junction. At higher voltages, the slope of the V–I characteristics decreases. Experiments were performed also with n^+-p diodes provided with field-limiting rings. The p-type silicon substrate resistivity was 450–650 Ωcm. The V–I characteristics were measured again under carefully controlled ambient conditions and are plotted in Figure 15.18, which shows the division of the voltage among the main junction and all the three rings. There is a very rapid increase of the leakage current at very low applied potentials. When compared to similar p^+-n diodes, the leakage currents of the n^+-p diode at 10 V are four orders of magnitude higher. These results were shown to be due to the formation of an inversion layer at the p-type silicon surface.

The rings may still be able to perform their function after the inversion layer is pinched off, but the high leakage will remain objectionable. It is, therefore, necessary to use all possible precautions to avoid the oxide contamination and channel formation.

Figure 15.18 Voltage–current characteristics of n⁺-p diode. (After Kao and Wolley [15.10].)

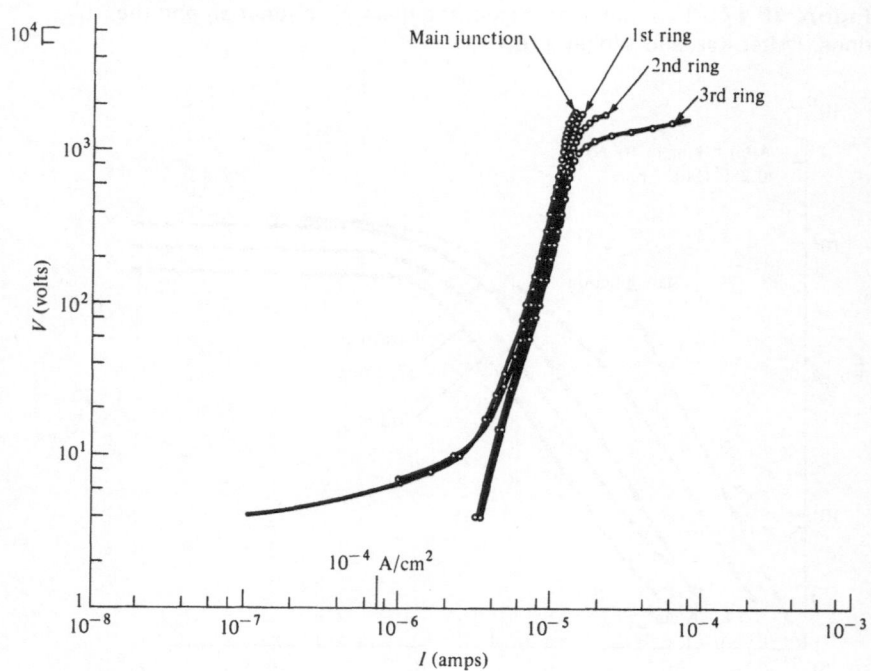

15.5.2
Field calculations with field-limiting rings

Adler and Ferro [15.12] have presented a theoretical and experimental study of planar diffused diodes with field-limiting rings in the presence of a surface charge. Circular planar diodes were fabricated to test the result of the computation performed for reverse-biased junctions by an approach similar to that used by Davies and Gentry [15.13] and by Cornu [15.14]. The method involves the solution of the two- and three-dimensional Poisson equation, relying on the assumption that the electric field in the space-charge (depletion) region vanishes at the region's boundaries and that charge neutrality exists outside the space-charge region. An additional criterion introduced by Adler and Ferro states that once the fields have reached a steady-state condition upon the application of the reverse potential, there must be a zero net current flowing into the ring. Figure 15.19 shows the cross-section of the circular-type diode under consideration. To satisfy the criteria of the zero net current into the ring, the ring junction cannot be either completely reverse or heavily forward biased. To satisfy this requirement, the potential of the ring must be equal to the lowest potential along the metallurgical ring-substrate junction (V_{low}) minus the built-in potential of about 0.8 V. Any voltage different from this value would result in a net forward or reverse current.

The shapes of the depletion-region boundaries for the case of a

Figure 15.19 Circular diode showing the location of the depletion region for voltages at and above the punch-through voltage for a device with positive surface charge. (After Adler and Ferro [15.12].)

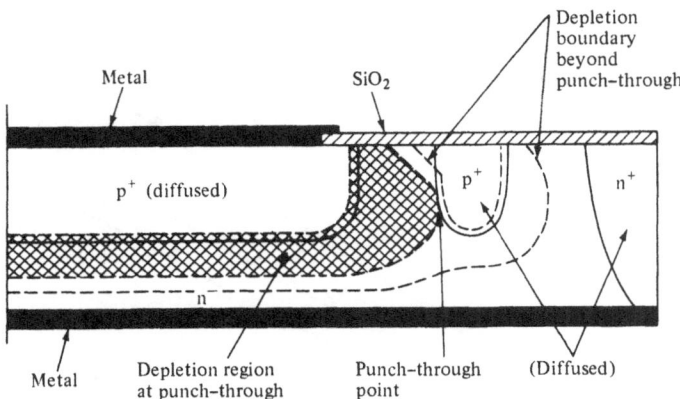

positive surface charge in the SiO_2 are shown in Figure 15.19. The point at which the punch-through takes place is much below the surface. It can be seen that because of the inversion layer (channel) presence, there will be an undepleted semiconductor region of a triangular shape even when the potential is increased above the punch-through voltage value. This triangular region does not exist when the oxide charges are either negative or nonexistent. The presence of the nondepleted triangular region and its effect on the voltage-dividing capability of the structure is not yet fully understood. Theoretical computations and experimental data [15.12] indicate (Figure 15.20) that the field ring is not 100-percent effective in taking over the excess voltage above the punch-through voltage. The discrepancy between the expected ring behavior and the calculated performance is due to the multidimensional nature of the problem. As a consequence, there is a larger voltage drop across the main junction at the surface than expected from the one-dimensional theory, and as a result a smaller potential appears across the ring junction.

The spacing between the main junction and the ring has a substantial effect on the ring's voltage-dividing ability. The larger the ratio of the junction depth to the spacing between the junction and the ring, the better the ring performance. The voltage-dividing property is also affected by the presence of the surface charges, and the silicon-doping level. Positive charge which induces a channel on the p-type substrate reduces significantly the field-limiting ring effect.

Figure 15.20 Experimental and theoretical dependences of the field-ring voltage on the applied voltage to the main p^+-n diode. (After Adler and Ferro [15.12].)

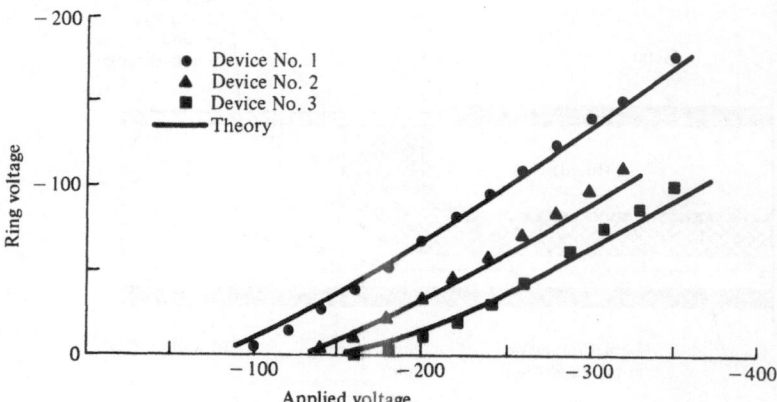

15.5.3
Devices with ultrahigh breakdown voltages

Matsushita *et al.* [15.15] suggested and developed a novel structure utilizing field-limiting rings capable of breakdown voltages in
excess of 10000 V in a single device. The new structure has been
applied to thyristors (gate-controlled switches) and transistors. As
can be seen from Figure 15.21, the concentric field rings are used on
both the top and the bottom surfaces of the silicon pellet. The contacts to both n- and p-type regions are made on the top surface. By
a judicial choice of electrode spacing, the breakdown of the main
junction (according to [15.15]) can be avoided and shifted to the
most exterior ring junction and it increases with the number of the
rings used. This result seems to contradict, to some extent, the more
pessimistic predictions of Adler and Ferro [15.11]. The principle of
operation of the multiple-ringed device relies on the pellet being thin
enough so that the depletion region reaches through to the bottom
of the pellet at the comparatively low voltage of a few hundred volts.
After that, further spreading of the depletion region may occur only

Figure 15.21 Cross section of a structure with the field-limiting
ring junctions on both the top and the bottom surfaces of the
substrate. (After Matsushita *et al.* [15.15].)

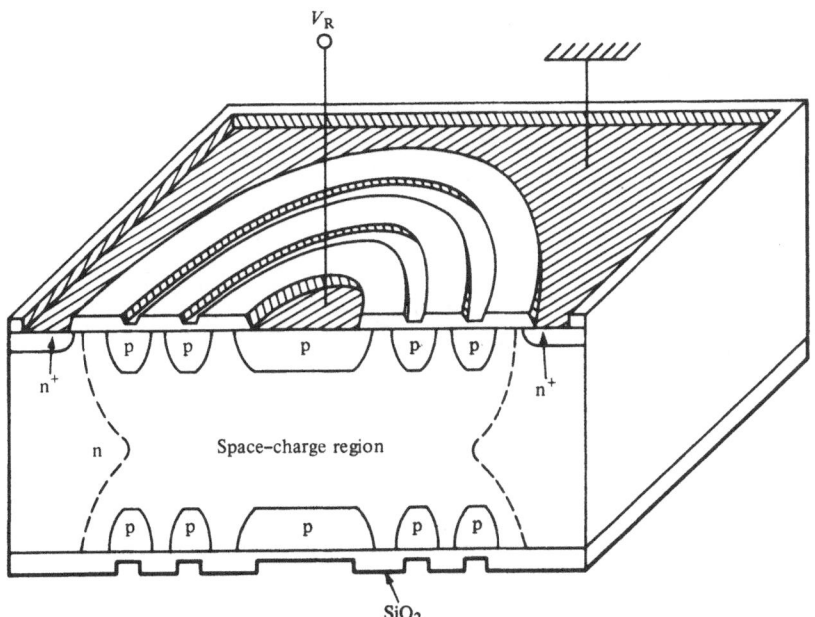

laterally to the right and to the left of the device center. The higher the voltage, the more lateral spreading will take place, and the applied potential will be divided between the main junction and the ring junctions at the top surface. A similar situation will exist at the bottom surface where there are almost an equal number of rings present so that the applied voltage will be distributed the same way as on the top surface. Figure 15.22 compares the breakdown voltage of the diode with field rings to the mesa and to the standard planar-type diodes. The breakdown voltage of the new structure does not depend on the resistivity of the substrate, so that relatively low resistivities yield very high breakdown. Figure 15.23 is the cross-sectional view of a thyristor utilizing the double-ring principle. The thyristor had a pellet size of 4 × 7 mm, the n-type resistivity was 60 Ωcm, and the total number of field-limiting rings was 10 on the top and 11 on the bottom surface. The n-emitter depth was 10 μm, the p-base depth was 25 μm, and the pellet thickness 200 μm. Maximum current $I_{T_{max}} = 100$ mA and the turn-off gain $G_{off} = 10$, with a gate current of 1 mA. The maximum blocking voltage achieved was 5000 V.

Due to the lateral rather than standard vertical structure of the thyristor, the main current flow is away from the cathode and the

Figure 15.22 Breakdown voltage of diodes with various structures as a function of resistivity. (After Matsushita *et al.* [15.15].)

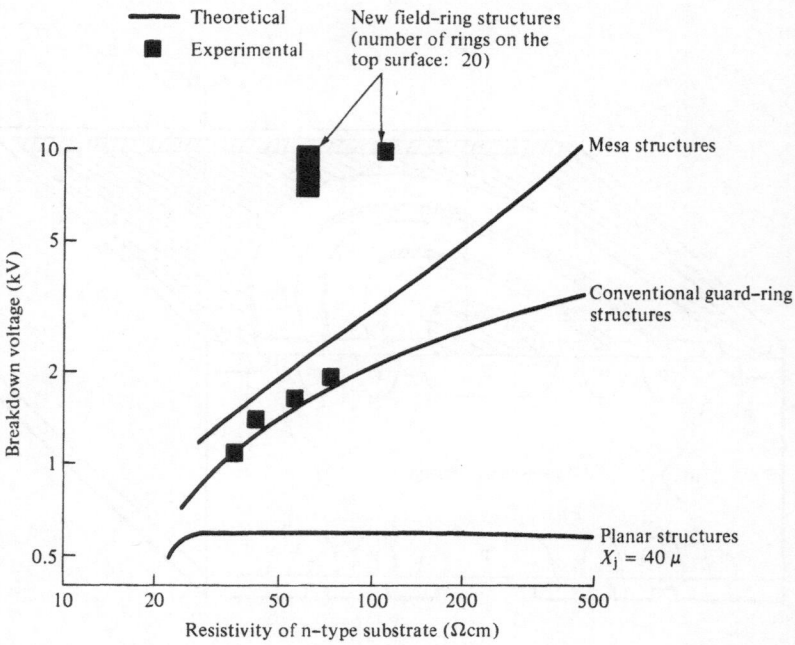

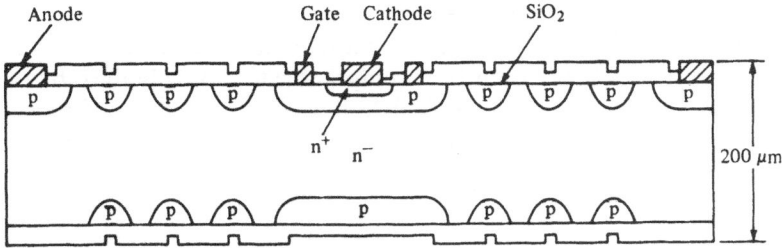

Figure 15.23 Cross section of an SCR with the double field-limiting ring junctions. (After Matsushita *et al.* [15.15].)

p-base radially in the n-base toward the p-type anode. The width of the n-base is, therefore, obviously much longer than that obtained with a standard SCR construction. As a result, a larger voltage drop will occur in the n-base of the new structure than in the ordinary SCR for the same current level. This obviously puts a limit on the maximum current one can use without running into very large dissipation problems.

15.6
Etch-contoured planar junctions

Temple and Adler [15.3] achieved high avalanche breakdown voltages in planar p-n junctions by extending the heavily doped side of the junction beyond the metallic contact and by etching away part of the heavily doped side as shown in Figure 15.24. (This is the same approach as described for the plane (mesa) structure in Section 14.6) As with mesa devices, the success in obtaining high-voltage breakdowns in a consistent way depends on the ability to carefully control the chemical etching, an operation which usually presents serious difficulties (the breakdown is a sensitive function of the etching depth).

Figure 15.24 Planar junction with partially etched highly doped p+ region. (After Temple and Adler [15.3].)

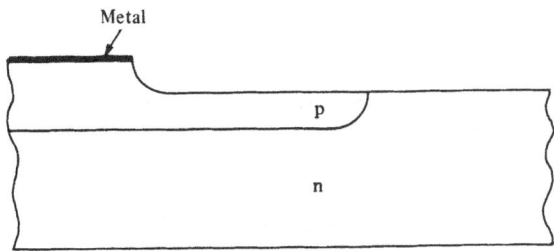

239

Temple and Adler [15.3] have been able to obtain breakdown voltages very close to those they computed theoretically. The results are somewhat better or at least comparable to those obtained by the authors when they used single field-limiting rings or etched a deep moat into the planar structure. The devices had to be surface passivated.

References

15.1 S. M. Sze and G. Gibbons. Effect of junction curvature on breakdown voltages in semiconductors. *Solid State Electron.*, *9*: 831, 1966.

15.2 O. Leistiko, Jr. and A. S. Grove. Breakdown voltage of planar silicon junctions. *Solid State Electron.*, *9*: 847–852, 1966.

15.3 V. A. K. Temple and M. S. Adler. Calculation of the diffusion curvature related avalanche breakdown in high voltage planar p-n junctions. *IEEE Trans. Electron Devices*, *ED-22*: 910–916, 1975.

15.4 D. S. Zoroglu and L. E. Clark. Design considerations for high voltage overlay annular diodes. *IEEE Trans. Electron Devices*, *ED-19* (1): 4–8, 1972.

15.5 L. E. Clark and D. S. Zoroglu. Enhancement of breakdown properties of overlay annular diodes by field shaping resistive films. *Solid State Electronics.*, *15*: 653–657, 1972.

15.6 T. Aoki, T. Matsushita, H. Yamoto, H. Hayashi, M. Okayama, and Y. Kawana. Oxygen-doped polycrystalline silicon films applied to surface passivation. *J. Electrochem. Soc. Technical Digest*, March 1975, pp. 167–170.

15.7 G. Spaden and P. L. Hower. A new high voltage planar transistor technology. International Electron Device Meeting, Washington, D.C., October 1971.

15.8 E. Muñoz, J. M. Boix, J. Liabrés, J. P. Honico, and P. Piqueras. Electronic properties of undoped polycrystalline silicon. *Solid State Electron.*, *17*: 439–446, 1974.

15.9 T. Matsushita, T. Aoki, T. Otsu, H. Yomoto, H. Hayashi, M. Okayama, and Y. Kawana. Highly reliable high voltage transistors by use of the Sipos process. *IEDM Technical Digest*, Washington, December 1975, pp. 167–170.

15.10 Y. C. Kao and E. D. Wolley. High-voltage planar p-n junctions. *Proc. IEEE*, *55* (8): 1409–1414, 1957.

15.11 C. T. Sah, R. N. Noyce, and W. Shockley. Carrier generation and recombination in p-n junction characteristics. *Proc. IRE*, *45*: 1228–1243, 1957.

15.12 M. S. Adler and A. P. Ferro. Field calculations of planar diffused semiconductor field limiting rings. *IEDM*, December 1973.

15.13 R. L. Davies and F. E. Gentry. Control of electric field at the surface of p–n junctions. *IEEE Trans. Electron Devices, ED-11*: 313–323, 1964.

15.14 J. Cornu. Field distribution near the surface of beveled p-n junctions in high voltage devices. *IEEE Trans. Electron Devices, ED-20*: 347, 1973.

15.15 T. Matsushita, H. Hayashi, and H. Yagi. New semiconductor devices of ultra-high breakdown voltage. *IEDM*, December 1973.

16

Thyristor thermal response

Summary

Heat is generated in a thyristor when it is in either the blocking or the conducting mode. In the first case, the heat generation takes place primarily at the reverse-biased blocking junction. For the case of forward conduction, it is customary to assume (for simplification) that the heat is generated in the thyristor in the plane passing through the device center and parallel to the center junction.

The maximum allowable thyristor dissipation is determined by the maximum permissible junction temperature and the device's thermal resistance. A generalized thermal impedance concept has been introduced in order to establish maximum dissipation limits during such transient conditions as the turn on, turn off, and current surges.

16.1
Introduction

One of the most important thyristor characteristics is the maximum temperature at which the device can still operate reliably under steady-state or transient conditions such as those existing during current surges.[1]

When a semiconductor device is in operation for any period of time, heat is generated as long as there is a current flow either in the forward-conducting or blocking state.

In the blocking state the current flow is due to generation-recombination phenomena and surface leakage. At room temperature this current is on the order of microamps for good devices. At

---

[1] By USA industry definition a surge (nonrepetitive) on-state current is a current of short time duration and specified waveshape.

242

elevated temperatures this type of current is very much increased so that power dissipation, particularly for high-voltage devices, may be quite appreciable. This current does not cause any appreciable Joule heating in the device body so that the heat generation is limited to the space-charge region of the blocking junction, where the generated hole–electron pairs undergo many collisions with the semiconductor crystal lattice while being accelerated across the depletion region by the electric field, increasing the lattice temperature. If the junction reaches the avalanche breakdown condition and the current level is high, there is always a possibility of current localization in a very small region due to the junction nonuniformities caused by impurity precipitants, junction irregularities, etc. Because of the highly localized nature of the current, the heat will be confined to a very small area and, as a result, the device may be destroyed if the maximum allowable temperature is exceeded.

When a thyristor is in the conducting state, the number of places where the heat is generated is much increased. In the on state, potential drops develop due to the majority carrier flow in the body of the device and at the contacts; thus Joule heat is generated. In the on state, all three junctions are forward biased and minority carriers are injected to some degree in all device regions. This gives rise not only to heating but to cooling of some device regions, as discussed below. Because of this complicated thermal situation, it is customary (for the sake of simplicity) to assume that the heat in a thyristor is generated in a plane passing through the device center and parallel to the center junction. This simplification is permissible when the device is operated for a time longer than the time needed for the heat to go through the silicon pellet [16.1, 16.2].

16.2
Heat generation and absorption in the forward-biased thyristor

The flow of current through a forward-biased p-n junction involves (1) the generation of a hole at the p-silicon–metal ohmic contact; (2) the generation of an electron at the n-silicon–metal ohmic contact; (3) recombination of hole–electron pairs in the silicon between the contacts; (4) disappearance, i.e., the infinite recombination rate of electrons at the p-silicon–metal contact; and (5) disappearance of holes at the n-silicon–metal contact. The region where most of the recombination takes place is determined by the diode dimensions in relation to minority-carrier diffusion lengths. For poor lifetimes, when the p-region and n-region thicknesses are much larger than the respective L_p and L_n diffusion lengths, the forward current will be determined by the recombination of hole–electron pairs in or near the space-charge region. On the other hand, for perfectly good lifetimes little recombination will take place in the semiconductor and almost all minority carriers will disappear at the ohmic contacts. For intermediate values of the lifetime, the current will be determined both by the carriers' recombination in the neutral region of the semiconductor and by their vanishing at the ohmic contacts. For simplicity, it is expedient to analyze the forward current in the two limiting cases of the very poor and very good lifetimes.

16.2.1
Case of a poor lifetime

This situation was studied by Stafeev [16.3] and by Bullis [16.4]. In this case the majority carriers are generated at the ohmic contacts, the heat is absorbed from the crystal lattice by the charge carriers, and the metal–silicon interfaces are cooled. The generated carriers move toward the junction and recombine in its vicinity, liberating the energy in the form of heat. The amount of heat liberated less the amount of heat absorbed per second is equal to $I_F V_A$, the product of the forward current and the applied potential. The crystal will *heat up* in the junction vicinity.

16.2.2
Case of a good lifetime

If the minority carrier lifetime is good, we have $d_p \ll L_n$ and $d_n \ll L_p$ (Figure 16.1). This condition has been studied by Stafeev [16.3], Bullis [16.4] and Hall [16.5]. In this case, cooling takes place at the p-n junction; the minority carriers are continually injected across the internal potential barrier ψ and, by their kinetic energy, carry heat away from the junction. The electrons are injected from the n into the p region, overcoming the barrier height ψ while the holes proceed in the opposite direction. The heat being carried away is given by

$$P_j = (I_n + I_p)\psi \tag{16.1}$$

where I_n and I_p are the electron and hole current, respectively. Since we have assumed a very good lifetime, almost all the minority carriers reach the ohmic contacts. The Fermi levels of the metal contacts and of the silicon are aligned at the metal–silicon interfaces since, ideally, no voltage drop occurs at these contacts. If ϕ_n and ϕ_p are energy barrier heights at the ohmic contacts, then the electrons will have to climb the energy barrier ϕ_n while coming from the metal into the

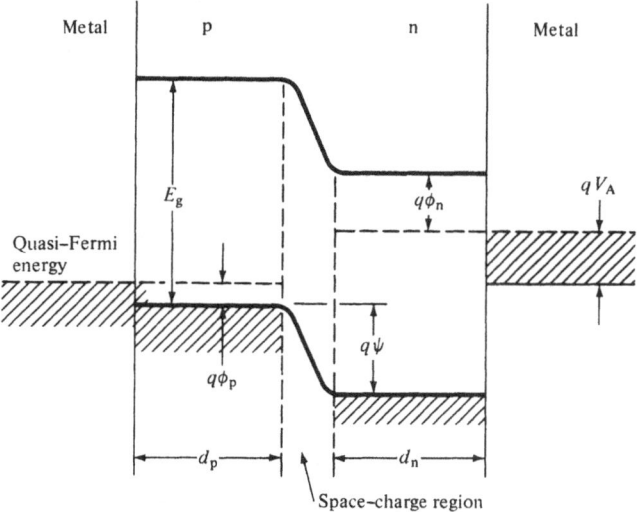

Figure 16.1 Forward-biased p-n structure with ohmic metal contacts [16.4].

semiconductor; by doing so, they will absorb $I_n \phi_n$ watts at the n-metal contact. The interface will be cooled. Assuming again no significant recombination, the same electrons will release $I_n[(E_g/q) - \phi_p]$ watts at the p-metal contact since they will drop from a higher to a lower energy level. By adding algebraically all these components, we obtain the total electron energy

$$P_n = I_n\left(\frac{E_g}{q} - \phi_p - \phi_n\right) \tag{16.2}$$

Analogously we will have for holes

$$P_p = I_p(I_g - \phi_p - \phi_n) \tag{16.3}$$

Therefore, the total power dissipation is

$$P = P_n + P_p + P_j = \left(\frac{E_g}{q} - \phi_p - \phi_n - \psi\right)I_F \tag{16.4}$$

since $I_n + I_p = I_F$. However, we can see from Figure 16.1 that $E_g/q = \phi_n + \phi_p + \psi + V_A$; therefore, the total power dissipation is

$$P = V_A I_F \tag{16.5}$$

Regardless then of the device carrier lifetime and the device dimensions, the dissipated power will always be equal to the $V_A I_F$ product.

245

For poor lifetimes, due to the recombination, the heat is liberated in the vicinity of the forward-biased junction; for good lifetimes, cooling occurs at the p-n junction because the minority carriers carry their heat away from the junction and the energy of electrons is released (heat is generated) at the p-metal contact while the energy of holes is released at the n-type contact.

When the device dimensions are comparable to the diffusion lengths of the corresponding minority carriers, the situation becomes less clearcut since the p-n junction is cooled by injection but, at the same time, is heated by the energy of recombination released in the junction vicinity.

16.2.3
p-n-p-n structure

The results of the foregoing can be applied now to the case of a p-n-p-n structure in the on state, when all three device junctions J_1, J_2, and J_3 (Figure 16.2) are forward biased [16.4]. For the sake of simplicity it is assumed that a negligible amount of holes flows through the J_3 junction (n emitter) and that the same is true for the electrons flowing through the J_1 junction (p emitter). This implies that the hole–electron recombination must take place in the region between the J_1 and J_3 junctions.

If I_{p_2} is the hole current going from region 2 into region 3 and I_{n_2} is the electron current going from the region 3 into region 2, then

Figure 16.2 Forward-biased p-n-p-n structure with ohmic metal contacts [16.4].

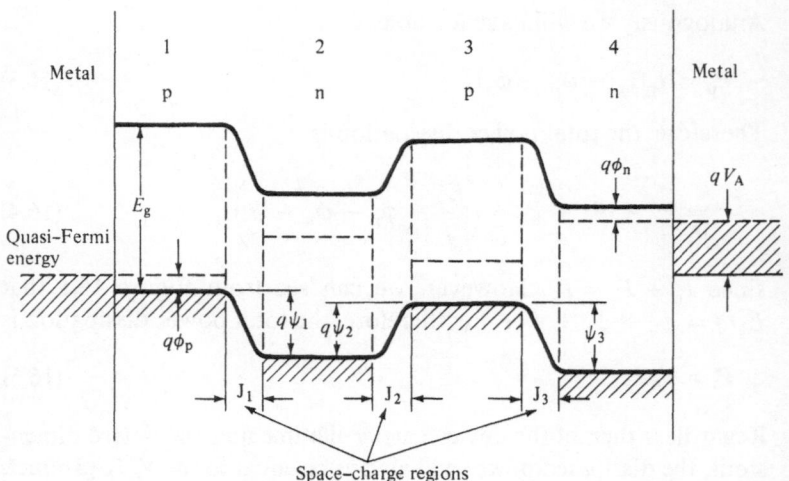

the total current collected by the J_2 junction is

$$I_F = I_{p_2} + I_{n_2} \qquad (16.6)$$

Since, according to our assumption, the current is carried by holes only in p region 1 and by electrons only in n region 4, the hole current reaching the p contact and the electron current reaching the n contact are both equal to I_F. The equilibrium hole density in region 1 is maintained by electrons flowing *into* the metal, which is equivalent to the injection of holes into the semiconductor with an absorption of heat (cooling) equal to $I_F \phi_p$.

On the right-hand side of the device the electrons are injected over a barrier ϕ_n in order to maintain the equilibrium electron density in region 4 with the amount of heat absorbed equal to $I\phi_n$.

Holes injected across the J_1 junction extract $I_F \psi_1$ watts, and the electrons injected across the J_3 junction absorb $I_F \psi_3$. At the junction J_2, a total of $I_F \psi_2$ watts is liberated. In addition, the carrier recombination process liberates $I_F E_g/q$ watts in regions 2 and 3. Adding algebraically all these heats absorbed and generated we obtain

$$P = I_F\left(\psi_2 - \phi_p - \phi_n - \psi_1 - \psi_3 + \frac{E_g}{q}\right) \qquad (16.7)$$

But since (from Figure 16.2)

$$\psi_2 + \frac{E_g}{q} - \phi_p - \phi_n - \psi_1 - \psi_3 = V_A \qquad (16.8)$$

we have again

$$P = I_F V_A \qquad (16.9)$$

The same result would be obtained if we assumed that the recombination occured in regions 1 and 4 rather than in 2 and 3 in all cases; the heat is liberated (heating) at the center junction J_2.

It is obvious by now that the heat in the p-n-p-n structure is generated and absorbed in several places and any accurate thermal mapping or computation would require a thorough identification of heat sources and sinks. For approximate calculation, however, it is customary to assume for the p-n-p-n structure that the heat is generated in a plane going through the device and parallel to the center junction. This approximation appears to be adequate as long as the heat transit time through the device is short compared to the duration of the applied current pulse. Since the heat diffusivity in silicon is about 0.9 cm^2/sec, the transit time through a 0.025-cm-thick silicon pellet is less than 1 μsec.

It is common to call the temperature at the center plane the junction temperature.

16.3
Determination of junction temperature

16.3.1
Maximum allowable junction temperature

Power is dissipated and heat is generated in a thyristor during all phases of the device operation, i.e., during the turn on, the conduction period, the turn off, and the blocking period. Power also is dissipated in the gate during the time the gate potential is applied. The major part of the dissipation, however, is due to the power loss when the device is in the conducting state. In some cases, when special care is not taken and the di/dt rate of the current rise is very high and the current is very much localized, the power dissipation may become excessively high. Also, at high operating frequencies the turn-on losses may exceed the on-state losses. Due to the power dissipation, the device temperature rises until the thermal equilibrium is reached, as determined by the rate of the device cooling. The maximum allowed temperature under these circumstances is usually much below the temperatures which may be deleterious to the device because of surface degradation, solder melting, etc. The upper limit is determined by the capability of the device to function within the specified ratings. In thyristors the maximum temperature is usually allowed to rise to 125°C since above this temperature excessive leakage current may cause the device to lose the gate-control capability.

If the heating of the device proceeds at a faster rate than its cooling and the thermal feedback mechanism is present, then a thermal runaway situation may arise, leading to the device destruction. This may happen, e.g., when the reverse-blocking losses generated in the junction increase with junction temperature at a faster rate than the dissipation capability of the heat sink. Figure 16.3 illustrates a thermal condition for which the junction temperature increases at a rate greater than the dissipation capability of the heat sink [16.6]. The figure indicates that some level of power dissipation (5 W) is associated with conduction in the on state. The thermal resistance of the heat sink is such that the junction temperature increases with, e.g., dissipation at a rate of 10°C/W. A device operating in this condition operates stably as a thermal system at point A with some temperature rise. If now the heat sink is made less effective in device cooling, the rate of increase of the junction temperature rises and the slope changes, e.g., from 10 to 20°C/W so that the operating point of the device heat-sink system shifts from point A to point B, which is unstable; a thermal runaway then occurs and the device temperature increases without control until the device is degraded or destroyed.

In the case of power surges of very short duration, it is possible

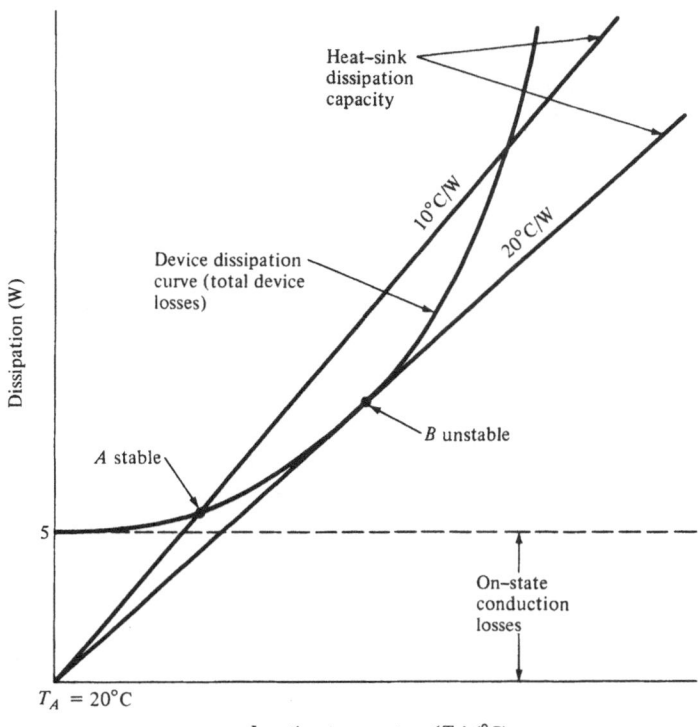

Figure 16.3 Stable and unstable thermal situation [16.6].

for the temperature to rise much above the limit imposed by the device ratings. Although device control may be temporarily lost, the device is not going to be damaged if the temperature rise is of very short duration.

During a portion of the turn-on time while the current has not yet time to spread along the cathode, the localized temperatures also may rise much above the maximum rated value, as was shown in Chapter 10.

16.3.2
Thermal resistance

In order to determine the device temperature rise theoretically, it is necessary to know both the maximum power dissipated at a given switching state and the device thermal resistance. The temperature rise is determined from the product of the dissipated power and the thermal resistance

$$\Delta T = P\theta \tag{16.10}$$

249

where ΔT is the temperature rise above ambient (or heat-sink temperature), P is the dissipated power, and θ is the device thermal resistance. Equation (16.10) holds both in the steady state and in the transient condition. However, in this last case, both ΔT and θ become time dependent. The transient θ is always smaller than the steady-state value.

16.3.3
Triac's thermal resistance

For a bidirectional p-n-p-n switch (triac) there are two junction temperatures, each associated with one of the two operating quadrants. In the ideal cases the thermal characteristics for both quadrants are nearly the same. More often, though, there is a considerable asymmetry in the thermal resistances of the device due to the different heat-flow paths and the forward voltage drops; it is, therefore, necessary to specify which quadrant is under consideration. The thermal resistance of a triac can be represented by a Y model [16.7, 16.8] shown in Figure 16.4. Branches A and B represent the thermal resistances of the two halves of the device. The common leg C represents the thermal resistance of the path through the package to the heat sink with reference temperature T_C.

16.3.4
Power dissipation in the on state

The forward voltage drop of a thyristor in the on state can be expressed (see Chapter 7) as

$$V_T = V_j + k_1 J_T^{1/2} + R J_T \tag{16.11}$$

where V_j is the J_3-junction voltage drop, k_1 is a constant dependent on the device physical parameters, J_T is the current density, and R is the ohmic resistance per unit area.

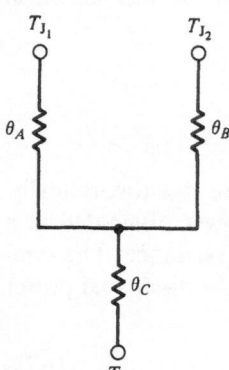

Figure 16.4 Triac thermal resistance equivalent circuit.

For a sinusoidal waveform, the instantaneous current may be expressed as

$$i_F = I_{TM} \sin \Theta \tag{16.12}$$

with $\Theta = \omega t$, where ω is the angular frequency, t is the time, and I_{TM} is the current peak value. The instantaneous dissipated power is

$$V_T i_F \tag{16.13}$$

Therefore, the average dissipated power can be computed from the integral

$$P_{(av)} = \frac{1}{2\pi} \int_{\Theta_1}^{\Theta_2} V_T i_F \, d\theta \tag{16.14}$$

where $\Theta_2 - \Theta_1 = \omega(t_2 - t_1)$ corresponds to the conduction interval $t_2 - t_1$.

16.3.5
Power dissipation in the reverse- or forward-blocking state

If the instantaneous reverse potential is $V_R = V_{RM} \sin \Theta$ and the instantaneous reverse current is i_R, the instantaneous reverse power is

$$P_R = V_{RM} i_R \sin \Theta \tag{16.15}$$

i_R is normally a function of time and of the potential V_R and may be represented [16.1] roughly as

$$i_R = I_0 + k_2 V_{RM} \sin \Theta \tag{16.16}$$

where I_0 is the intercept current with no gate current present, k_2 is the slope of the I–V characteristic, and $V_{RM} \sin \Theta$ is the instantaneous reverse potential (Figure 16.5). The average power dissipated will be

$$P_{R(av)} = \frac{1}{2\pi} \int_{\Theta_3}^{\Theta_4} V_{RM} i_R \sin \Theta \, d\Theta \tag{16.17}$$

where $\Theta_4 - \Theta_3 = \omega(t_4 - t_3)$, with $t_4 - t_3$ being the blocking interval of time.

The dissipation in the blocking mode is increased significantly if the gate of a thyristor is forward biased. In the forward-blocking mode the equivalent gain of the n-p-n-p hook transistor is equal to

$$\beta_F = \frac{\alpha_N}{1 - \alpha_N - \alpha_P} \tag{16.18a}$$

251

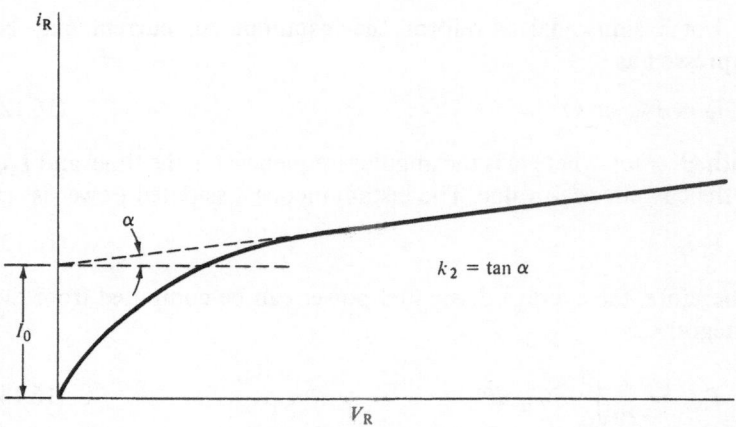

Figure 16.5   Instantaneous reverse current versus voltage [16.1].

so that if i_g is the gate current, the reverse leakage current of equation (16.16) should be increased by an additional term

$$\frac{\alpha_N i_g}{1 - \alpha_N - \alpha_P} \tag{16.18b}$$

In the reverse-blocking mode the equivalent gain is

$$\beta_R = \frac{\alpha_N}{1 - \alpha_{PI}} \tag{16.19a}$$

where α_{PI} is the inverse DC alpha of the p-n-p transistor section, and the leakage current of Equation (16.16) should be increased by an additional term of

$$\frac{\alpha_N i_g}{1 - \alpha_{PI}} \tag{16.19b}$$

16.3.6
Power dissipation during the turn-off

When a thyristor is switched off by a sudden application of a reverse potential as in the circuit-commutated turn off, one of the transistor sections always recovers before the other one (see Chapter 8). The high reverse current is limited at the beginning by the load impedance only, since the device still continues to present a small resistance to the current flow. This large current traverses one junction that is in avalanche breakdown. Under these conditions the power dissipated may reach an unacceptable level in some specific cases.

16.3.7
Power dissipation for the nonrecurrent surges

16.3.7.1 Step-like current surge. We are considering now the case when the thyristor is in the conducting state and either the applied voltage or the load impedance is changed in a steplike fashion. Under these conditions, the computation of the forward drop as per Equation (16.11) is not applicable since the expression was derived for the steady-state situation. When the load impedance is suddenly lowered, the current may rise only if the holes and electrons which were forward biasing the center collector junction are removed in order to supply the current carriers. The withdrawal of these charges takes the center junction out of saturation so that it becomes temporarily reverse biased and the device starts forward blocking for a short period of time. However, this situation does not persist, since the still forward-biased n- and p-type emitters inject enough electrons and holes to supplement the withdrawn ones and the center junction then becomes very rapidly rebiased in the forward direction. Thus, at the beginning of the transient the forward voltage drop becomes very large when the center starts blocking the current [16.1], but as soon as the carriers begin recharging the center junction the voltage drop begins to decay exponentially until it reaches its steady-state value V_T (Figure 16.6). The current, on the other hand, increases exponentially until it reaches its steady-state value. The power dissipated during this kind of a transient occurring due to the change in the load impedance or of the applied anode potential is distributed

Figure 16.6 Voltage transient due to a current surge [16.1].

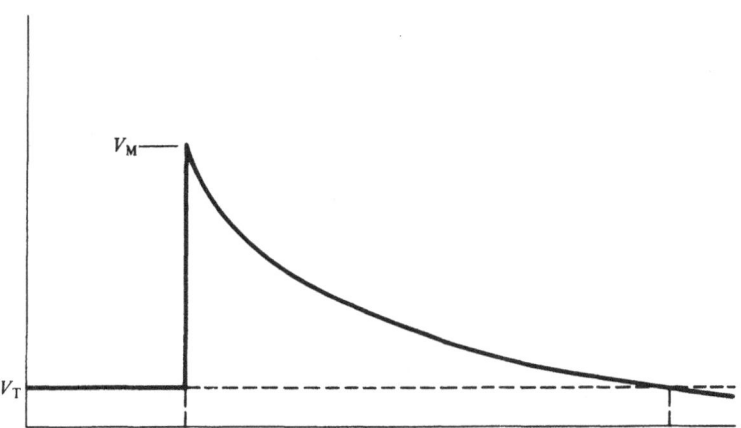

uniformly over the entire device area. If the transient is caused, however, by the gate triggering, then the dissipation will be highly localized.

The average power dissipated during the transient may be obtained from the following approximate expression [16.1], which can be derived for the case of a small circuit impedance:

$$P_{avg} \approx \frac{V_M I_M}{4.4} \frac{t_r}{t_s} + V_T I_M \tag{16.20}$$

where I_M is the steady-state value of current after transient; V_M is the instantaneous maximum voltage across the thyristor at the start of the transient, $V_M \gg V_T$ is assumed; t_r is the rise time for current from 10 to 90 percent of its final value; t_s is the duration of the surge period; and V_T is the forward voltage drop under steady-state conditions.

Expression (16.20) assumes that the current prior to the surge is much smaller than the current I_M after the surge and the current rise time is much smaller than the surge duration.

16.3.7.2 Gradual current surge. If the onset of the thermal dissipation is not instantaneous but gradual, then the power dissipated can be calculated using the steady-state voltage drop appropriate for the current level considered. For current surges considerably exceeding current densities of 1000 A/cm^2, Equation (16.11) may be approximated by

$$V_T = V_j + RJ \tag{16.21a}$$

where V_j is the voltage drop across the n-emitter junction at the current density J, and R is an equivalent resistance per unit area. The heat generated is, therefore,

$$P = V_j J + J^2 R \tag{16.21b}$$

per unit area of the device. In order to simplify computations, Gentry [16.2] made the following assumption in the case of a p-n diode: the power $V_j J$ is dissipated in a *single plane* dividing in two parts an infinitely long silicon bar; the $J^2 R$ heat is generated *uniformly* on one side only of that plane. With these assumptions the heat flow equation yields an expression for the temperature rise ΔT above the steady-state junction temperature before the occurrence of the surge:

$$\Delta T = aI^2 t + bV_j I t^{1/2} \tag{16.22a}$$

where a and b are constants and t is time. The first term of the right-hand side is due to the Joule heat and the second to the power dissipated in the junction. The maximum value of ΔT is just the temperature above which the device is degraded or destroyed and is equal to some constant value C, so that Equation (16.22a) becomes

$$aI^2T + bIt^{1/2} = C \qquad (16.22b)$$

where we assumed additionally that $V_j \approx 1$ V. The constant C could be computed, but it is much easier to obtain its value experimentally for a given device. It was found experimentally that Equation (16.22b) can be very well approximated by

$$I^n t = C \qquad (16.22c)$$

where $n \approx 2.7$ [16.9, 16.10]. For a pure resistor, like a fuse, the exponent $n = 2$, since the $V_j I$ term in Equation (16.22a) is non-existent. Since a thyristor during a surge can be represented by a simple diode, the above conclusions are valid for both types of these devices. It appears also that the relation (16.22c) is adequate for times considerably shorter and considerably longer than the duration of one-half cycle of a 60-cycle AC current.

In the event of a nonrecurrent surge, the rated junction temperature can be exceeded for a brief instant. For this short time the expected thyristor characteristics such as dv/dt, off-state blocking capability, will not be valid until the junction temperature drops down to its maximum rated value.

16.4
Generalized concept of thermal impedance

16.4.1
Heat flow

The typical method of thyristor cooling is to mount it on a metal heat sink which in turn may be cooled by convection, forced air, or water flow. In this discussion we limit ourselves to the review of the situation in which heat from the device is carried away *by conduction only* to a heat sink. This is the most common situation for power devices; therefore, device cooling by convection and radiation from the device enclosure may be normally neglected. Figure 16.7 illustrates schematically, and in one dimension only, how the heat is carried away from the central plane of a thyristor whose temperature we designate by T_j. The heat flows through the silicon pellet to the header and to the heat sink. The flow of the heat from the header to the air is negligible. The time-dependent heat-flow equation in one dimension [16.11] in a homogeneous medium can be written as

$$\frac{\partial T}{\partial t} = \alpha \frac{d^2 T}{dx^2} + \frac{q}{\rho c} \qquad (16.23)$$

255

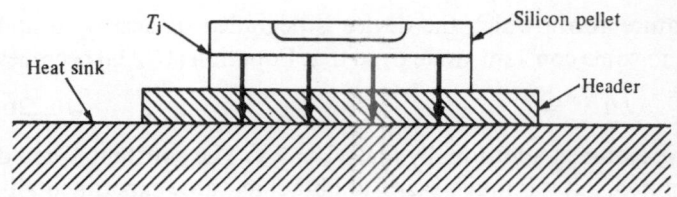

Figure 16.7 One-dimensional heat flow.

where c is the specific heat, k is the thermal conductivity, ρ is the material density, and

$$\alpha = \frac{k}{\rho c} \tag{16.24}$$

The parameter α is called the thermal diffusivity and is analogous to the charge-carrier diffusivity (diffusion constant); q is the heat generated per unit volume. In the steady state, Equation (16.23) becomes

$$-\frac{d^2 T}{dx^2} = \frac{q}{k}$$

The heat-flow equation in the steady state, therefore, reduces to the Poisson equation.

16.4.2
Thermal resistance and capacitance electric analog

According to Fourrier–Biot law the heat flow per unit time from a certain body can be expressed as

$$P = k \frac{dT}{dx} A \tag{16.25}$$

where A is the cross-sectional area available to the heat flow. If this law is applied to the steady-state heat conduction across some path of the length L, the above equation takes the form of

$$P = k \frac{\Delta T}{L} A \tag{16.26}$$

where ΔT is the temperature difference between the two ends of L. This can be rewritten in analogy to the Ohm's law as

$$P = \frac{\Delta T}{\theta} \tag{16.27}$$

where θ is the thermal resistance

$$\theta = \frac{L}{Ak} \qquad (16.28)$$

The temperature difference is analogous to the voltage; the heat flow, to the current; and θ, to the ohmic resistance. k and θ are temperature dependent.

The heat capacity C_T is the product of the material's mass m and the specific heat c

$$C_T = cm \qquad (16.29)$$

When a given sample absorbs a quantity of heat Q, its temperature is changed by an increment δT and the thermal capacitance, expressed in watt-seconds per degree Celsius, can be determined from the following expression

$$C_T = \frac{Q}{\delta T} \qquad (16.30)$$

16.4.3
Electrical transmission line analog

Very often the analytical solutions of the heat-flow equation are either difficult to obtain or closed-form solutions do not exist. A great simplification for the heat-flow problem is the possibility of using analogous electrical methods to simulate the thermal power flow. The heat flow can be well represented by the current flow in an electric transmission line approximated by an equivalent lumped circuit consisting of resistances and capacitances.

Table 16.1 gives the comparison of the various electrical and thermal quantities, while Figure 16.8 shows the thermal equivalent circuit for a transistor mounted on a heat sink.

Figure 16.8 Thermal equivalent circuit for a transistor mounted on a heat sink.

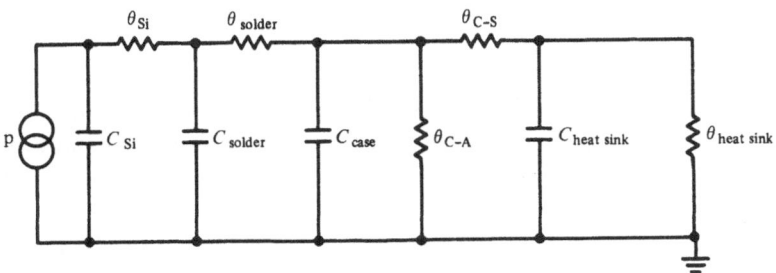

257

Table 16.1
Various thermal quantities and the corresponding electrical quantities

Thermal	Electrical
Heat generator (silicon crystal)	Current generator
Thermal resistance (θ) °C/watt	Resistance (R)
Thermal capacitance (C_T) watt-second/°C	Capacitance (C) coulomb/volt
Temperature difference ($T_1 - T_2$)	Potential difference ($V_1 - V_2$)
Power dissipation (P) watt	Current (I) amp
Thermal impedance °C/watt	Impedance (Z) volt/amp

The power dissipated within the silicon crystal results in a flow of the heat P in analogy to the flow of a constant current I (Figure 16.8). θ_{Si} is the thermal resistance of the thermal path in the silicon; θ_{solder}, in the solder and between the silicon pellet and the device case; θ_{C-S}, the thermal resistance from the case to the heat sink; θ_{C-A}, from the case to air; and $\theta_{heat\ sink}$, the thermal resistance from the heat sink to the ambient. The capacitances are marked accordingly. It was assumed in the above that the heat propagates in one direction only. In some cases, a double heat sink may be used; the device is mounted, then, in between the two heat sinks for a reduction of the overall thermal resistance.

16.4.4
Transient thermal impedance

If the device dissipates P watts and its total thermal resistance to the heat sink is θ, then the junction temperature rise above ambient will be

$$\Delta T = P\theta \tag{16.31}$$

ΔT is expressed in degrees Celsius, P in watts, and θ in degrees Celsius per watt. A similar equation holds for a transient condition with the proviso that ΔT and θ are now time dependent, so that Equation (16.31) becomes

$$\Delta T(t) = P\theta(t) \tag{16.32}$$

$\theta(t)$ is analogous to an impedance consisting of the combination of the resistances and capacitances. The thermal capacitance of the silicon pellet is obviously rather small because of its small mass; the thermal capacitance of the heat sink is generally much larger. For short-duration overloads, even the small thermal capacity may be large enough to prevent rapid device heating. The presence of a large thermal capacity will be helpful in the thyristor applications where large surge currents are present.

If a step function of power is applied to the device (Figure 16.9), the temperature difference between the junction and the case, taken as a reference point, rises as shown in the figure and approaches the steady-state temperature T_1 asymptotically. If the device stops dissipating power at the same time t_2 after the steady state was reached, the device will start cooling off as per Figure 16.9. The heating and the cooling curves (T_H, T_C) are conjugates of each other [16.12]. The ratio of the instantaneous temperature rise (curve T_H) to dissipated P is the device instantaneous thermal impedance $\theta(t)$. Figure 16.10 represents a typical plot of transient thermal impedance of a silicon-controlled rectifier in function of time. The thermal impedance shown for a silicon crystal mounted on an infinite heat sink approaches assympotically the steady-state thermal resistance with increasing time. At the other end, the thermal impedance curve approaches assympotically the impedance value for extremely short pulses.

It is of importance to know the magnitude of the temperature rise of the device for surge conditions which occur in switching such loads as electric bulbs or electric motors when the in-rush currents exceed many times the steady-state current. This temperature may

Figure 16.9 Response of junction temperature to the step input of dissipated power.

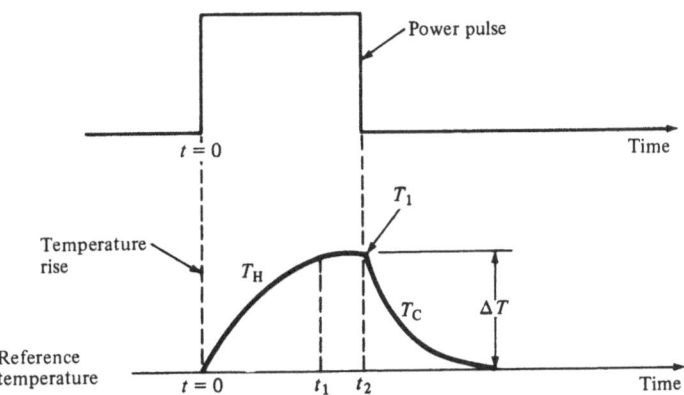

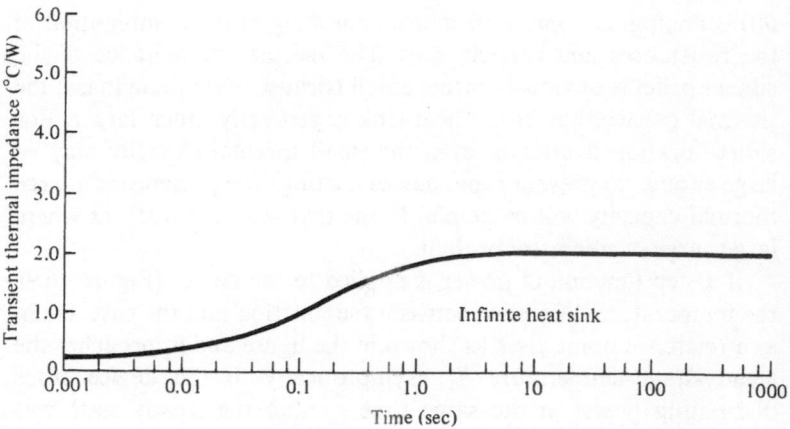

Figure 16.10   Typical transient thermal impedance of a silicon-controlled rectifier.

be determined if the transient thermal impedance is known, on the basis of measurements, and by using the superposition principle [16.12], to be discussed in Section 16.5.

16.4.4.1 Effect of the heat sink on the transient thermal impedance. The geometrical dimensions of the heat sink effect significantly the transient thermal impedance curve. In most cases, the size of the heat sink by the final user is not known; therefore, it may be necessary to make some simple computation in order to arrive at the θ-versus-time curve appropriate for the heat sink used. As a departure point, the curve for the device with the infinite heat sink can be used, to which the thermal transient impedance of the heat sink is added, yielding the transient impedance of the entire system [16.1, 16.11, 16.12].

16.5
Device temperature response to an arbitrary power waveshape

16.5.1
Single-pulse response

Curve T_H in Figure 16.9 represents the device temperature response to a sudden application of a load. After some time t_1, the instantaneous device temperature above the reference temperature will be

$$T_H = P\theta(t) \tag{16.33}$$

Once the input power is removed (time $= t_2$), the device starts cooling off following the T_C branch of the temperature curve. During this cooling phase the instantaneous device temperature above the reference point will be

$$T_C = P[\theta_s - \theta(t)] \tag{16.34}$$

where θ_s is $\theta(t)$ as $t \to \infty$, in other words, the steady-state thermal resistance. Equation (16.34) is valid only if the junction has reached its steady-state temperature before the input power is removed. Equations (16.33) and (16.34) permit the computation of the junction temperature at any time during the heating and cooling phases if the steady-state and transient thermal impedances are known.

16.5.2
Superposition approach for multiple pulses [16.12]

Quite often the heating pulses follow each other with very short time intervals so that the device cannot heat up to its steady-state value and cool off completely between the pulses. When this happens, the temperature increases after each successful pulse. An equilibrium is reached only when the temperature rise at each cycle becomes the same as the temperature drop, so that, eventually, successive peak temperatures are identical. This is the steady-state condition for that type of input. Superposition offers a numerical method of using the transient impedance curve to compute the temperature rise despite the fact that the steady-state condition implicit in Equations (16.33) and (16.34) is not met. This is achieved by expressing the input waveform by the superimposition of the positive and the negative step functions, with each of them extending to infinity in order to satisfy the steady-state requirement. The unit step is a function of time t and has the value equal to zero for all negative values of t but a value of unity for all positive values of time [16.13]. The function has, therefore, a discontinuity at $t = 0$. The unit step function is designated by $U(t)$. A similar function, but with the discontinuity at $t = t_1$ is, therefore, $U(t - t_1)$ (Figure 16.11). Figure 16.12a shows the heat input to the device in the form of rectangular pulses of diminishing amplitudes. These pulses can be replaced by an equivalent series of positive and negative step functions as per Figure 16.12b, which shows how the input pulses are constructed by superimposition of positive and negative step functions. By using the transient impedance curve of $\theta(t)$ one can obtain the individual temperature responses as per Figure 16.12c for each step function. Finally, Figure 16.12d represents the algebraic sum of the individual responses.

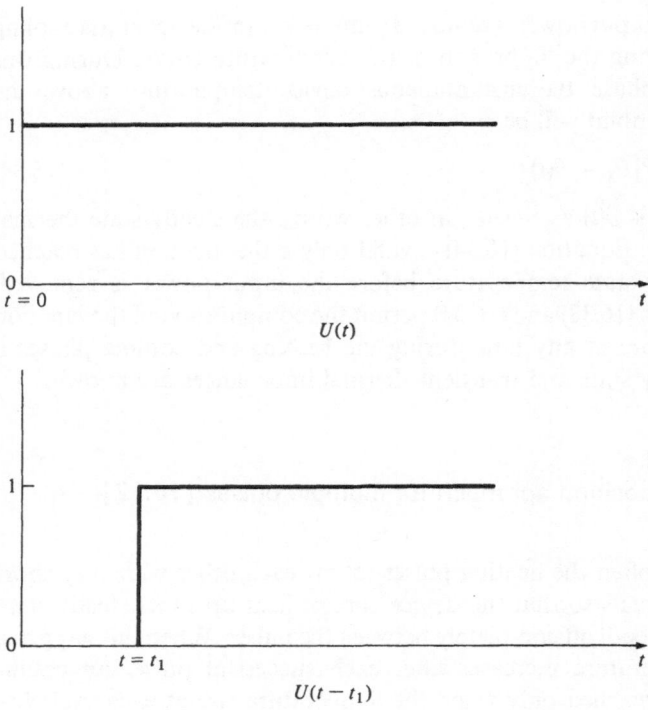

Figure 16.11 Unit step functions.

The superposition effects produced by the various steps at the times t_1, t_2, ... t_n of interest give a series of equations for the temperature rises of T_n above the reference temperature (heat sink temperature):

$$T_1 = P_0 \theta_1 \tag{16.35}$$

$$T_2 = P_0 \theta_2 - P_0 \theta_{(2-1)} \tag{16.36}$$

$$T_3 = P_0 \theta_3 - P_0 \theta_{(3-1)} + P_2 \theta_{(3-2)} \tag{16.37}$$

and so on. In these expressions T_n is the temperature rise at the time t_n, $\theta_{(2-1)}$ is the transient impedance at the time $(t_2 - t_1)$, and so on. By following this procedure it is possible to determine the temperature at any desired time for any number of rectangular pulses. The irregularly shaped pulses should be first approximated by rectangular pulses with the same averaged power and the same peak power (Figure 16.13) by altering the pulse duration by a constant N to maintain the peak-to-average relationship. This is the "worst-case" approximation since a rectangular power pulse will raise the junction temperature higher than any other waveshape with the same peak

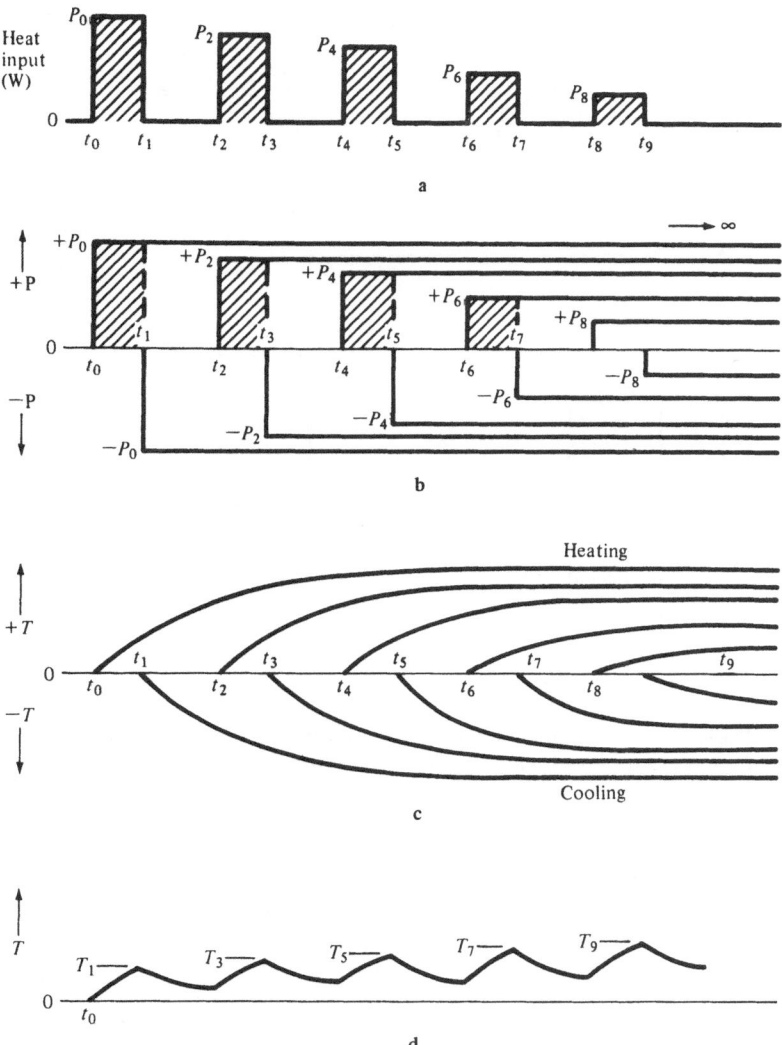

Figure 16.12 Use of superposition to determine peak junction temperature. (After Gutzwiller and Sylvan [16.12].)

and average values. A rectangular pulse concentrates the heating effect into a shorter time interval, minimizing the cooling during the pulse duration. Half- or full-wave sinusoidal inputs can be handled the same way as any other irregularly shaped pulses.

16.5.2.1 Superposition integral for exact temperature calculation.
If we allow the number of step inputs in Figure 16.12b to increase indefinitely and make all time intervals between the subsequent step

263

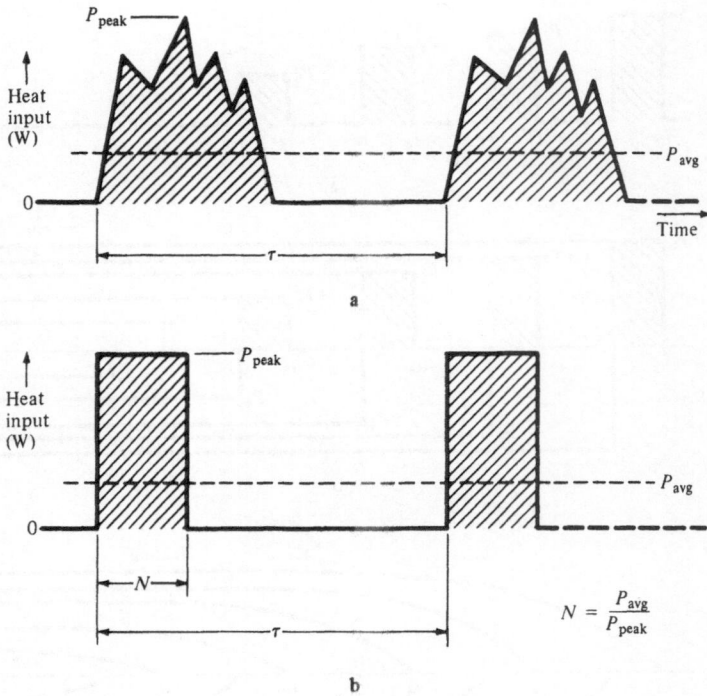

Figure 16.13 Approximating an irregularly shaped power pulse by a rectangular waveshape. (a) Actual heat input; (b) equivalent heat input. (After Gutzwiller and Sylvan [16.12].)

inputs $\Delta\tau$ approach zero, then we can express the power input in function of time in the following way [16.12, 16.13]:

$$P(t) = P(0)U(t) + \int_0^t \frac{\partial P}{\partial \tau} U(t - \tau) \, d\tau \tag{16.38}$$

The temperature rise $T(t)$ in function of time may be obtained from one of the following expressions:

$$\Delta T(t) = P(0)\theta(t) + \int_0^t \frac{\partial P}{\partial \tau} \theta(t - \tau) \, d\tau \tag{16.39a}$$

or

$$\Delta T(t) = P(t)\theta(0) + \int_0^t P(\tau) \frac{\partial \theta(t - \tau)}{\partial \tau} \, d\tau \tag{16.39b}$$

$P(0)$ is the applied power at $t = 0$, $\theta(0)$ is the transient thermal impedance at $t = 0$, and τ is the general time variable of integration which is used in order to avoid the confusion with one of the limits of integration which is equal to t.

264

The evaluation of the above integrals may require the application of numerical methods, and therefore it may in some cases be more laborious than the application of the approximate superimposition method.

16.6
Thyristor thermal impedance measurement

The rate of change of the forward voltage drop of a thyristor with temperature for *small conduction currents* is given by

$$\left.\frac{dV_F}{dT}\right|_{I\,=\,\text{constant}} = -b \tag{16.40}$$

where b is a parameter varying very slowly with temperature and thus can be considered a constant with a comparatively wide temperature range [16.14]. The value of b is about 1.5 mV/°C in the temperature range 20–125°C. The forward characteristic of a p-n-p-n device can serve, therefore, for the determination of the p-n-p-n device temperature operated in either the AC or DC mode. To obtain the greatest accuracy of the forward voltage drop method, V_F must be measured at the same low current level as used for the calibration. The circuitry can be simplified if the cooling rather than the heating of the junction can be observed. A high-level DC or AC current is passed through the device until the junction reaches its steady-state value. The current is then switched from the high level down to the low measurement level and the voltage drop is monitored as the device cools. If the device is mounted on a good heat sink, if the thermal properties of the material in the device are linear, and if radiation and convection effects can be neglected [16.12, 16.15], then it is possible to convert the junction temperature during the cooling into the complementary junction temperature during the heating, since the two temperatures are conjugate to each other. As a result, the junction temperature above the reference (heat sink) temperature may be expressed as follows:

$$T(t) = \frac{V_F(t) - V_F(0)}{b} \tag{16.41}$$

where $V_F(t)$ is the low-current voltage drop as measured at the time t during the cooling cycle and $V_F(0)$ is the extrapolated low-current forward drop at $t = 0$, i.e., at the time when the high-level current which was used to heat the device was turned off. The transient thermal impedance of the p-n-p-n device with reference to the heat-sink temperature can be obtained from

$$\theta(t) = \frac{T(t)}{P} \tag{16.42}$$

265

where P is the average power dissipated in the devices prior to the turn-off of the high-level current. As an alternative method of junction-temperature measurement, the threshold gate-trigger current of an AC-operating triac may be used [16.15], since this current is a very sensitive, exponential function of temperature.

16.7
Computer-aided thermal analysis

The solution of the heat-flow equation for power devices may, in some cases, be handled purely analytically. This approach may provide accurate answers when the geometries involved are simple and material properties are linear in nature. Most cases, however, are not in this category, so that computerized numerical solutions are to be sought, particularly for the multidimensional problems.

The most commonly used approach to the solution of the thermal problems is the finite-difference method [16.16] based on the concept of lumped parameters. Devices to be analyzed are divided into small regions called nodes, For the steady-state problems the heat balance requires that the heat that enters each node must equal the heat that leaves the node. For transients, the heat that enters the node less the heat that goes out of the node during a small time increment must equal the product of the node heat capacity and the node temperature rise during the increment of time. In addition, all boundary conditions pertinent to the problem at hand must be satisfied.

An example of the finite-difference method for the one-dimensional case is shown in Figure 16.14 [16.16]. The boundary temperatures are T_A and T_B, and the heat is generated internally in each node. The length of the time increment is Δt and the heat generated in the nth node per unit volume is designated by q_n. The temperature of the nth node at the start of the time increment is T_n and at the end of the time increment it is T_n'. The finite-difference formulation proposed by Liebman [16.17] gives the heat balance on the nth node

Figure 16.14.   Three-node thermal model. (After Wenthen [16.16].)

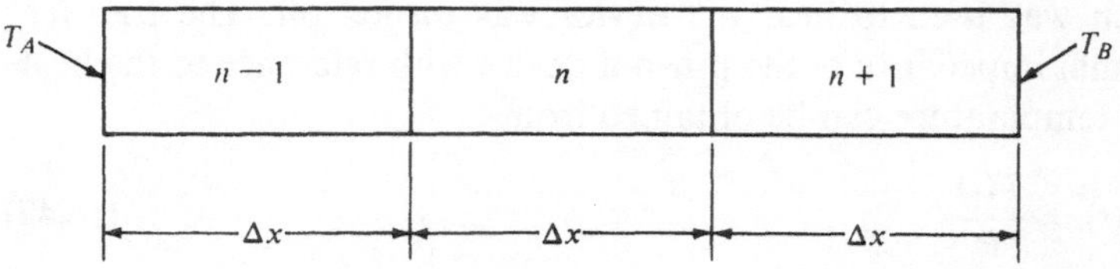

in relation to the neighboring node temperatures of the end of the time increment:

$$T'_{(n+1)} - 2T'_n + T'_{(n-1)} = M(T'_n - T_n) - \frac{(\Delta x)^2 q_n}{k} \qquad (16.43)$$

where

$$M = \frac{(\Delta x)^2 \rho c}{k \Delta t} \qquad (16.44)$$

when k is thermal conductivity, ρ is density, c is specific heat, and Δx is node thickness. Similar expressions can be obtained for other nodes. Since the neighboring node temperatures at the end of the time step are also unknown, the method requires simultaneous solution of the heat balances on all the nodes. The main advantage of Liebman's method is its stability for any time increment chosen, whereas other methods may oscillate (in the mathematical sense) for time intervals longer than the thermal time constant of the smallest node. This leads to the use of an excessive number of computations. In some cases, it is possible to reduce the number of equations to be solved by using shorter segments (Δx) near the heat source where rapid variations of temperature are expected with respect to the distance of heat travel and/or time, and longer segments for more remote sections of the heat-flow path, where the temperature varies more slowly [16.18].

Since using a very large number of nodes increases the computation time and cost, simplifying approaches have to be used to minimize the number of nodes, i.e., taking advantage of existing symmetries, making only gradual transitions from large segments to small segments [16.11].

References

16.1 F. E. Gentry, F. W. Gutzwiller, N. Holonyak, Jr., and E. E. Von Zastrow. *Semiconductor Controlled Rectifiers*. Englewood Cliffs, N.J.: Prentice-Hall, 1964.

16.2 F. E. Gentry. Forward current surge failure in semiconductor rectifiers. *AIEE Trans.*, *77* (1): 746–750, 1958.

16.3 V. I. Stafeev. Injection heat transfer. *Soviet Phys.—Solid State*, 2: 406–412, 1960.

16.4 W. M. Bullis. Minority carrier thermoelectric cooling. *J. Appl. Phys.*, *34* (6): 1648–1649, 1963.

16.5 R. N. Hall. An analysis of the performance of thermoelectric devices made from long lifetime semiconductors. *Solid State Electron.*, 2: 115–122, 1961.

16.6 RCA Solid State Power Circuits Technical Series, SP-52. RCA Solid State Division, Somerville, N.J., 1971.

16.7 General Electric SCR Manual, 5th ed., Syracuse, N.Y.: General Electric, 1972.

16.8 J. LePonner and J. M. Peter. Resistance et impedance thermique des triacs. *Rev. Gen. d'Electricité, 81* (11): 711–719, 1972.

16.9 J. Neilson. RCA Solid State Division, private communication, 1972.

16.10 W. E. Newell. Dissipation in solid state devices—The magic of I^{1+N} *PESC Record*, 1974, pp. 162–173.

16.11 D. Baker and W. O. Fleckenst. In *Physical Design of Electronic Systems* (D. C. Koehler, C. E. Roden, and R. Sabis, eds.). Bell Telephone Laboratories, Design Technology Vol. 1. Englewood Cliffs, N.J.: Prentice-Hall, 1970.

16.12 F. W. Gutzwiller and T. P. Sylvan. Power semiconductor ratings under transient and intermittent loads. *Trans. AIEE, 79*: 699–706, 1960.

16.13 S. Goldman. *Transformation Calculus and Electrical Transients.* Englewood Cliffs, N.J.: Prentice-Hall, 1949.

16.14 Hua Quen Tserng and Hubert R. Plumlee. The forward voltage technique to measure junction temperatures of AC operating triacs. *IEEE Trans Electron Devices, ED-17* (9): 755–761, 1970.

16.15 Hua Quen Tserng and Hubert R. Plumlee. Temperature measurements of AC operating triac using a gate trigger current technique. *IEEE Trans. Electron Devices, ED-17* (9): 761–765, 1970.

16.16 F. T. Wenthen. Computer-aided thermal analysis of power semiconductor devices. *IEEE Trans. Electron Devices, ED-17* (9): 765–770, 1970.

16.17 G. Liebman. The solution of transient heat flow and heat transfer problems by relaxation. *Br. J. Appl. Phys., 6*: 129–135, 1955.

16.18 W. H. Parker. Computed transient temperatures for silicon diodes. RCA-Electronic Components, Lancaster, Internal Report, April 1972.

Also of interest

W. E. Newell. Transient thermal analysis of solid state power devices—Making a dreaded process easy. *PESC Record*, 1975, pp. 761–770.

Thyristor circuits basics

Summary

Phase control and zero-voltage switching are the basic power control methods used in the SCR and triac circuits. Phase control is achieved by resistive or resistive–capacitive networks or by triggering pulses derived from such devices as diacs, unijunction transistors, two-transistor trigger circuits, and neon bulbs.

In order to turn off the thyristor, it is necessary to reduce the main current below the holding-current level. In an AC circuit, the current passes through zero every half cycle so the turn off of the thyristor also is assured every half cycle. The commutation of a thyristor in a DC circuit requires, however, the use of additional circuitry.

To avoid either false triggering or damage to a thyristor due to transients, some simple protective networks are usually recommended.

Practical applications of SCRs and/or triacs are shown in the circuits of light dimmers, universal motor controllers, heat controllers, and DC -to- AC inverters.

17.1
Introduction

In recent years the use of thyristors in power control or power conversion circuits has become quite common. The currents which are typically controlled by thyristors range from fractions of an ampere up to several thousand amperes.

The discussion in this chapter is limited essentially to the simplest circuits using either SCRs or triacs and to a few special gate-triggering devices. As rectifying unidirectional devices, SCRs are suited to control all kinds of DC circuits such as inverters (i.e., electrical systems that change DC to AC power), or DC power supplies. For the control of AC power, on the other hand, triacs are most desirable from the economy point of view.

269

Thyristors, as switches, have two possible states: the high-impedance off state and the low-impedance on state. To trigger a thyristor from the off to the on state, it is necessary to apply to the thyristor gate a current of sufficient direction and amplitude. When the device's main current exceeds the latching current, the gate-triggering current can be removed and the device will remain in conduction.

In order to turn off the thyristor, it is necessary to reduce the main current below the holding-current level, which is normally on the order of several milliamps for medium-power devices. In the AC circuit, the current passes through zero every half cycle so the turn-off also is assured every half cycle. To turn off a DC circuit, however, it is necessary to use additional circuitry to reduce the main current below the holding current. Thyristor turn off is referred to frequently as commutation.

By appropriate thyristor triggering it is possible to achieve load-voltage waveforms illustrated in Figure 17.1. The AC load-voltage

Figure 17.1   Load voltage waveforms.

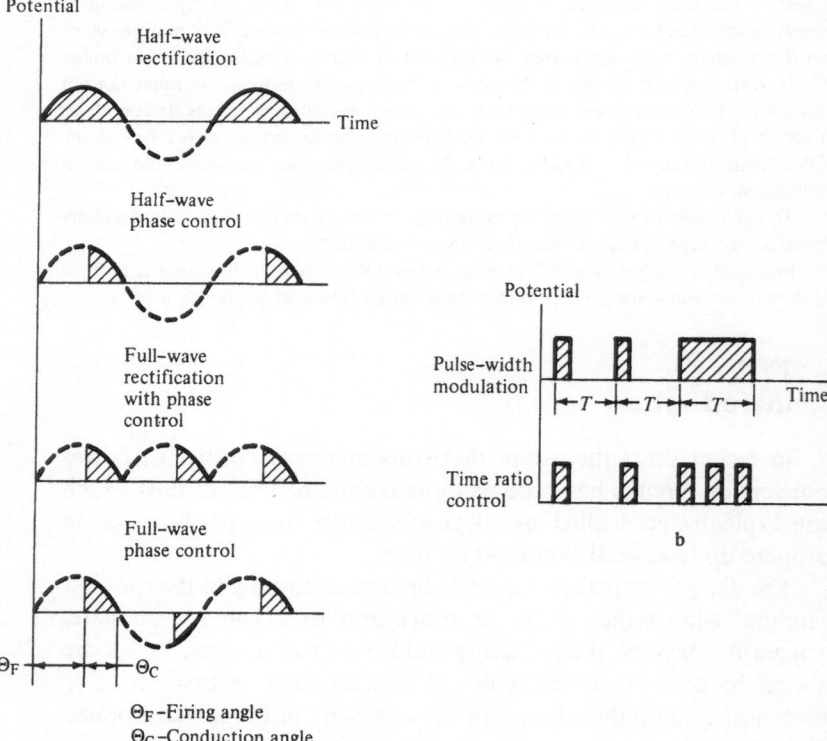

Θ_F –Firing angle
Θ_C –Conduction angle

270

waveforms illustrate, essentially, the so-called phase control which allows the load current to flow during some definite portion of the AC cycle. The average load power or voltage can be varied broadly by varying the time of switch closure.

In Figure 17.1b the current has the form of rectangular pulses. The pulse technique consists of chopping the load voltage, and the ratio of the on to the off time determines the average power to the load. Two modes of operation are shown in the figure: pulse-width modulation and the time-ratio control. In the first case, the average value of the voltage is modified by varying the pulse width from zero to some maximum value T. In the second, all the pulses have the same constant duration but the repetition rate varies.

17.2
SCR and triac control and triggering methods

17.2.1
Phase control

In many power-control thyristor circuits partial cycles of the applied AC voltage are switched to the load, as is shown in Figure 17.1a. Phase control, as this method is called, is achieved by varying the electrical angle of the applied AC voltage waveform at which the thyristor current is initiated, i.e., by varying the firing angle Θ_F (Figure 17.2). The conduction angle, Θ_C, is the number of electrical degrees of the applied AC voltage waveform during which the thyristor is in conduction. In a half-wave circuit $\Theta_F + \Theta_C = 180$ degrees; in a full-wave circuit, using a triac, $2(\Theta_F + \Theta_C) = 360$ degrees.

The triggering signal can be obtained either from the same power source used for the load or from a completely separate source, if this additional expense is warranted by the circuit flexibility. The use of the same source is the most commonly used approach.

Triggering can be accomplished by DC or AC signals or by short-duration pulses. This last technique provides a firing system less dependent on temperature variation and the spread in the gate turn-on current from device to device. A short pulse may have a large overdrive amplitude—much bigger than the required gate turn-on current—and result in less power dissipation because of its short duration. Figure 17.3 represents the simplest type of SCR triggering by a trigger signal derived from a common DC power source.

The gate contact is connected by a variable resistor R_g to the thyristor anode; in this way a potential of right polarity is developed

271

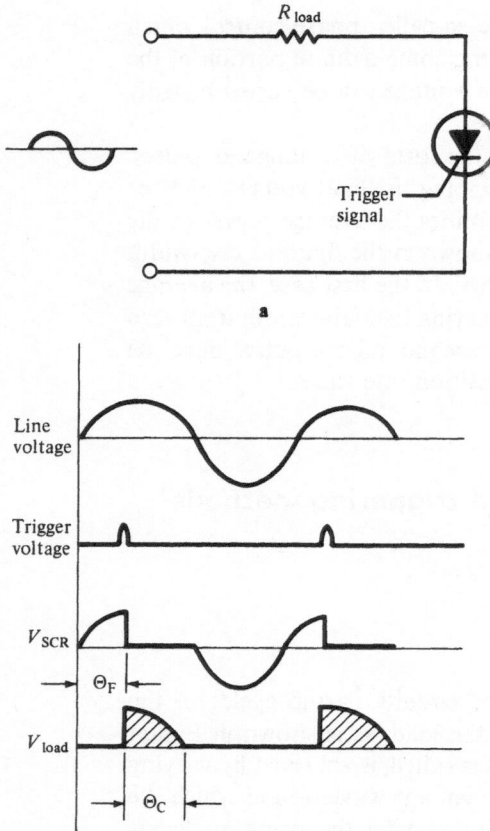

Figure 17.2 Firing (Θ_F) and conduction (Θ_C) angles for half-wave controlled rectification.

between the gate and the cathode. The magnitude of the gate current is limited by the presence of R_g, and for some sufficiently small R_g, when the switch S is closed, a high enough gate current is flowing to turn the SCR on. After the SCR is already turned on, the anode-to-cathode voltage drops down to the forward-voltage drop of the device V_T (1–2 V). At this point, almost the entire line voltage appears across the load. For an AC operation of an SCR, it is necessary to add to the gate circuit a diode (Figure 17.3b) which blocks the current when the potential of the anode becomes negative. When a triac is used with an AC power supply, the diode is not necessary since the device works at both current alternations.

For an AC operation, a circuit like that shown in Figure 17.3b can be used. The gate current is derived here from the common AC current source. In the AC case, a diode is added in series with the variable resistance R_g in order to prevent the application of the negative-gate bias when the SCR anode becomes negative. As R_g is

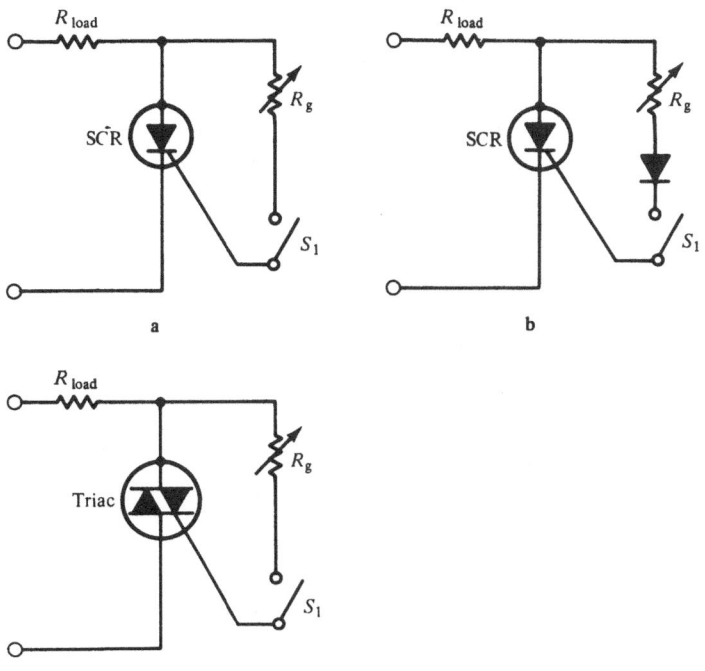

Figure 17.3 Simple thyristor triggering methods. (a) Resistance triggering of SCR; (b) resistance-diode triggering of SCR; (c) resistance triggering of triac.

decreased, the gate current increases and the SCR triggers close to the beginning of the positive cycle. The earlier in the cycle the SCR starts conducting, the larger will be the average potential across the load. The circuit of Figure 17.3b can be turned on only between an angle of zero and 90 degrees. Because of the SCR latching it cannot be turned off between 90 and 180 degrees. As a result, only about 50 percent of the positive half cycle is controlled (Figure 17.4a). For a triac, the situation is very similar except that the conduction takes place for parts of both the positive and negative half-cycles. Figure 17.4a shows the situation when the firing angle Θ_F has reached its maximum 90-degree value. For lower values of R_g, Θ_F will be less than 90 degrees and the conduction angle Θ_C will move toward the beginning of the cycle, as indicated by the arrows.

In order to obtain a firing-angle delay greater than 90 degrees, one may use a resistance-capacitance network in place of a simple resistor triggering as shown in Figure 17.5. The addition of the capacitor C_1 provides a built-in lag in the gate voltage so that triggering can be delayed to angles greater than 90 degrees. The phase

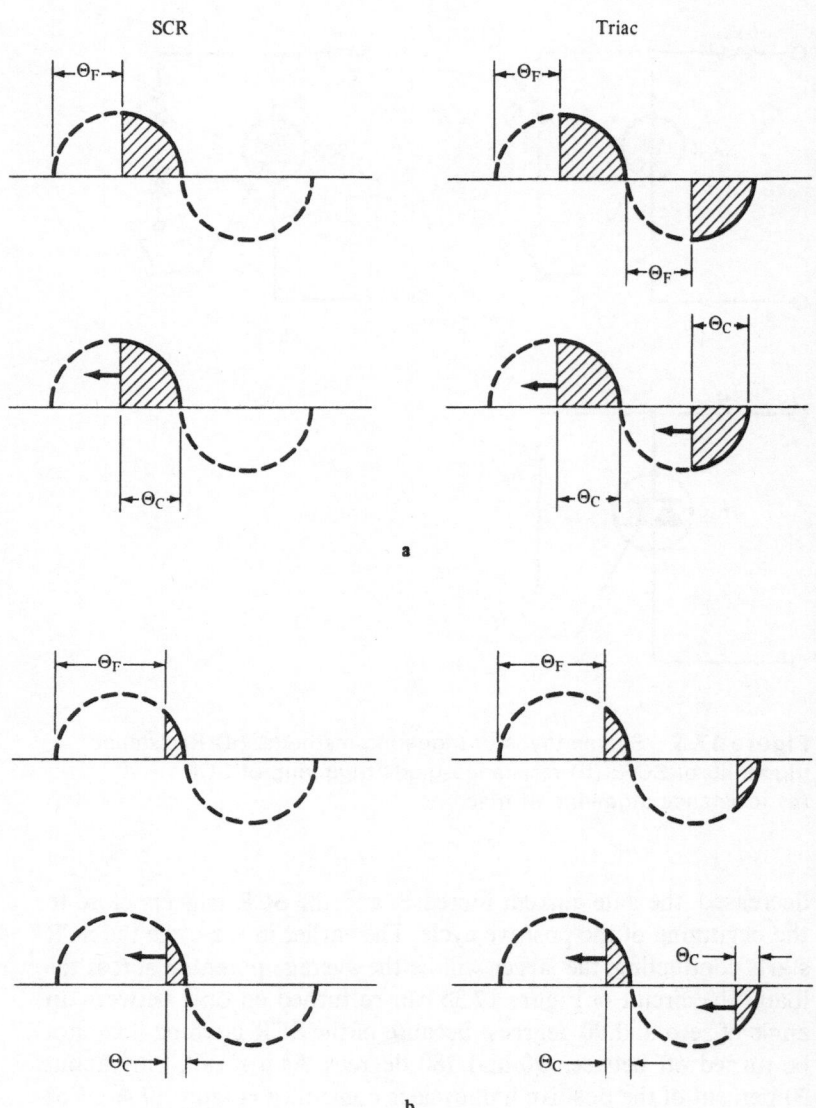

Figure 17.4 Degree of control over conductive angles with (a) a resistive network; (b) a resistive–capacitive network.

lag is illustrated by the vector diagram of Figure 17.6 which shows that the lag angle Θ is dependent on the relative values of the resistance R_1 and the reactance of the capacitor C_1.

Figure 17.4b illustrates the situation when the resistive–capacitance (RC) network is used. The firing angle Θ_F may become now greater than 90 degrees and the conduction angle Θ_C may vary from

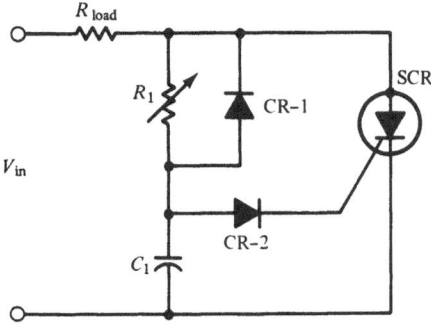

Figure 17.5 Resistive–capacitive triggering network and diodes for phase-control triggering.

a value almost equal to 180 degrees down to almost 0 degree, as indicated by the arrows.

Figure 17.7 shows a typical transfer characteristic, i.e., the percentage of available power versus potentiometer setting for a simple half-wave RC-diode phase control.

Since the triggering current must be supplied by the line voltage through the resistor, the capacitor must be selected such that its charging current is high compared with the gate-triggering current I_{gt} at the instant of the largest firing angle.

Resistance and resistance–capacitance trigger circuits have one great disadvantage: the gate voltage rises slowly to the triggering level. Because of the wide variation in gate characteristics among thyristors the control potentiometer resistance setting may yield a different conduction angle for different thyristors or temperature conditions. Power level in the control circuit is high because the

Figure 17.6 Vector diagram for phase control.

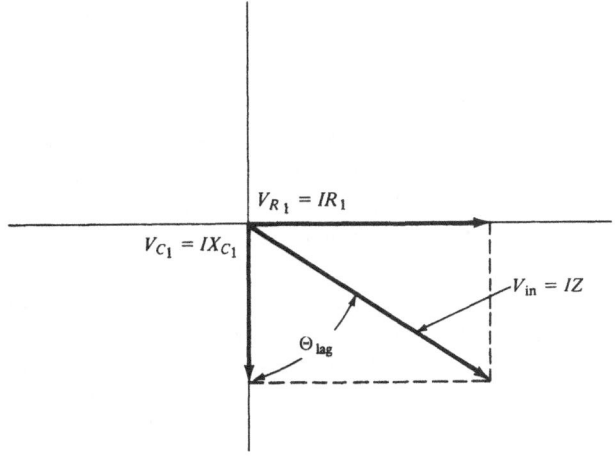

275

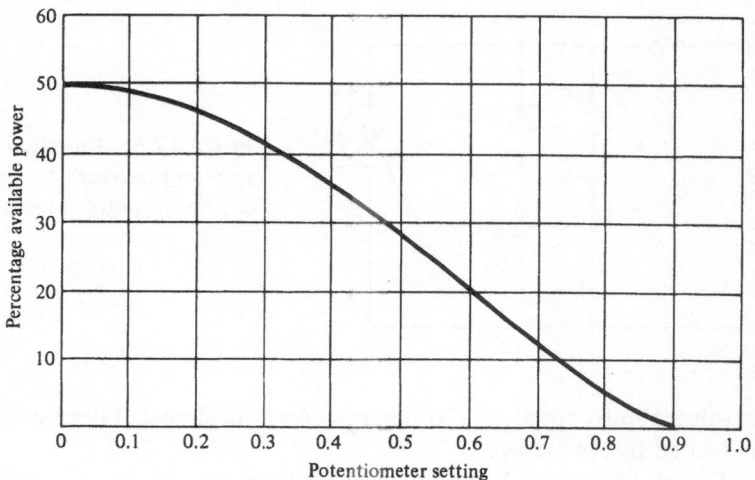

Figure 17.7 Transfer function of a simple half-wave *RC*-diode phase control.

entire triggering current must flow through the resistive part of the network. Also, the *RC* triggering circuits do not lend themselves readily to the automatic, self-programmed, or feedback-control systems.

The performance of the circuit is improved somewhat by use of a double *RC* section (see Figure 17.12). This technique increases the rate of rise of gate voltage in the vicinity of the triggering potential and, therefore, minimizes the effects of gate differences on the conduction angle.

17.2.2
Zero-voltage switching

Power control in a thyristor circuit also can be achieved by switching the thyristor on and off for some desired number of complete half or full cycles. This type of control is usually referred to as zero-voltage switching, zero-crossover switching, or integral-cycle switching. Figure 17.8 illustrates the relationship between the line and load voltages for both SCR (half-wave) and triac (full-wave) power-control circuits for zero-voltage switching.

Zero-control switching is of great importance in minimizing radio interference and in the reduction of transient current in inductive circuits. In zero-voltage switching only two levels of input power are delivered to the load; the load receives the full amount of power for a period of time and zero power for another period of time.

276

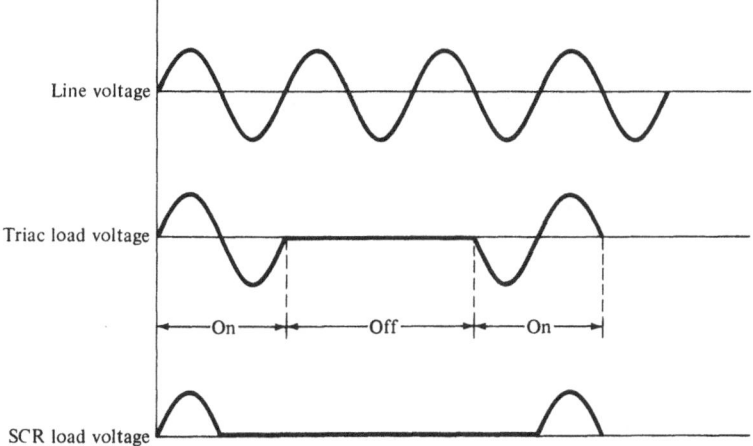

Figure 17.8   Typical zero-voltage switching waveforms.

The average power delivered to the load depends, consequently, on the ratio of the "power-on" interval to the "power-off" interval.

The zero-voltage switching is advantageous in the control of high-inertia systems such as electric furnaces. The resistance heaters have a thermal response much slower than the 60-Hz line; therefore, the control during some small part of a cycle is not needed and it is sufficient to switch entire cycles of power and do the switching near the zero-voltage crossing. In essence, this technique involves applying bursts of complete sine waves to the load, separated by periods of zero voltage.

17.2.3
Pulse triggering

The disadvantages of the *RC* triggering networks can be largely overcome by the use of two-terminal triggering devices exhibiting negative-resistance characteristics that provide pulses to trigger the thyristor. These triggering devices have a narrower range of characteristics and are not so temperature-sensitive. Pulse triggering can accommodate wide tolerances in triggering characteristics by over-driving the gate. The power level in pulse-control circuits may be quite low since the required triggering energy, i.e., the product I_{gt} $V_{gt}t$, can be stored slowly, then discharged rapidly at the desired triggering instant. The use of pulse triggering enables low-power, signal-type components to control large high-current thyristors.

Some of the triggering devices used are the neon gas discharge lamp, diac, unijunction transistor (UJT), two-transistor switches,

277

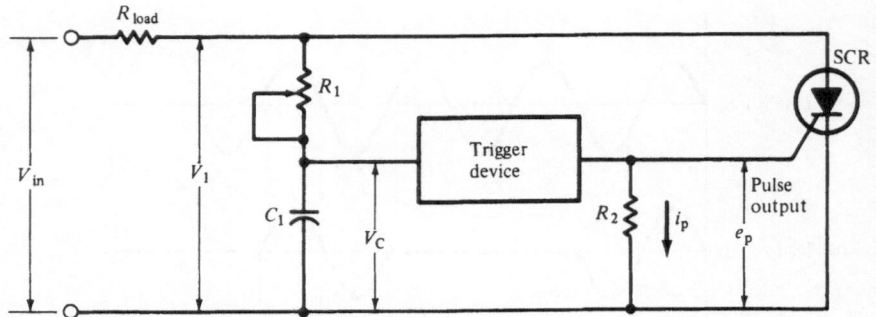

Figure 17.9 Thyristor power control with a trigger device.

Schockley diodes, special types of thyristors, and pulse transformers. The Zener diode is not included since its reverse-avalanche breakdown characteristic does not exhibit negative incremental resistance.

The trigger-pulse–producing devices usually operate by discharging a capacitor into the thyristor gate [17.1, 17.2]; their circuits function essentially in a relaxation oscillation mode. Typical equivalent circuits for thyristor power control and the trigger device voltage–current (V–I) characteristic are shown in Figures 17.9 and 17.10, respectively. R_1 is a variable resistance and the resistance R_2 *includes the thyristor gate resistance.* Between points 1 and 2 the V–I curve (Figure 17.10) exhibits an incremental negative charac-

Figure 17.10 Trigger device as a relaxation oscillator.

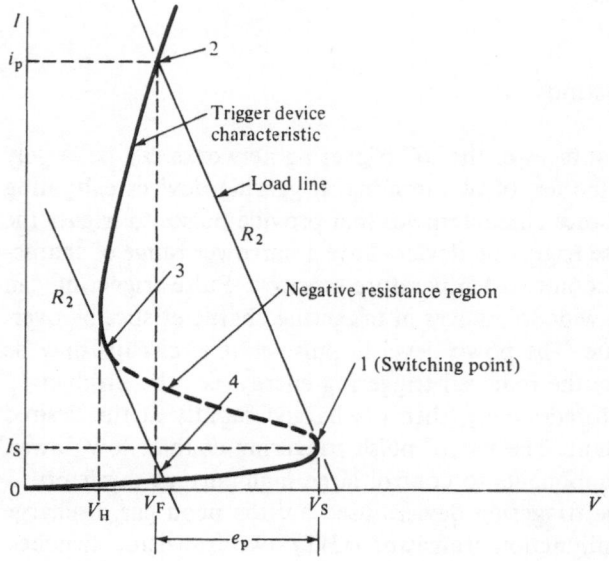

278

teristic; an increase in voltage results in a decrease of current. The trigger device switches from its blocking state to the conducting state after the applied voltage exceeds the switching voltage V_S (point V_S-I_S of the characteristic). Lines with the slope R_2 represent the load lines for the R_2 resistor.

The resistance R_1 is varied so that the input voltage to the resistance R_2 changes and the load line R_2 shifts until it intersects the triggering device curve at point 1, at which the negative-resistance slope of the devices characteristic is equal to the slope of the load line R_2. When this condition is reached the operating point shifts from point 1 to point 2, and the capacitor C discharges very rapidly

Figure 17.11   (a) Voltage–current characteristic of triggering devices. (b) Typical gate-current waveform.

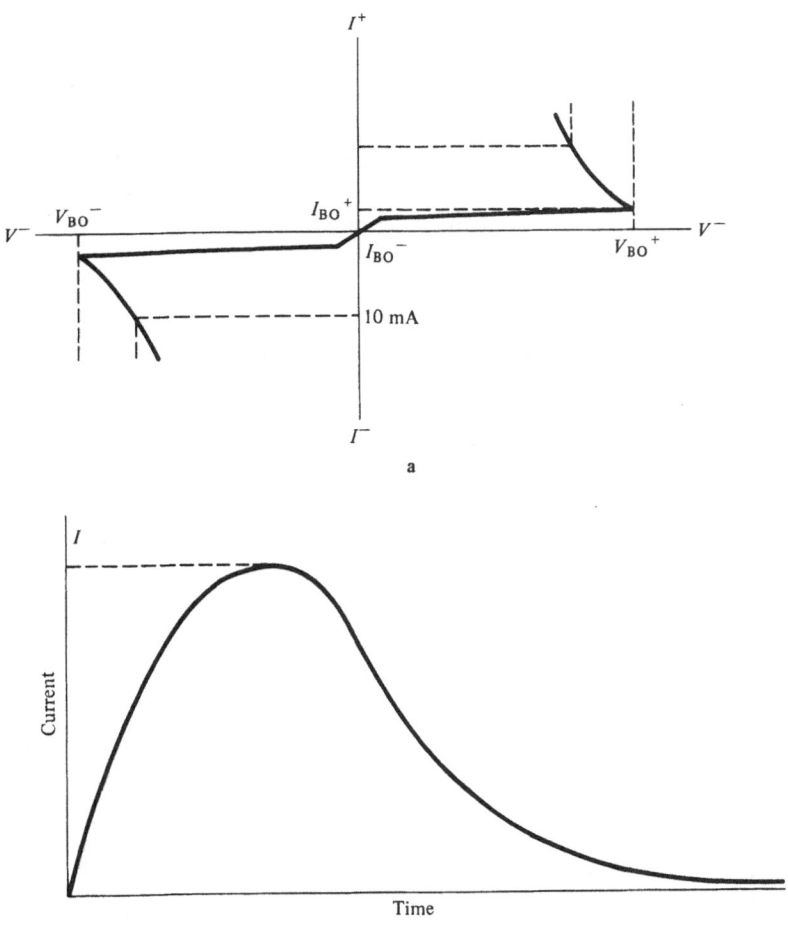

a

b

developing a voltage pulse across R_2 ($e_p = R_2 i_p$) of the magnitude sufficient to trigger the thyristor. During the discharge the operating point follows the trigger device characteristic from point 2 to point 3 until at point 3 the negative-resistance slope becomes again equal to the R_2 slope. Beyond point 3 the capacitor continues discharging along the R_2 load line (far to the left of the diagram) until it reaches point 4; the capacitor recharges then through R_1 (Figure 17.9) and the cycle repeats itself, i.e., the trigger device oscillates in the relaxation mode.

The trigger device needs, of course, some definite time for the turn-on. If this switching time is much smaller than the circuit time constant, then the peak voltage e_p is just the difference between the switching voltage V_S and the triggering device forward-voltage drop V_F. If, however, the opposite is true, then the pulse voltage e_p and, consequently, the pulse current i_p are reduced by the effective trigger-device resistance. Figure 17.11b shows the typical shape of the gate-current pulse that is produced. The time needed to reach the peak gate current is a function of the speed at which the triggering device is switched from its high-impedance to its low-impedance state. The magnitude and duration of the gate pulse produced by the triggering device and the capacitor must be adequate to fire the thyristor. A curve of turn-on time as a function of the gate-pulse magnitude, provided in manufacturing company's published data on the thyristor, defines the minimum requirements.

17.3
Thyristor triggering devices

We are limiting our discussion here to only the most commonly used devices.

17.3.1
Neon lamp

One of the simplest and least expensive of the triggering devices is the neon gas discharge lamp. Prior to triggering, such a lamp exhibits an internal resistance of several thousand ohms and a breakdown voltage between 50 and 100 V. When the discharge is initiated, the current through the lamp is limited only by the external resistance, and the device exhibits a negative-resistance characteristic similar to the one represented by Figure 17.11a. A typical current pulse resulting from a capacitor discharging through a neon bulb and a thyristor gate is very similar in shape to that shown in Figure 17.11b. The bidirectional characteristic of the neon lamp allows it to trigger

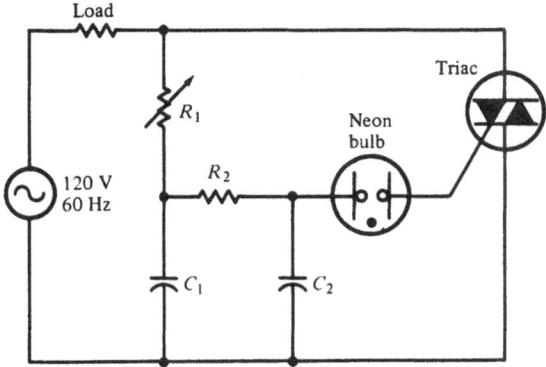

Figure 17.12 Neon bulb as thyristor triggering device [17.1].

both SCRs and triacs. Figure 17.12 illustrates the use of a neon bulb for the SCR power-control circuits.

The neon lamp makes an excellent trigger for circuits operated from 115 V or higher. There are a few disadvantages of the neon bulb. The spread of firing voltages greater than 20 V from unit to unit in nonselected batches is quite common. When a neon bulb is used to trigger from a 120-V line, the voltage loss as great as 10 percent can occur at the load due to the relatively high breakdown voltage of the device. The breakdown voltage of a neon bulb is sensitive to radiation, so when precise control is required it may be necessary to shield the bulb or to obtain bulbs specially treated to minimize the effect of radiation. The GE type 5AH neon lamp is radioactive isotope stabilized and is now being used in many low-cost SCR control circuits.

17.3.2
Trigger diodes (diacs)

The three-layer trigger diode is similar in construction to a bipolar transistor but differs from it in that the doping concentrations at the two junctions are approximately the same and there is no contact made to the base region. The equal doping levels result in a symmetrical, bidirectional characteristic similar to the one represented by Figure 17.11b. Diacs require smaller firing voltages than neon bulbs, they have a higher pulse-current capability, and, being solid-state, they have a longer life. The solid-state diodes have avalanche breakdown voltages in the range of 27 to 37 V and are particularly useful for triggering bidirectional thyristors (triacs). These diodes exhibit leakage currents on the order of 50 μA before

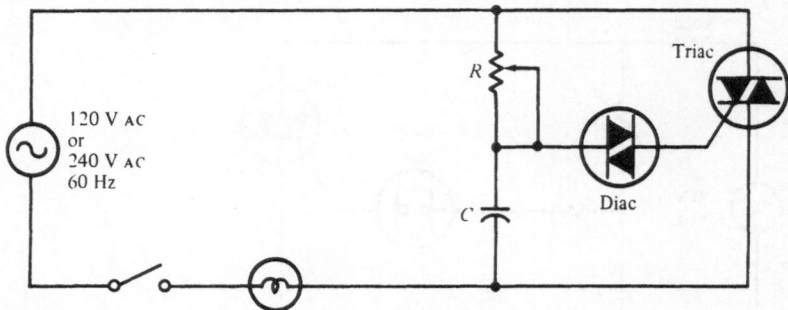

Figure 17.13 Basic triac–diac light-control circuit. (After Neilson [17.3].)

the actual breakdown is reached. The breakdown is of the same nature as the one at the open base transistor.

When an increasing positive or negative voltage is applied across the terminals of the diac, a leakage current I_{BO} flows through the device until the voltage reaches the breakover point V_{BO}. The reverse-biased junction then undergoes avalanche breakdown. The breakover point exhibits negative-resistance characteristics.

Figure 17.13 shows an AC circuit with a diac controlling the firing of a Triac in a light-control circuit (dimmer) [17.3].

The magnitude and the direction of the gate current pulse are determined by the value of the phase-shift capacitance, the dynamic diac's impedance, and the thyristor-gate impedance.

The interaction of the RC network and the trigger diode results in a hysteresis effect when the triac is initially triggered at a small conduction angle. The hysteresis effect is characterized by a differ-

Figure 17.14 Charging cycle of the capacitor–diac network. (After Neilson [17.3].)

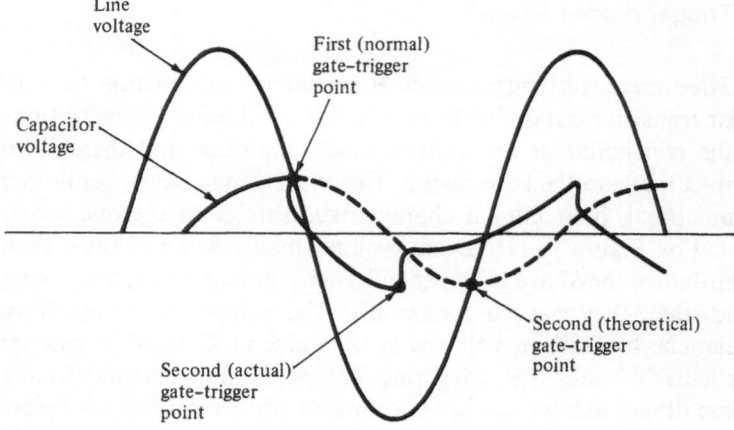

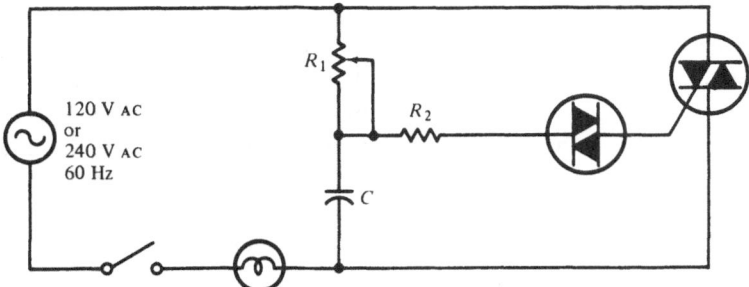

Figure 17.15 Light-control circuit incorporating a resistor in series with the diac. (After Neilson [17.3].)

ence in the control potentiometer setting when the triac is first triggered and when it is turned off. The resistance value to "start" and to "stop" the load current will not be the same. Hysteresis is caused by an abrupt decrease in the capacitor voltage when triggering begins. Figure 17.14 shows the charging cycle of the capacitor–diac circuit. The large AC sine wave represents the line voltage; the smaller AC sine wave represents the normal charging cycle of the capacitor. Gate triggering occurs at the first point of intersection of the two waves. At this point, however, there is an abrupt decrease in the capacitor voltage (solid line). As a result, the capacitor begins to charge during the next half-cycle at a lower voltage and reaches the trigger voltage in the opposite direction earlier in the cycle [second (actual) gate-trigger point].

The hysteresis effect can be reduced by use of a resistor in series with the trigger diode and gate, as shown in Figure 17.15. The series resistor slows down the discharge of the capacitor through the trigger diode. Consequently, the capacitor does not lose as much charge while triggering the triac and produces a smaller hysteresis effect. The "double–time-constant" circuit (Figure 17.16) also

Figure 17.16 "Double-time-constant" light-control circuit. (After Neilson [17.3].)

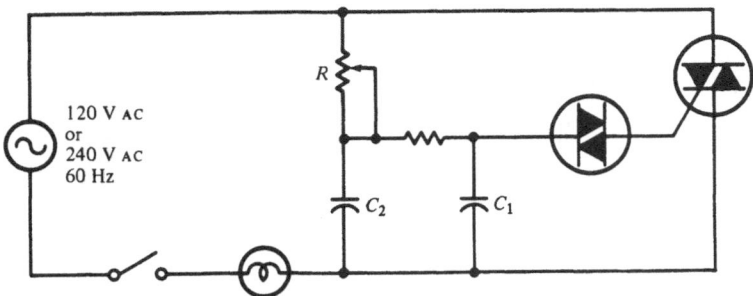

improves the performance of the single–time-constant-control circuit. The additional RC network not only extends the control phase angle, as mentioned before, but also minimizes the hysteresis effect. The added capacitor C_2 reduces hysteresis by charging to a higher voltage than C_L and maintaining some voltage on C_1 after triggering. As gate triggering occurs, C_1 discharges to form the gate-current pulse. However, because of the longer C_2R time constant, C_2 restores some of the charge removed from C_1 by the gate-current pulse.

17.3.3
Unijunction transistor

A unijunction transistor (UJT) is a three-terminal, two-layer device formed by a p-emitter placed asymmetrically in an n-type base with two base contacts B_1 and B_2 (Figure 17.17). Between B_1 and B_2 the unijunction transistor has the characteristics of an ordinary resistance. This interbase resistance R_{BB} has values in the range from 4.7 K to 9.1 K at 25°C. The device switches into conduction between the emitter E and base contact B_1, with a negative resistance characteristic, whenever its emitter voltage, V_E, exceeds a specified fraction (the intrinsic stand-off ratio η) of the interbase voltage V_{BB}.

When V_E is made sufficiently positive the emitter diode becomes heavily forward biased and injects holes that flow into the n-base toward the ground; the R_{B_1} becomes conductivity modulated and very small so that the emitter–base 1 voltage drops from a high value to a very low value. Since the voltage drops with the increase of current, the emitter–base 1 V–I characteristic exhibits a negative resistance region.

Figure 17.17 (a) Unijunction transistor symbol; (b) equivalent circuit.

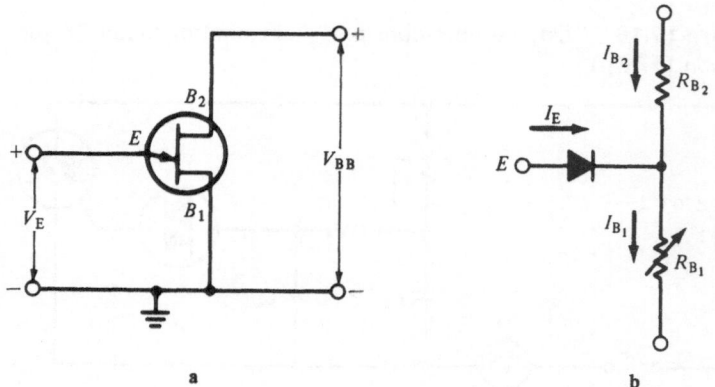

a

b

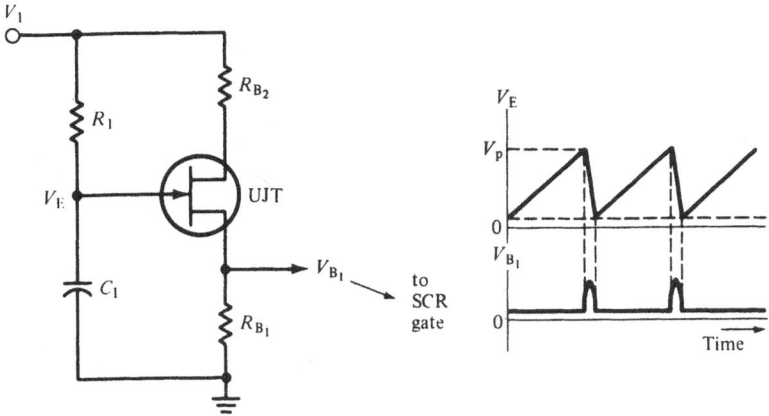

Figure 17.18 Basic UJT relaxation oscillator trigger circuit with V_E and V_{B_1} waveforms [17.2].

The emitter voltage at which triggering takes place, V_p (the peak-point voltage), is given by $V_p = \eta V_{BB} + V_D$, where V_D is the voltage drop of the emitter to base-one diode (about 0.5 V) and the value of η lies between 0.51 and 0.82 [17.2]. The basic UJT pulse-trigger circuit is shown in Figure 17.18. The capacitor, C_1, charges through R_1 until the emitter voltage reaches the peak-point voltage, whereupon UJT turns on and discharges C_1 through R_{B_1}. When the emitter voltage reaches a value of about 2 V, the UJT turns off and the cycle is repeated. The period of oscillation is given by $T \approx R_1 C_1 \ln(1/1 - \eta)$ [17.2, 17.4]. The waveform of the voltages V_E and V_{B_1} are as shown in Figure 17.18. V_{B_1} may be coupled through a resistor or diode or through a transformer to the gate of a thyristor. Manual control of the frequency of repetition may be achieved by the variation of R_1.

The disadvantage of the unijunction device is that it is unilateral with regard to the current flow and requires a DC voltage. As a result, diodes must be used to assure that no reverse voltage appears across the device when it is used in AC circuits. The output pulses are positive-going and can be used to trigger SCRs directly. For triacs, a transformer of capacitive coupling is required.

17.3.4
Programmable unijunction transistor

The UJT has its period of oscillation determined by the relationship

$$T \approx R_1 C_1 \ln\left(\frac{1}{1 - \eta}\right) \tag{17.1}$$

285

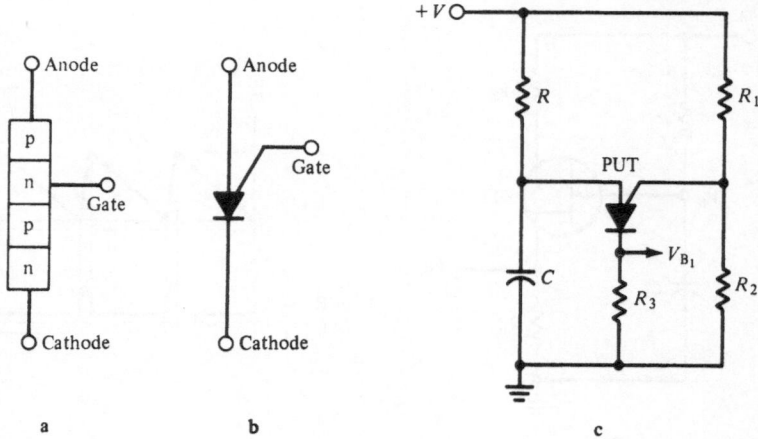

Figure 17.19 Programmable unijunction transistor: (a) structure;
(b) symbol; (c) typical circuit.

where the intrinsic stand-off ratio η is fixed by the manufacturer.
The programmable unijunction transistor (PUT), on the other hand,
can be programmed for various values of η and consequently has a
wider frequency range than the UJT.

The PUT consists basically of a p-n-p-n structure as shown in
Figure 17.19. The gate contact is made here to the *n-base*. The
equivalent circuit is the same as of an SCR except for the presence of
an n gate in place of a p gate. Once turned on, the gate loses its control
and to turn off the PUT it is necessary either to reduce the voltage
across it to zero or to reduce the current to below the holding current
level.

A typical circuit utilizing the PUT is shown in Figure 17.19 and
the waveforms observed are very similar to the ones shown in Figure
17.18. The period of oscillation is given by the same equation as for
the UJT, but with

$$\eta = \frac{R_2}{R_1 + R_2} \tag{17.2}$$

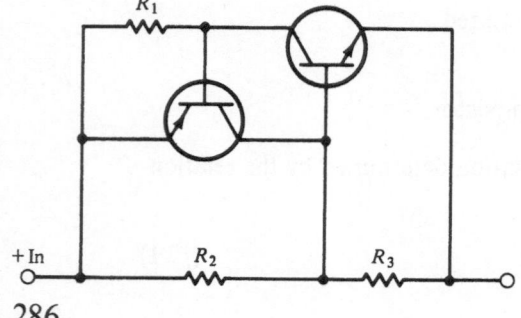

Figure 17.20 Two-transistor
switch.

where the R_1 and R_2 are the resistances of the voltage divider as per Figure 17.19. The frequency of oscillations can, therefore, be controlled by the R_C product and/or by R_1 and R_2. This device can be made less sensitive to the temperature variations than the UJT by using suitable temperature compensation components.

17.3.5
Two-transistor trigger circuit

A two-transistor trigger circuit that has characteristics similar to those of a diac is shown in Figure 17.20. The circuit is very similar to an equivalent circuit of an SCR, and when in operation it regeneratively switches from the high-impedance to the low-impedance state and its V–I characteristic is similar to that of a diac. Proper biasing of this circuit yields triggering voltages of 15 V or less. Figure 17.21

Figure 17.21 Half-wave motor control with no regulation. (After Yonushka [17.5].)

Courtesy RCA Solid State Division, Somerville, N.J.

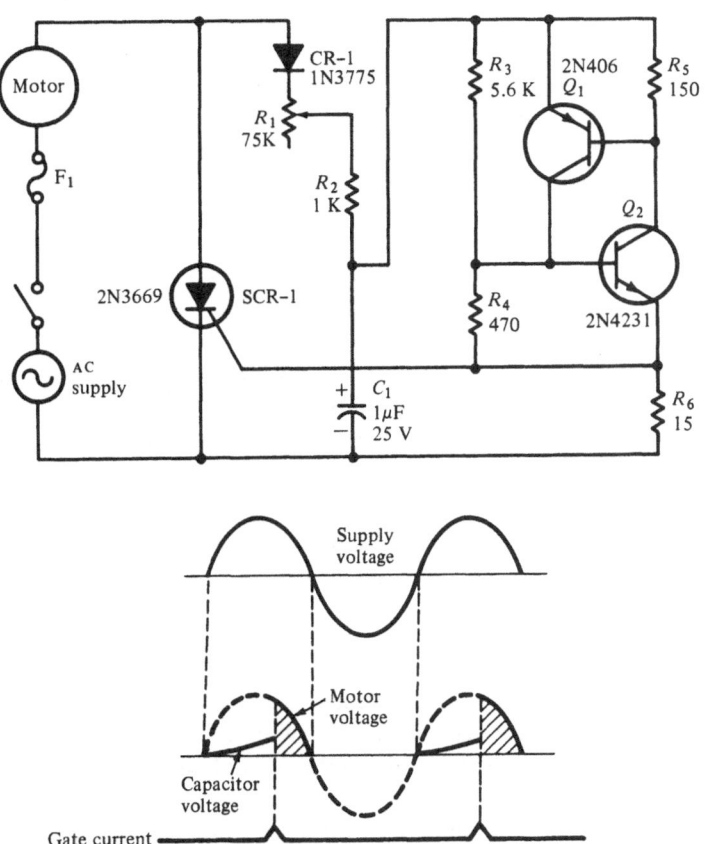

287

shows an SCR circuit that uses a two-transistor regenerative-trigger network with a motor load [17.5]

17.3.6
Silicon unilateral switch

The silicon unilateral switch (SUS) is similar to the PUT and also has an anode gate in place of the usual p-type cathode gate. It is a small SCR with a built-in avalanche (Zener) diode between the anode and cathode. The device turns on when the anode voltage exceeds that of the avalanche diode and is used in the relaxation circuit as described in Section 17.2.3 and Figure 17.18.

In contrast to a UJT, the SUS switches at a fixed voltage and its turn-on current is much higher than that of a UJT. As a result, the lower and upper limits of the time delays obtainable with the SUS are more restricted.

17.3.7
Silicon bilateral switch

The silicon bilateral switch (SBS) consists of two SUS connected back-to-back and integrated on the same silicon chip. It has $V–I$ characteristics in the first and third quadrant; it operates as a switch with both polarities of the applied voltage. The SBS is very useful for triggering of triacs with alternate positive and negative gate pulses [17.2].

17.4
Commutation

The thyristor turn-off is usually referred to as commutation. In an AC circuit an SCR or a triac will be turned off every time the AC line sine wave goes through zero. In the case of AC and of an SCR, some definitive time interval must elapse after the anode current passes through zero before it is capable of blocking forward voltage without turning on again. This limits the maximum frequency at which the device can be used. In the triac the device must turn off at the instant the AC current passes through zero, since essentially a triac is an integrated circuit consisting of two SCRs in inverse parallel connection, and there is a strong interaction between the two devices. In particular, inductive loads can cause a turn-off problem. For instance, if the triac is in series with an inductive load and the triac will attempt to turn off at the instant the current passes through zero. The supply voltage, because of the inductance, will be leading the current and, therefore, will have a nonzero value at this

288

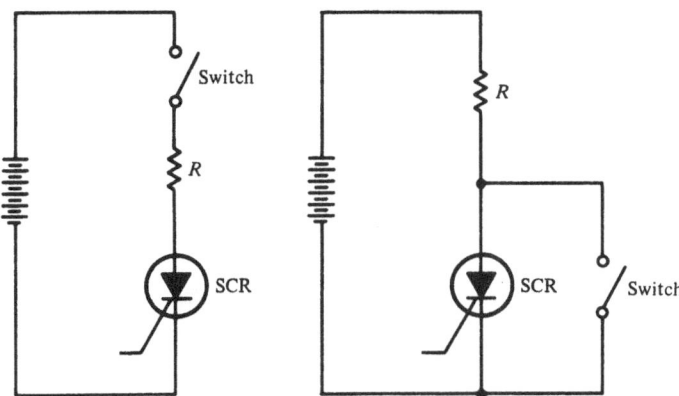

Figure 17.22 Commutation by current interruption.

instant. This voltage will appear across the triac, and the rate of rise dv/dt will often be large enough to prevent the triac from turning off. This can be usually prevented by the use of a resistance-capacitance circuit connected between the main terminals.

Many applications require the anode–cathode current flow to stop at other than the natural AC zero crossing; also in the DC circuits the voltage does not go through zero. Since the gate has no pronounced control over SCR turn off,[1] some external commutation means must be employed. There are two basic SCR commutation methods. One consists of interrupting the anode current, as shown schematically in Figure 17.22. The interruption of the current by some kind of a mechanical switch is limited to very few applications. One of its disadvantages is the very fast voltage rise (dv/dt) seen by an SCR.

A more acceptable approach is to use the forced commutation method as depicted by Figure 17.23. The commutation is performed by the capacitor C_1 and an additional SCR-1. When SCR-2 is in the on state, the current flows through the load and the capacitor C_1 charges. When the additional SCR-2 is triggered on to the low-resistance state, C_1 is effectively connected in parallel to the SCR-1 since the SCR-2 has a very low on resistance. The charge on C_1 is then opposite to the SCRs forward voltage; therefore, the SCR-1 is turned off and the current is transferred to the R_1–SCR-2 path. The current rating of the SCR-2 can be much smaller than that of the SCR-1 since it must be only on for a time equal to t_q, i.e., usually for just a few microseconds. The value of the commutating capacitor for a resistive load must be sufficient to supply reverse voltage to SCR-1 *until* it turns off. This leads to the use of usually expensive large and

---

[1] GTO is excluded from this discussion.

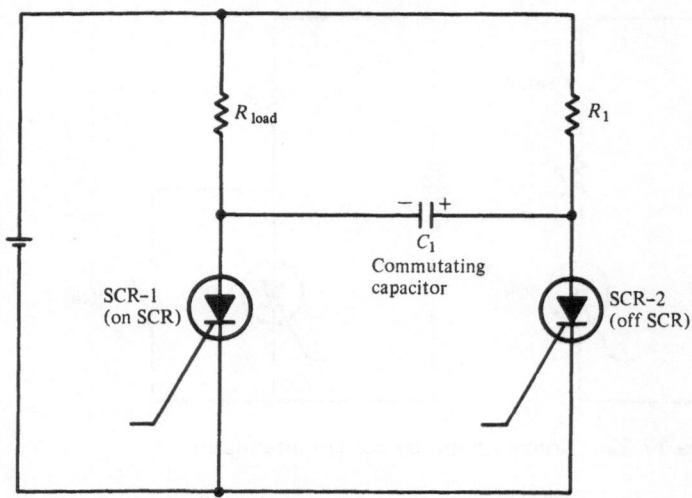

Figure 17.23 SCR commutation by commutating-capacitor technique.

nonpolarized capacitors because of the bidirectional current flow through C_1. Figure 17.24 demonstrates a turn-off technique which momentarily deprives the SCR of its forward current and voltage. In this case, the saturated bipolar transistor (collector and emitter forward biased) provides essentially a momentary short. A low-power device is quite satisfactory since the transistor will have to operate only a few microseconds until the thyristor turns off.

17.4.1
Self-commutation by an LC circuit

Before the gate pulse is applied the capacitor charges up to the polarity indicated in Figure 17.25. When the SCR-1 is triggered, current flows in two directions. The load current I_R flows through R

Figure 17.24 SCR commutation by momentary-shorting technique.

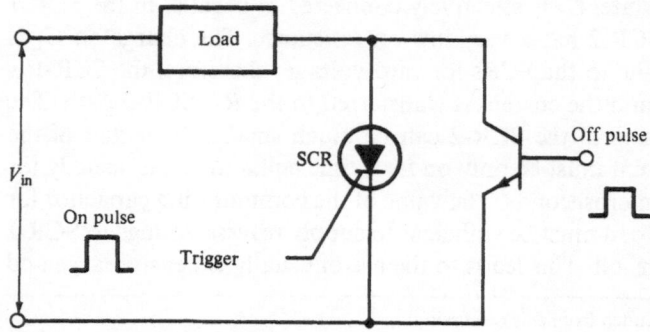

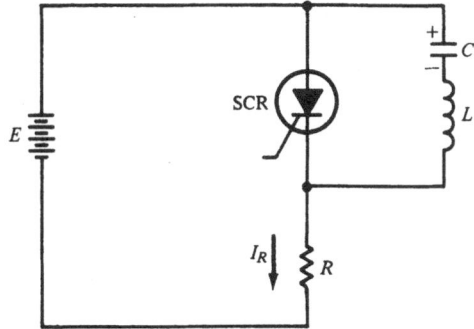

Figure 17.25 Self-commutation by an LC circuit.

and a pulse current flows through the resonant circuit LC. The pulse current charges C up in the *reverse* polarity. The resonant-circuit current will then reverse and attempt to flow through the SCR in opposition to the load current. The SCR will turn off when the reverse resonant-circuit current is greater than the load current.

17.5
Effect of an Impedance between the gate and cathode of a thyristor

Let us consider first the consequences of adding a resistance between the gate and cathode, as per Figure 17.26. This is similar to the so-called shorted-emitter configuration, in which the cathode

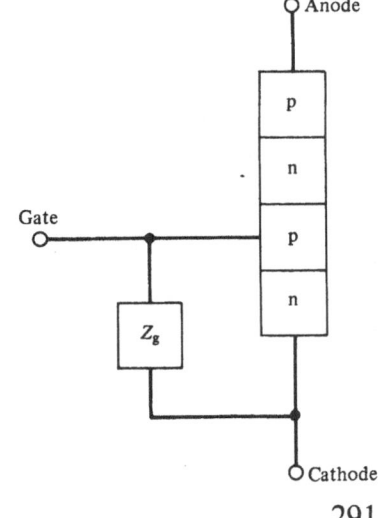

Figure 17.26 Resistance or reactance between gate and cathode.

291

metallization extends beyond the cathode and on to the p-gate region. Adding a parallel resistance provides a path other than the n-p cathode-gate junction for the thermally generated reverse-leakage current, or the dv/dt-induced anode current. The addition of a resistance causes the SCR to be less temperature and dv/dt sensitive, analogously to a shorted emitter. A capacitor connected in parallel to the gate-cathode circuit will decrease the gate sensitivity to high-frequency noise, but the low-frequency sensitivity (60 Hz, for instance) will be unaffected. The disadvantages of the gate-to-cathode capacitance are that it causes the rise time of the gate current to increase and may cause failure to turn off or commutate because the capacitor continues to supply energy to the gate after the holding current is reached. An inductance from the gate to the cathode reduces sensitivity to slow changes while maintaining high sensitivity to the sudden changes. The effect on turn-off time is just the opposite of a capacitor; t_{off} can be significantly reduced due to the reverse (negative) gate current induced in the gate inductance during turn off.

17.6
Thyristor protection circuits

Some of the precautionary measures which are taken to protect thyristors from either false triggering or damage are summarized in Figure 17.27. If the anode voltage is applied too rapidly, the dv/dt rating may be exceeded and the thyristor will turn on despite the fact that no gate-triggering current is applied. A protective circuit consisting of a resistor R_S and capacitor C_S connected across the thyristor will prevent the occurrence of this type of false triggering. The computation of the R_S and C_S values can be performed using [17.6].

A very large rate of current rise (di/dt) may be destructive to the device. The introduction of an inductance in series with the load may bring the di/dt to safe limits Figure 17.27b.

Very often, transient suppressors are added between the line and the thyristor, as shown in Figure 17.27c. The transient suppressor may consist of a neon glow lamp, metal-oxide varistors, or selenium thyrectors [17.2].

Because triac switching from the high-impedance to the comparatively low-impedance state can occur in less than 1 μsec, the current in the load increases extremely rapidly. This rapid rise of load current produces radio-frequency interference (RFI) extending into the range of several megahertz. Although this effect is much less pronounced than the RFI caused by other devices such as AC/DC brush-

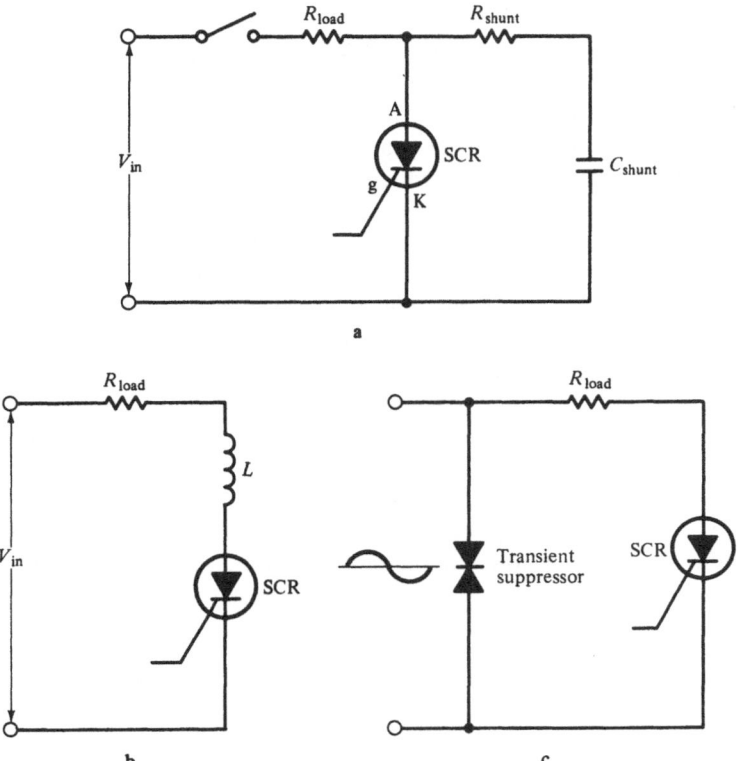

Figure 17.27 Thyristor protection circuits. (a) dv/dt suppression; (b) di/dt suppression; (c) transient suppression.

type motors, some means of RFI suppression are generally required. The most commonly used approach is that shown in Figure 17.28. An inductor is connected in series with the triac control circuit to restrict the current rate of rise, and a filter capacitor is used in parallel with the entire network to bypass high-frequency signals.

Figure 17.28 RFI-suppression network.

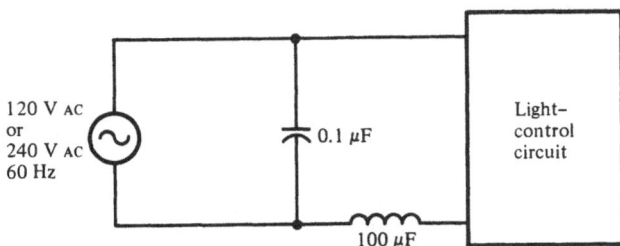

293

17.7
Some thyristor applications

SCRs and triacs are successfully used in a very large number of power circuits which can be very roughly categorized as follows: (1) static switching circuits which include such applications as replacement for electromechanical relays in battery-charging regulators, flasher circuits, current-limiting circuit breakers, electronic "crowbar" protection, time-delay circuits, and electroluminescent panel drivers; (2) power control of electric motors, light dimmers, DC power supplies; (3) power-conversion circuits to change DC to AC; (4) frequency-control circuits such as DC choppers, power inverters (DC to AC), and cycloconverters (one frequency to another); (5) temperature and air conditioning controls with phase- or zero-voltage switching; and (6) TV deflection circuits. This is, of course, not a comprehensive list.

The detailed description of all these circuits is much beyond the scope of this discussion and, therefore, we limit ourselves here to the simplified description of only a few representative circuits. For further reading see [17.1, 17.2, 17.8–17.10].

17.7.1
Triac light dimmer

This application was already discussion above in connection with the application of a diac and double-time–constant light control. Figure 17.29 [17.11] shows a typical phase-control circuit for lamp dimming, heat control, and universal-motor speed control, with values of various components. This circuit includes some of the protective network discussed above.

17.7.2
Universal-motor control

Fractional horsepower motors are, in most cases, series wound and capable of operation from either AC or DC power sources. The field winding of a universal motor is in series with the armature and external circuit, as shown in Figure 17.30. The current through the field winding produces a magnetic field which cuts across the armature conductors. The interaction of this field with the magnetic field developed by the armature current results in the armature torque and rotation. High starting torque, adjustable speed characteristics, and small size are distinct advantages of a universal motor over a comparably rated single-phase induction motor. One of the simplest and most efficient means of varying the impressed voltage, and thus

294

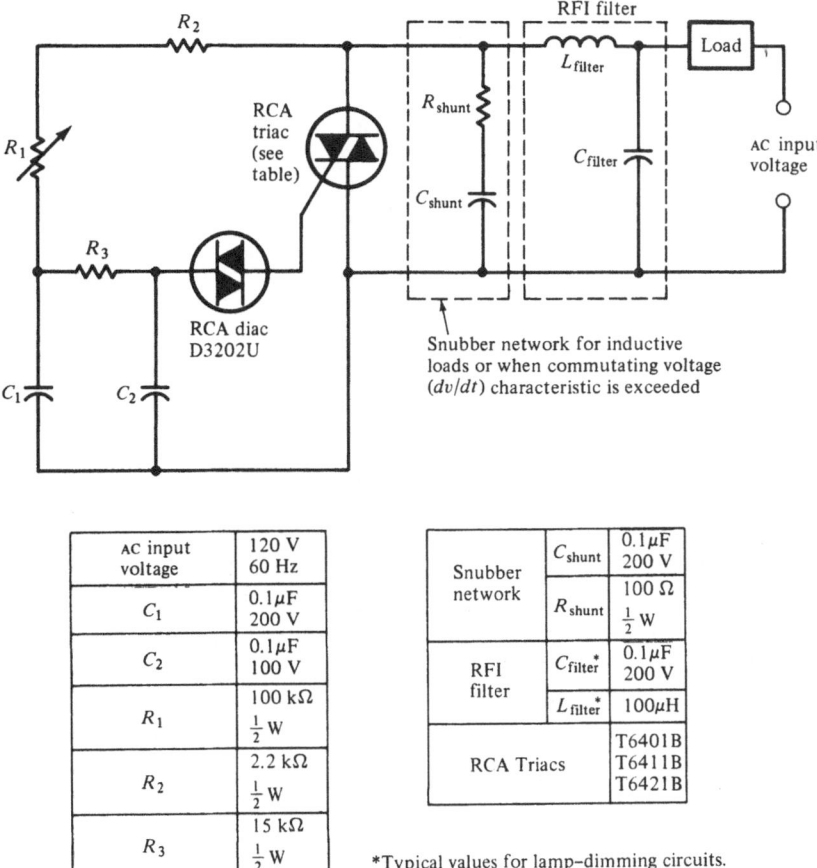

AC input voltage	120 V 60 Hz
C_1	0.1μF 200 V
C_2	0.1μF 100 V
R_1	100 kΩ $\frac{1}{2}$ W
R_2	2.2 kΩ $\frac{1}{2}$ W
R_3	15 kΩ $\frac{1}{2}$ W

Snubber network	C_{shunt}	0.1μF 200 V
	R_{shunt}	100 Ω $\frac{1}{2}$ W
RFI filter	C_{filter}^*	0.1μF 200 V
	L_{filter}^*	100μH
RCA Triacs		T6401B T6411B T6421B

*Typical values for lamp–dimming circuits.

Figure 17.29 Typical phase-control circuit for lamp dimming, heat control, and universal-motor speed control. (After RCA Solid State Data Book [17.11].)

Courtesy RCA Solid State Division, Somerville, N.J.

the speed of a universal motor, is by the control of the conduction angle of a thyristor placed in series with the load. Figure 17.31 shows the variation of the motor speed with the SCR conduction angle for both half-wave and full-wave line voltages [17.5]. The motor control circuits are divided into two classes: regulating and nonregulating. In this context regulation means load sensing and the existence of a feedback network to prevent the motor speed changes.

Figure 17.21 shows a half-wave motor control circuit without regulation [17.5]. The phase shift is provided by a network consisting of R_1, R_2, and C_1. When the voltage across C_1 becomes large enough during the positive half cycle, a sufficiently high current

295

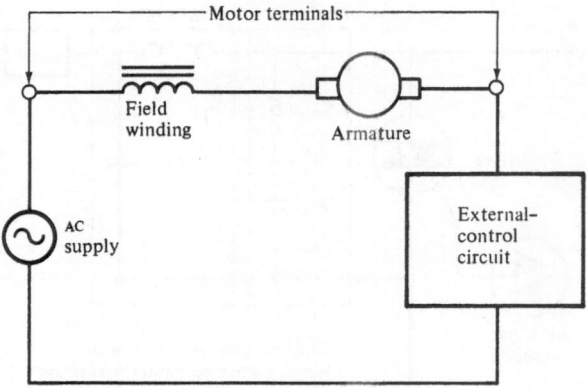

Figure 17.30 Schematic diagram for a series-wound universal motor.

flows in the base of the transistor Q_1 in order to turn it on. Q_1 supplies, in turn, the base current to Q_2 so that Q_2 turns on and supplies more base current to Q_1, and so on. The circuit is regenerative and leads to a very rapid saturation of transistors Q_1 and Q_2. The capacitor C_1 discharges through the saturated (very low series resistance) transistors into the SCR gate. When the SCR fires, the

Figure 17.31 Comparison of full- and half-wave control of univeral motors [17.5]. (After Yonushka [17.5].)

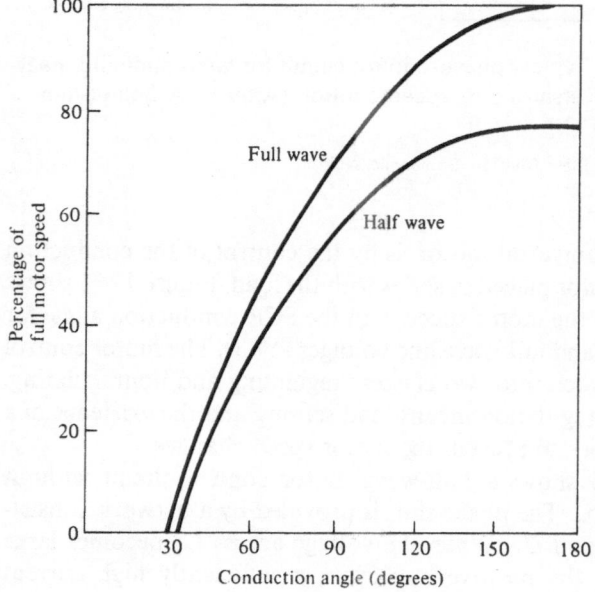

remaining portion of the positive half cycle of AC power is applied to the motor.

Speed control is accomplished by adjustment of the potentiometer R_1. With the components' values as shown (Figure 17.21), the threshold voltage is approximately 8 V and the maximum conduction angle is about 170 degrees. It should be pointed out that due to the universal-motor characteristics, half-wave control offers almost as much as full-wave (Figure 17.31).

17.7.3
Heat control

Phase-control circuits are very effective for electric heat control, but they require expensive filtering due to severe RFI. An on–off circuit (Figure 17.32) is more economical and provides synchronous

Figure 17.32 Synchronous switching on–off heat controller. (After RCA Solid State Data Book [17.11].)

Courtesy RCA Solid State Division, Somerville, N.J.

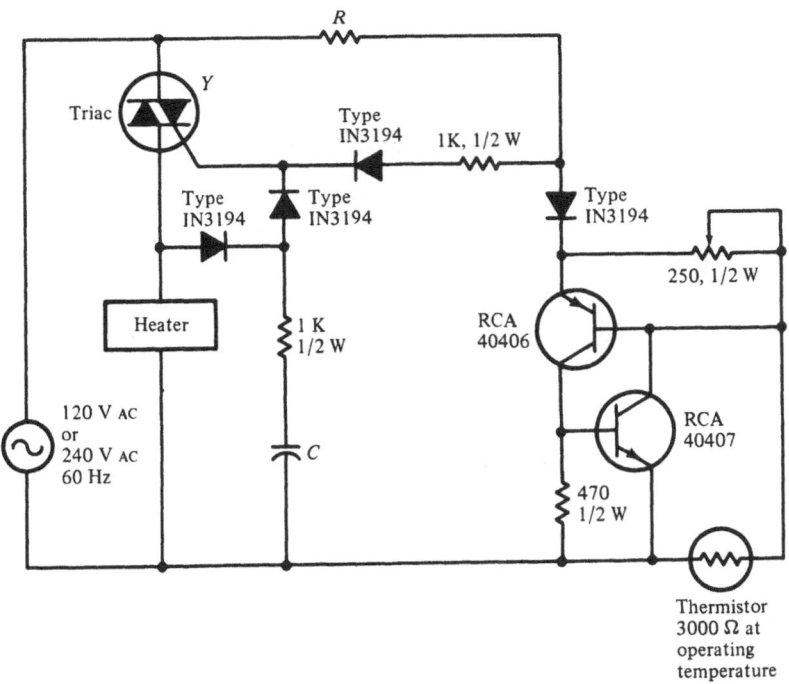

	120 V AC, 60 Hz	240 V AC, 60 Hz
R	2.2 kΩ, 5 W	3.9 kΩ, 5 W
C	0.5μF, 200 V	0.5μF, 400 V
Y	T4700B	T4700D

297

switching close to the zero-input voltage crossing to minimize the RFI. A temperature-sensing thermistor activates the two-transistor regenerative switch controlling the triac. When the temperature is low and the thermistor resistance is high, the two-transistor switch remains open, the triac is triggered on the line-positive half cycles, a voltage is applied to the heater, and the capacitor C is charged to the line peak voltage. The discharging of C through the triac gate triggers the triac on the opposite half cycle. The purpose of the diode–resistor–capacitor network is to trigger the triac on the line-negative half cycles after it is triggered on the positive half cycles providing integral AC voltage cycles to the heater.

As soon as the desired temperature is reached, the thermistor resistance drops and activates the two-transistor switch, shunting the trigger current away from the triac gate, and the triac stops conducting. When the temperature falls below the desired level, the two-transistor switch turns off and the triac is turned on again on the positive half cycle, and so on. A superior circuit to the one described includes the so called proportional integral cycle control which avoids thermal over- and undershoots due to the on–off action [17.1, 17.11].

Figure 17.33 Inverter configurations. (After Mapham [17.12].)

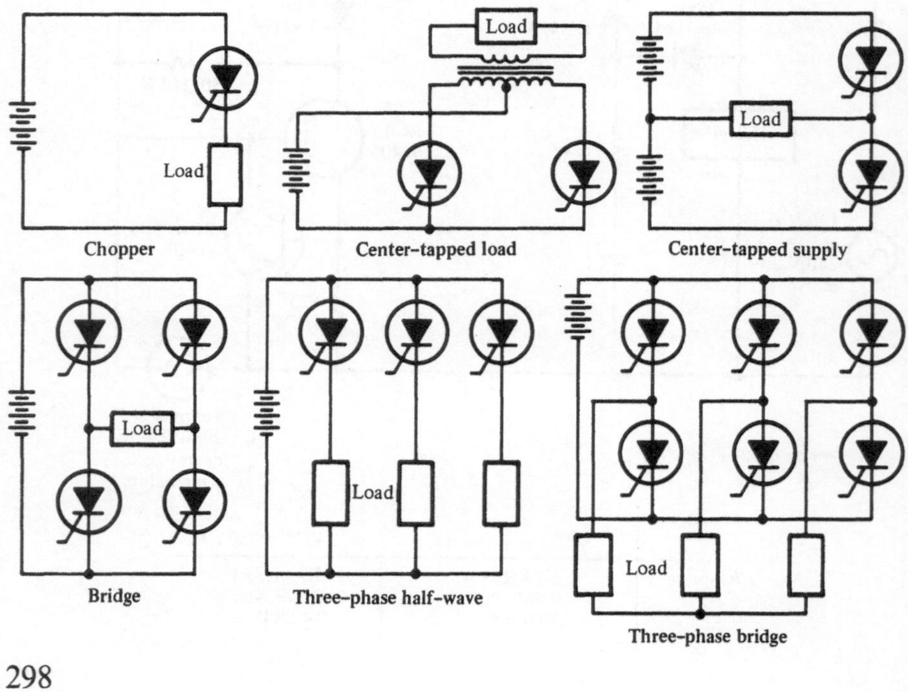

Chopper   Center–tapped load   Center–tapped supply

Bridge   Three–phase half–wave

Three–phase bridge

17.7.4
Inverters

There are several classes of inverter circuits transforming DC to AC [17.12] such as class A, self-commutated by resonating the load; class B, self-commutated by an LC circuit; class C, C or LC switched by a load-carrying SCR; class D, L or LC switched by an auxiliary SCR; class E with external pulse source for commutation; and class F, AC line commutated. Each of these classes has some advantages or disadvantages for some specific applications.

Inverter circuits may occur in several configurations in a way analogous to rectifier circuits (Figure 17.33). We are limiting our discussion here to the class C inverter, also known as McMurray–Bedford inverter [7.12, 7.13, 17.14]. This type of circuit is useful at frequencies below about 1000 Hz. External means must be used for regulation.

The circuit represented by Figure 17.34 is commutated by a capacitor in conjunction with an inductor and a center-tapped transformer. When SCR-1 is triggered, current starts flowing from the battery E_B through the inductor and into one half of the primary winding of the transformer. The center tap D is now at voltage E_B with respect to point A. By transformer action, a voltage is induced in the second half of the transformer and point B is at a voltage E_B in respect to the center tap D. Consequently, a potential equal to $2E_B$ appears across the capacitors C. When the SCR-2 is triggered, point B is in effect connected to ground and the capacitor voltage

Figure 17.34 Parallel inverter circuit with capacitive commutation.

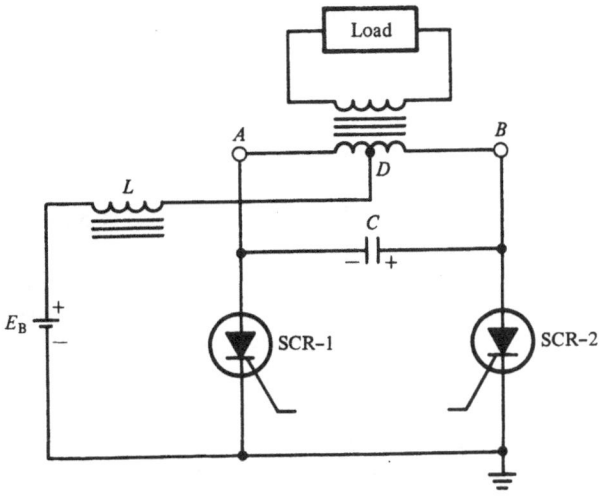

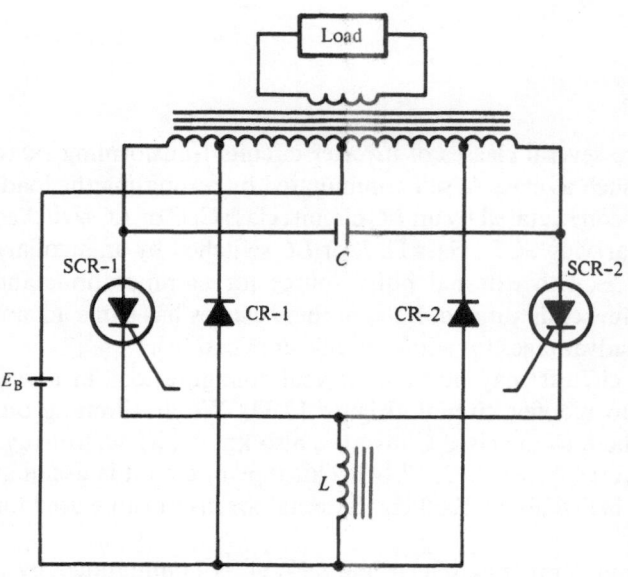

Figure 17.35   McMurray–Bedford inverter. (After Mapham [17.12].)

appears now across SCR-1. Due to the reverse polarity applied to the SCR-1 anode, SCR-1 turns off. Similar commutation takes place when SCR-1 is triggered on again and SCR-2 turns off. The voltage across the load is essentially a square wave. During commutation, the inductor L_1 and the capacitor C must supply the load current. The shorter the turn-off time, the smaller these components can be. When the load is inductive, the capacitor and the inductor must carry much higher current than for the case of a purely resistive load, and the volt–ampere rating of the capacitor must exceed that of the load.

The McMurray–Bedford inverter is an improvement over the circuit just described (Figure 17.35). The operation is similar; the circuit includes, however, two diodes CR-1 and CR-2 which are tapped into the primary of the transformer at about 10–15 percent from the end of the windings. They feed back to the DC supply the reactive power stored in the two capacitive or inductive loads. In the first case, the energy stored in the load at the end of a half cycle of an AC voltage is returned to the supply at the beginning of the next half cycle; in the second case, the energy stored in the capacitive load in the beginning of a half cycle is returned to the supply at the end of that half cycle.

The load voltage obtained approximates a square wave, but it can be converted to a sine wave by using a suitable filter.

300

References

17.1 RCA. Solid state power circuits. Technical Series SP-52, RCA Solid State Division, 1971.

17.2 *GE SCR Manual*, 5th ed., Syracuse, N.Y.: General Electric, 1972.

17.3 J. M. Neilson. Light dimmers using Triacs. RCA Application Note AN-3778.

17.4 F. E. Gentry, F. W. Gutzwiller, N. Holonyak, Jr., E. E. Von Zastrow. *Semiconductor Controlled Rectifiers*. New York: Prentice-Hall, 1964.

17.5 J. V. Yonushka. Application of RCA silicon controlled rectifiers to the control of universal motors. RCA Thyristor Application Note AN-3469, Somerville, N.J., 1968.

17.6 J. V. Yonushka. Triac power control application. RCA Thyristor Application Note AN-3697.

17.7 J. E. Wojslawowicz. Analysis and design of snubber networks for dv/dt suppression in thyristor circuits. RCA Application Note AN-4745.

17.8 John D. Harnden, Jr. and Forest B. Golden, Eds. *Power Semiconductor Applications*. New York: John Wiley, 1972.

17.9 B. D. Bedford and G. R. Hoft. *Principles of Inverter Circuits*. New York: John Wiley, 1964.

17.10 W. McMurray. *The Theory and Design of Cycloconverters*. Cambridge, Mass.: MIT Press, 1972.

17.11 RCA Solid State, 1975 Thyristors/Rectifiers Data Book, Somerville, N.J., SSD-206C.

17.12 N. W. Mapham. The classification of SCR inverter circuits. IEEE International Convention Record, Part 4, pp. 99–105, 1964.

17.13 W. McMurray. SCR inverter commutated by an auxiliary impulse. 1964 Proceedings of the Intermag Conference.

17.14 N. W. Mapham. An SCR inverter with good regulation and sine wave output. *IEEE Trans. Industr. Gen. Applic.*, *IGA-3*: 176–187, 1967.

Also of interest

G. J. Deboo and C. Burrows. *Integrated Circuits and Semiconductor Devices*. New York: McGraw-Hill, 1971.

P. Atkinson. *Thyristors and Their Applications*. London: Mills & Boon, 1972.

F. F. Mazda. *Thyristor Control*. New York: John Wiley, 1973.

Index